PROTEOMIC APPLICATIONS IN CANCER DETECTION AND DISCOVERY

PROTEOMIC APPLICATIONS IN CANCER DETECTION AND DISCOVERY

Timothy D. Veenstra

Laboratory of Proteomics and Analytical Technologies
Advanced Technologies Program
SAIC-Frederick, Inc.
Frederick National Laboratory for Cancer Research
Frederick, Maryland 21702, USA

Published by John Wiley & Sons, Inc., Hoboken, New Jersey.
Published simultaneously in Canada.

For general information on our other products and services or for technical support, please contact our Customer Care Department within the United States at (800) 762-2974, outside the United States at (317) 572-3993 or fax (317) 572-4002.

Wiley also publishes its books in a variety of electronic formats. Some content that appears in print may not be available in electronic books. For more information about Wiley products, visit our web site at www.wiley.com.

Library of Congress Cataloging-in-Publication Data:

Veenstra, Timothy Daniel, 1966–
  Proteomic applications in cancer detection and discovery / Timothy D. Veenstra, Laboratory of Proteomics and Analytical Technologies, Advanced Technologies Program, SAIC-Frederick, Inc., Frederick National Laboratory for Cancer Research.
       pages cm
    Includes index.
    ISBN 978-0-471-72406-3 (hardback)
  1. Cancer–Genetic aspects.   2. Proteomics.   3. Biochemical markers.   I. Title.
    RC268.4.V44 2013
    616.99′4042–dc23

                                                                                    2012049877

Printed in the United States of America

10 9 8 7 6 5 4 3 2 1

CONTENTS

PREFACE

Almost everyone on this planet has been affected one way or another by cancer. Either you have suffered from this disease yourself, or know a family member or friend who has. Cancer continues to be the second leading cause of death in the United States behind heart disease. The billions of dollars spent annually on cancer research has produced better detection and treatment strategies resulting in a continuing increase in the overall survival rates for patients. Unfortunately, statistics are not very comforting to an individual who has to watch their loved one suffer and die from this horrific disease. Ultimately, what everyone desires is a cure for cancer, making this disease no more difficult to deal with than something like appendicitis. Unfortunately, cancer is a very complex disease with each type containing many stages and subtypes. Cancer is also a very personalized disease. Individuals with the same type of cancer may not respond equally well to the same treatment, suggesting unique differences exist at the molecular level of each patient.

This era of science is able to explore these molecular differences with much greater detail than ever before. Technical advances in genomics have provided us a blueprint of human cells and the ability to recognize aberrations that give rise to cancer. This genetic blueprint, however, only gives part of the story. A major part of the story continues is told by the proteins that are expressed from this genetic blueprint, along with the modifications that "decorate" these proteins. The various functions carried out by these proteins define the state of the cell. Obviously, these functions are not exclusively determined from the genetic blueprint or the human body would not contain such a diversity of different cell types.

For this reason, proteomics is critical in advancing our knowledge of cancer. The advances made in proteomic technologies in the past decade have been enormous. Thousands of proteins can be analyzed in a single experiment allowing global differences between normal and cancerous cells to be measured. While the amount of data being generated has provided a number of exciting insights into cancer, it has also showed us our inadequacies in characterizing and comparing cellular proteomes. Beyond its complexity, the proteome of any cell, tissue, and organism is in a constant state of flux. Identifying specific differences in two proteomes is essentially trying to hit the bull's eye of two moving targets with a single arrow. Extrapolating this scenario to multiple samples quickly illustrates the challenges faced by proteomic scientists.

While this challenge seems enormous, we can be encouraged by the fact that the present state of proteomics allows us to even consider analyzing cancer at a global, or systems biology, view. Obviously, technology has driven proteomics up to its present state, but public health needs will be the main driver of proteomics in the future. The impact of proteomics, like genomics, will continue to grow as physicians, clinicians, and other health providers become more aware of the potential of this field in diagnosing and treating cancer patients.

One of the goals of this book is to help continue to bridge the knowledge gap that exists between scientists that develop and apply proteomics technologies and oncologists who focus on understanding the biological basis behind cancer manifestation and progression. Proteomic technologies are becoming widespread in their use, but they are still sequestered in specialized laboratories that possess the necessary expertise to optimize their use. While it is not critical for every scientist who uses proteomics to become an expert in the technology, it is important that he/she at least understand how the data are acquired and analyzed. Understanding these steps enables the scientist to determine his/her level of confidence in the final results. Proteomic studies, particularly global analysis, can produce data that overwhelm most investigators, and a greater understanding of the potential value, and pitfalls, of the results is necessary. Only through the combined questions and knowledge of proteomic and oncology laboratories will proteomics reach its potential impact on the diagnosis and treatment of cancer.

ACKNOWLEDGMENTS

Writing this book is as close to completing a marathon as I will ever come. Much like running a marathon, writing this book required inspiration, dedication, and perseverance. I obtained these attributes from my wonderful wife, Christine, and our three children, Jacob, Prairie, and Benjamin. Christine, your kind words of encouragement and love provide the incentive I need to always keep moving forward. Jacob, Prairie, and Benjamin, I want this book to be a reminder to each of you that you can complete anything you set your mind to. I also want to thank my parents, Peter and Shirley Veenstra, who taught me the importance of keeping my word and completing what I agreed to do. I am grateful to John Wiley & Sons for their patience. More than anyone, I want to thank Jesus Christ who gives me the grace I need daily.

This project has been funded in whole or in part with federal funds from the National Cancer Institute, National Institutes of Health, under contract #HHSN261200800001E. The content of this publication does not necessarily reflect the views or policies of the Department of Health and Human Services, nor does mention of trade names, commercial products, or organization imply endorsement by the U.S. Government.

1

SYSTEMS BIOLOGY

1.1 INTRODUCTION

In the classic, simplistic view of biomolecules within a cell, DNA, RNA, proteins, and metabolites are intimately connected. DNA acts as a template to produce RNA, which then serves as a template for protein production. Proteins then act on metabolites, converting them into whatever nutrients the cell requires while acting back upon the DNA and RNA from which it was created. This model is an oversimplification; however, it serves to illustrate that these classes of biomolecules are interconnected. Not only are these different classes of biomolecules interconnected, but so are molecules within the same order. For instance, a later chapter will discuss protein interactions in greater detail and why deciphering them is so important in understanding cancer. All of these biomolecular connections and interactions are necessary for a cell or organism to function. Just like an engine must be connected to the transmission in an automobile, if these biomolecules do not interact in some fashion, no work is produced within cells.

A new paradigm has emerged within the life sciences in the past decade on how biomolecular connections are studied. In the past, most scientists studied a single gene or protein very intently, collecting a great deal about its sequence, structure, and/or function. Indeed, many investigators became associated with the gene or protein they

Proteomic Applications in Cancer Detection and Discovery, First Edition. Timothy D. Veenstra.
© 2013 John Wiley & Sons, Inc. Published 2013 by John Wiley & Sons, Inc.

studied. For example, if a word association game was being played and the name Dr. Bert Vogelstein was mentioned, p53 would be the immediate response (1). Max Perutz and John Kendrew would be associated with the crystal structures of hemoglobin and myoglobin, respectively (2, 3). Laboratories that are highly focused on a single (or small number of) biomolecule(s) have been the driving force in our present understanding of the mechanism of cancer development and will continue to play a significant role in the future. There are, however, an emerging group of laboratories that focus on obtaining global views of a cell's biological machinery and how it is perturbed in diseases such as cancer.

The past couple of decades have seen the emergence of discovery-driven science. In discovery-driven science, experiments are performed in a nonbiased manner, and inductive reasoning is used to explain the resulting observations. This approach is in stark contrast to hypothesis-driven science in which experiments are performed to answer a specific question, and deductive reasoning is used to reach a logical conclusion based on known principles. Hypothesis-driven studies obtain a large amount of specific information on a few biomolecules, whereas discovery-driven studies obtain sparse amounts of information on as many biomolecules as possible. In the hypothesis-driven mode, discrete and subtle changes in a particular biomolecule of interest are studied and used to develop further hypothesis on how that molecule may function in the context of the cell. For example, much of my first couple of years as a postdoc was spent identifying calcium-binding sites within calbindin D_{28K}. Calbindin D_{28K} contained six EF-hand Ca^{2+}-binding domains, and through a series of experiments using both fluorescence spectrophotometry and mass spectrometry (MS), we were able to show that it binds 4 mol of Ca^{2+}/mol of protein (4, 5). Once the binding stoichiometry was established, other studies using deletion mutants established which EF hands bound the Ca^{2+} ions.

In a discovery-driven mode, gross changes in a large number of biomolecules are measured in an attempt to deduce some conclusion concerning the overall biomolecular composition of the samples or the effects of a specific perturbation on its composition. Discovery-driven studies are not designed to gather a lot of information about a specific molecule, but a small amount of information about a lot of molecules. I think back even further to my graduate school days. My thesis involved using nuclear magnetic resonance (NMR) spectroscopy to conduct three-dimensional structural studies of two proteins, thymosin $\alpha 1$ and ribonuclease A (RNaseA). I had memorized the primary structure of thymosin $\alpha 1$ and could tell you the positions of the eight cysteinyl and four histidinyl residues within RNaseA, along with bond distances when uridine vanadate was bound within its active site (6). In most of the studies our laboratory performs nowadays, we are content to identify at least three peptides from any given protein. The move toward discovery-driven science has been driven by the development of technologies that have dramatically increased the numbers of biomolecules that can be studied in a single experiment. These technologies include next-generation DNA sequencers for sequencing entire genomes (7), mRNA arrays capable of measuring the expression levels of thousands of genes (8), and highly sensitive mass spectrometers capable of identifying thousands of proteins within a complex mixture (9). As discussed throughout this book, these new capabilities brought as many challenges as they have breakthroughs. Many of these challenges are a direct result of "data overload." In a typical

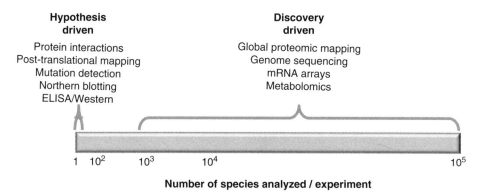

Figure 1.1. The move from hypothesis- to discovery-driven science. In hypothesis-driven studies, only a very small number (i.e., one to five) of molecules are studied per experiment. The move to discovery-driven studies, accelerated by the development of technology and software, has seen a leap to where hundreds and thousands of molecules are analyzed per experiment.

global quantitative proteomic study, significant abundance changes will be observed for 30–40% of the proteins identified. This percentage can equate to upward of 400 proteins. The natural tendency of scientists is to interpret the data so that every piece fits neatly. The unprecedented size of the datasets that are routinely accumulated combined with our rudimentary knowledge of cellular function, however, results in a frustrating inability to fit every piece logically together. This frustration can often cause the hypothesis-driven researcher to abandon discovery-driven technologies. For better or worse, it is going to require a combination of ideas from both hypothesis- and discovery-driven scientific fields for studies at the systems biology level to be successful.

Part of the challenge we encountered in moving from hypothesis- to discovery-driven studies was the size of the step. As illustrated in Figure 1.1, the move was anything but gradual. Basic research such as identifying protein interactions, post-translationally modified proteins, and mutations using hypothesis-driven technologies (Western blotting, ELISAs, Northern blotting, etc.) quickly morphed into global studies using discovery-driven technologies such as next-generation sequencing, mRNA arrays, high throughput MS, etc. Often it has seemed like making the leap from putting together a 10-piece puzzle of Winnie the Pooh to a 5000-piece flower garden jigsaw puzzle overnight.

1.2 WHAT IS SYSTEMS BIOLOGY?

I think one of the best, most understandable definitions of systems biology was put forth about a decade ago by Dr. Trey Ideker. His definition is summarized as the use of systematic genomic, proteomic, and metabolomic technologies to acquire data for the construction of models of complex biological systems and diseases (Figure 1.2) (10). Systematically determining the pieces (i.e., DNA, RNA, proteins, and metabolites) that

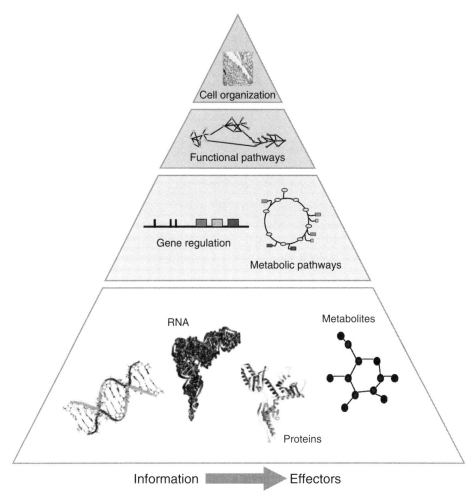

Figure 1.2. Conceptualization of the building blocks of systems biology. Systematically iden-tifying the key components of a cell is the first step in determining how these molecules function to regulate specific individual processes (e.g., gene regulation or metabolic path-ways). Associations between these processes are determined to identify functional pathways and networks. It is these pathways and networks that provide the synchronicity necessary for a cell or organism to survive and respond.

comprise a biological system is the foundation required for developing systems biology. By integrating our understanding of how different biological components function, systems biology aims to enhance our knowledge of the living systems and develop predictive models on how they behave when perturbed.

Why is systems biology important? Personally I believe it is because the human cell is an intricately designed machine in which all the parts need to act correctly in sync for its success and survival. A simple analogy is Roy Halladay. "Doc" Halladay is arguably the best pitcher in Major League Baseball. His success as a pitcher is directly linked to his hard two-seam sinking fastball. In trying to determine how he is able to throw the ball so hard, one must take a systems level view. It is not simply that his arm is strong enough to throw the ball so fast. The velocity that he generates is a product of how he plants his foot, performs his leg kick, turns his hips, rotates his shoulder, snaps his wrist, and even positions his head. Almost every part of his body works in a synchronized fashion to produce the end result. If he rotates his shoulder too early or does not kick his leg high enough, he will lose velocity and command of the pitch.

The human cell is no different, just a lot more complicated. Just consider cell division. The G1, S, G2, M, and C phases must occur in this exact sequence and errors during any phase can result in cell death or uncontrolled cell division (i.e., cancer). Also consider the release of energy from the hydrolysis of ATP, which is needed for the cell to function. Although we tend to just focus on the hydrolysis of ATP to ADP as the point of energy release, every major class of biomolecule in the cell was required to produce that energy reserve. DNA was required to provide the template that could be used by proteins to transcribe the messages required for translation of mRNA by proteins into other proteins that function to breakdown metabolites that result in the production of ATP. To completely understand how a cell or living organism functions, we are going to have to take a systems biology view.

1.3   WHAT SYSTEMS DO WE NEED TO STUDY?

Identifying which systems we need to study in systems biology is not that easy. The more we learn about the cell, the more factors need to be considered. When I was in college, I only had to think about three classes of RNA (mRNA, rRNA, and tRNA); now miRNA needs to be considered. The types of metabolites in the cell ranges from water-soluble metals (i.e., Ca^{2+}, Zn^{2+}, etc.) to high molecular weight, water-insoluble lipids. Since the cell is not a closed structure, a complete systems biology view needs to take into account its environment. In humans, the environment ranges from the effects from proximal to quite distal cells. To make the situation more complicated, we exist in a minimum four-dimensional universe that requires us to take changes over time into consideration.

Obviously, we presently do not have the technological capabilities or knowledge to provide the ultimate systems biology view of the human cell, however, that should not prevent scientists from making progress. To make progress, we initially need to take a very simplistic view of the cell. For the purpose of this book, we are going to consider the four major classes of biomolecules: DNA, RNA, proteins, and metabolites. Each of these classes can be further broken down into subclasses such as introns, exons, enhancers, promoters, and so forth for DNA or lipids, metals, sugars, and so forth for metabolites (Figure 1.3). In modern systems biology, the specific type of information gathered for each type of biomolecule may be different. For genomics (DNA), much of the focus is on mutation detection and gene copy number; for transcriptomics (RNA), the focus is on relative abundance and posttranscriptional modifications; for proteomics (proteins), the focus is relative abundance and posttranslational modifications (PTMs); and for

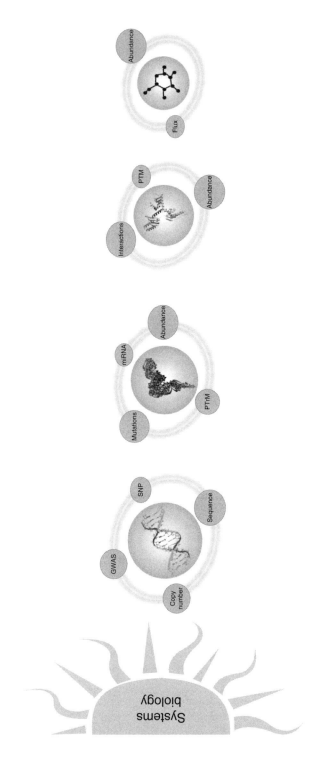

Figure 1.3. While DNA, RNA, proteins, and metabolites are the major components of systems biology, each major class of biomolecule has several related characteristics that need to be taken into account when pursuing a true systems biological view of the cell.

TABLE 1.1. List of Major "Omic" Technologies, the Molecules They Target, and What They Measure

Technology	Biomolecular Class	Parameters Measured
Genomics	DNA	Copy number
		Gene sequence
		Mutations
		Genome-wide associations
Transcriptomics	Messenger RNA	Relative abundance
		Mutations
		Posttranscriptional modifications
Proteomics	Proteins	Relative abundance
		Posttranslational modifications
		Protein interactions
Metabolomics	Metabolites	Abundance (relative and absolute)

metabolomics, the focus is again primarily relative abundance (Table 1.1). Since any cell requires all four classes of these biomolecules to act in concert for survival, it logically follows that a minimal systems biology view would incorporate information obtained from genomic, transcriptomics, proteomic, and metabolomic studies. To understand how a systematic view of the cell can be attempted and interpreted, it is necessary to examine the types of data acquired for each biomolecular class.

1.3.1 Genomics

Genomics is the study of genomes and the genes that are contained within. There are approximately three billion base pairs in the human genome that encode approximately 22,000 genes. According to the Human Genome Project, only 0.1% of bases vary between individuals (11). Since this book has a focus on technology and cancer, I am going to try to focus primarily on genomics in cancer. A cancer cell is a direct descendant of the fertilized egg from which the patient developed; however, its genome has accumulated a set of differences from its progenitor fertilized egg (12). These mutations, known as somatic, to distinguish them from parent to child inheritable germline mutations, may encompass several distinct changes in a DNA sequence. These changes include single-base substitutions, insertions, or deletions of small or large segments of DNA; rearrangements in which segments of DNA have broken and rejoined to DNA elsewhere in the genome; and increases or decreases in gene copy numbers from the two copies present in a normal diploid genome. Cells can also have obtained entirely new, foreign DNA sequences that contribute to carcinogenesis. Many of these foreign sequences arise from viruses such as human papilloma virus, Epstein–Barr virus, human herpes virus 8, hepatitis B (and C) virus, human T lymphotropic virus 1, and Merkel cell polyomavirus (13, 14). It is now known that seven human viruses cause 10–15% of human cancers worldwide (15).

Over the past several years, the sequencing of cancer genomes has revealed a large number of somatic mutations that occur across a multitude of genes. As of early 2011, the Sanger Institute's Cancer Genome Consortium identified 436 genes with causative mutations in cancer. A vast majority of these mutations were found in oncogenes and tumor suppressor genes that control signaling pathways that regulate functions such as cell growth and division (16). As of March 2011, the Catalogue of Somatic Mutations in Cancer database contained over 41,000 unique somatic mutations distributed across over 19,000 genes (http://www.sanger.ac.uk/genetics/CGP/cosmic/) (17). This database is a curation of experimentally determined somatic mutations published in the scientific literature. Obviously not all somatic mutations translate into cancer, but their frequency shows how dynamic a genome can be over the course of an individual's life. Somatic mutations can be classified as either driver or passenger mutations, depending on whether they are casually implicated in carcinogenesis (driver) or not (passenger). A key challenge to genome sequencing in the future will be differentiating driver and passenger somatic mutations.

In addition to mutations, epigenetics plays an important role in tumorigenesis and cancer progression (18–20). Epigenetics is the study of the regulation of gene expression that is independent of the DNA sequence of the gene. There are multiple mechanisms that affect the expression of a gene beyond its sequence including cytosine methylation, histone deacetylation and methylation, and chromatin remodeling. Like mutations, epigenetic patterns can be inherited (germ-line) and/or acquired (somatic) (21). Epigenetic events can also be influenced by selection pressures (22).

DNA cytosine methylation has been shown to regulate cell processes by influencing gene expression. This influence has been shown through studies that demonstrate distinct cell lines with variations in their DNA methylation patterns behave markedly different (23,24). Studies have shown that hypermethylation in promoter CpG islands of specific genes is associated with a variety of cancers such as colorectal, gastric, endometrial, lung, and prostate (25). Cytosine methylation is also strongly associated with histone modification and nucleosomal remodeling (26).

These epigenetic modifications occur on a genome-wide scale, and their importance is demonstrated by the ability to predict the risk of prostate cancer recurrence based on histone modification patterns (27). The mechanism of tumorigenesis is believed to involve both epigenetic changes and genetic mutations. Therefore, knowledge of both of these genomic characteristics is critical for completely understanding cancer on a personalized level.

The task of identifying driver and passenger somatic mutations, as well as epigenetic patterns within genomes, has been largely facilitated by the development of faster and more sensitive genome sequencing platforms (Table 1.2) (28). Next-generation (or second-generation) technologies have utilized improvements in PCR-based amplification that do not require *in vivo* cloning and coupled these developments with innovative sequencing chemistries and detection methods. The developments have resulted in longer read lengths and a dramatic decrease in the cost of genome sequencing.

Over the past several years, several next-generation sequencing platforms became commercially available. These systems include the 454 Genome Sequencer (GS) FLX (2005) (29), Illumina Genome Analyzer (2006) (30), and Applied Biosystems (AB)

TABLE 1.2. List of Major Gene Sequencing
Technologies and the Year They were Introduced

Sequencing Technology	Year of Introduction
454 GS FLX	2005
Illumina	2006
AB SOLiD	2007
Helicos HeliScope	2007
PacBio	2010
Ion Torrent	2010

SOLiD (2007) (31). These systems operated based on PCR amplification of DNA fragments, which was necessary to produce adequately strong signals for detection during the sequencing step.

Soon after these technologies came online, third-generation sequencing methods were developed. A different approach, based on single-DNA-molecule sequencing, was introduced by the Helicos HeliScope, which is based on single-DNA-molecule sequencing (32). These technologies rely on highly sensitive detection techniques and circumvent some of the limitations associated with PCR-based methods, but also introduce other problems that are discussed below.

The sensitivity of the HeliScope eliminated the need of a PCR-amplification step. DNA sequencing was based on the synthesis performed using a reduced-processivity DNA polymerase. As labeled nucleotides were added to the reaction sequentially one at a time, each was read as it was incorporated using a highly sensitive photon detection system called total internal reflection fluorescence. The read lengths were between 25 and 35 bp, with an output of 21–35 Gb per run (>1 Gb/h).

While the HeliScope was not widely adopted, continued innovations in the sequencing process continued to enhance the feasibility of single-DNA-molecule sequencing. The PacBio RS, which is based on single-molecule, real-time technology, was commercialized by Pacific Biosciences. The PacBio RS operates through nanometer-diameter aperture chambers that are created in a 100 nm metal film (33). These apertures, called zero-mode waveguides (ZMWs), allow the selective passage of short wavelengths. A single-DNA polymerase is attached to the supporting substrate of each ZMW. The gene is sequenced in real-time through the fluorescence signal that is emitted by each phospho-linked nucleotide as it is sequentially incorporated into the growing DNA strand. With an array of ZMWs, the PacBio RS can concurrently sequence 75,000 DNA molecules, producing read lengths of over 1000 with an upper limit of >10,000 bp. The throughput is an astounding <45 min per run (http://www.pacificbiosciences.com).

The latest technology in high throughput genome sequencing is the Ion Torrent (34). Detection is based on the measuring changes in pH, as nucleotides are added during DNA synthesis. Multiple DNA strands can be sequenced on a single chip. Ion Torrent technology produces read lengths of about 200 bp (with 400 bp lengths expected soon) in about 2 h of total sequencing time (http://www.iontorrent.com).

The obvious benefit gained with these high speed sequencing technologies is increased genome coverage at a severely reduced cost. As the cost of sequencing decreases, however, the cost of library construction, capital equipment, and particularly analysis and interpretation is dominating the field of genomics. As technologies result in an exponential increase in data, the burden shifts toward developing and applying tools to turn these data into useful information.

1.3.2   Transcriptomics

Since genes give rise to transcripts, transcriptomics was the next logical "omics" technology to pursue. Transcriptomics, also sometimes referred to as functional genomics, is the study of RNA transcripts produced by the genome (35). The major goals in this type of analysis are either to identify transcripts that are differentially abundant within different cell systems or to recognize patterns that are associated with a particular biological state. A wide range of different technologies such as DNA microarray analysis and serial analysis of gene expression is used to obtain this type of biological information.

In transcriptomics, a microarray is constructed that contains thousands of probes that are representative of individual genes bound to an inert substrate such as glass or plastic. The principle behind microarrays is the hybridization between two complementary DNA strands. Two strands that possess a high number of complementary base pairs will bind tightly and remain hybridized even after a series of washing steps. In many microarray experiments, fluorescently labeled cDNA is prepared from RNA extracted from the samples of interest. The labeled DNA is allowed to hybridize to the individual probes on the array with the expectation that each will hybridize to complementary gene-specific probes. Since the sample is fluorescently labeled, confocal laser scanning can be used to measure the relative fluorescence intensity of each gene-specific probe. The fluorescent intensity is used as a measure of the level of expression of each particular gene. Abundant RNA sequences will generate strong signals, whereas the signal from rare sequences will be weak.

Microarray data can be generated using either a single- or a dual-color array. In a single-color array format, each sample is labeled and individually incubated with an array. The array is washed to remove any nonhybridized material, and the level of expression of each gene is reported as single fluorescence intensity. In a dual-color array, two samples of RNA are labeled with a different dye (i.e., Cy3 and Cy5) (36). The two samples are mixed in a 1:1 ratio and introduced concurrently to the array and allowed to hybridize. The fluorescent intensities arising from the different dyes are measured, which provides a ratio of the amount of RNA that was isolated from the different samples. Regardless of whether a single- or dual-color array is used, the end result is a comparative measure of the expression for each gene in the two samples. As with most "omic" technologies, the trend in transcriptomics was "the more the merrier." The current version of Affymetrix GeneChips (Human Genome U133 Plus 2.0) permits the entire transcribed human genome to be measured on a single array (37). This array covers greater than 47,000 transcripts represented by more than one million distinct oligonucleotide entities.

Transcriptomics has become immensely popular in the past 15 years. These types of studies are now routinely conducted to determine which genes are expressed in a particular cell type or tissue as well as to compare their expression levels. There are a number of ways in which this type of information can be used. One of the most popular is to discover potential disease-specific biomarkers. In contrast to Northern blots that generally measured a single transcript per experiment, microarrays can analyze tens of thousands of transcripts, dramatically increasing the chances of finding genes that are uniquely expressed in particular samples. Microarrays can also be used to identify drug targets. For example, if the expression profiles obtained from a sample with a particular mutation are similar to that obtained from a drug treatment, the result may suggest that the drug inactivates the protein that is translated from the mutated gene.

Transcriptomics has also been used to classify diseases, particularly cancers. Cancer is a multifactorial disease that is not readily defined by a single aberration. By testing the expression profiles of a greater number of genes, cancers can be more accurately diagnosed. Many of these studies are showing that the gene expression profiles of cell types that were thought to be very similar can be quite disparate. For example, RNA amplification and Lymphochip cDNA microarrays were used in a recent study conducted by Dr. Louis Staudt's laboratory at the National Cancer Institute (USA) to profile hemopoietic stem cells, early B, pro-B, pre-B, and immature B cells with the aim of better characterizing normal human B cell development (38). Hierarchical clustering was conducted on 758 differentially expressed genes resulting in the clear separation of the gene expression profiles into five populations. Genes involved in VDJ recombination along with B-lineage-associated transcription factors (TCF3 [E2A], EBF, BCL11A, and PAX5) were activated in E-B cells, prior to CD19 acquisition. Interesting expression patterns were observed for several transcription factors with unknown roles in B lymphoid cells, such as ZCCHC7 and ZHX2. B cells had increased expression of 18 genes (including *IGJ*, *IL1RAP*, *BCL2*, and *CD62L*) compared with hemopoietic stem cells and pro-BB cells. In addition, the myeloid-associated genes *CD2*, *NOTCH1*, *CD99*, *PECAM1*, *TNFSF13B*, and *MPO* as well as T/natural killer lineage genes were also expressed by early B cells. The expression of these genes in the specific cell populations was validated and confirmed at the protein level. These results provide novel insight into the gene expression profiles of human B cell in the early stages of development. Hopefully being able to more precisely identify the developmental stages of B-cell development can lead to greater understanding of the cellular origin of precursor B-cell acute lymphoblastic leukemia.

While gene-expression profiling has been the major use of microarrays, other types of information can be also delineated. Affymetrix introduced the GeneChip Mapping Arrays for genotyping almost half-a-million single nucleotide polymorphisms (39). These arrays allow large-scale linkage analyses, association studies, and copy number studies to be performed on a large number of clinical samples in a high throughput fashion. Comparative genomic hybridization, which detects chromosomal copy number changes, can also be performed using the introduction of microarray-based comparative genomic hybridization (arrayCGH) (40, 41). More recent applications of microarray technology include chromatin immunoprecipitation (ChIP) with hybridization of microarrays (ChIP-on-chip) (42, 43). This technology partially bridges the fields of transcriptomics and proteomics by identifying sites of DNA–protein interaction across the whole genome, as well as the analysis of the methylation status of CpG islands in promoter regions.

1.3.3 Proteomics

Since this book is entitled "Proteomics and Cancer Discovery" many of the chapters are devoted intensely to different aspects of proteomics. Therefore, the description of proteomics found in this chapter will not be in-depth. Suffice it to say proteomics is the technological equivalent of genomics and transcriptomics at the protein level. The major foci of proteomics are protein identification, relative quantitation, and characterization of PTMs, especially phosphorylation. While these are the major foci, just about everything that used to be referred to as "protein science" now falls under the umbrella of proteomics (Figure 1.4). The ability to even consider some of the global surveys that are being conducted in proteomics today is a direct result of the rapid developments made in MS technology. While the function of a mass spectrometer is to detect the mass-to-charge (m/z) ratio of ions, it is ultimately its ability to manipulate these ions prior to detection, and the speed at which it does so, that makes it such a powerful tool. A more detailed description of MS technology is provided in the next chapter; therefore, this section will only briefly focus on its capabilities in proteomics.

The primary attribute that enables proteomics to contribute data to potentially enable a systems biological understanding of the cell is the ability to characterize thousands of proteins in a single study. While much of this capability is derived from MS technology, the advent of effective chromatographic separations of complex peptide mixtures prior to MS analysis has been just as critical. Among the most critical developments were the coupling of liquid chromatography (LC) with MS and the invention of multi-dimensional protein identification technology (known as MudPIT) (44). In 1999, John Yates III demonstrated the identification of over 2000 proteins in yeast by MS using MudPIT. While these results were spectacular at the time, it took less than 5 years for the identification of thousands of proteins in complex mixtures to become more or less commonplace.

The ability to identify thousands of proteins in complex mixtures was the first step in developing the ability to measure their relative abundance in different systems. The next step was to devise methods to be able to quantitatively compare protein abundances. Many different methods were quickly developed based on the incorporation of stable isotopes into proteins either metabolically or by chemical modification and recently through direct measure of peptide count or signal intensity. Regardless of the quantitative method used, the measure on a global scale is always the relative abundance of proteins, analogous to that being measured at the mRNA level in microarray studies.

While the progress made in proteomics has been nothing short of astounding in the past few years, it still has two major deficiencies when compared with genomics and transcriptomics. These comparative deficiencies are coverage and throughput. Current technology is capable of sequencing entire genomes at an ever-decreasing cost. Microarrays are capable of measuring the relative abundance of almost 50,000 gene products

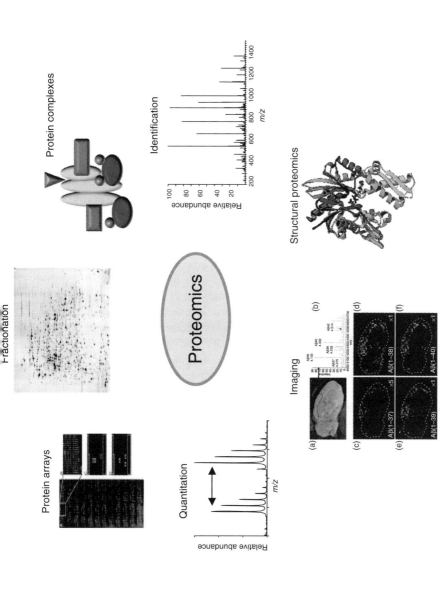

Figure 1.4. The myriad of different aspects related to proteomics. (Portions of figure reproduced with permission from Hudson ME, Pozdnyakova I, Haines K, Mor G, Snyder M. Identification of differentially expressed proteins in ovarian cancer using high density protein microarrays. *Proc. Natl. Acad. Sci. U.S.A.* 2007;104:17494–17499 and Seeley EH, Caprioli RM. Molecular imaging of proteins in tissues by mass spectrometry. *Proc. Natl. Acad. Sci. U.S.A.* 2008;105:18126–18131.)

13

through the use of over 1,000,000 oligonucleotide probes. Proteomics on the other hand is typically limited to a few thousand proteins, and acquiring and analyzing this level of data require on the order of 2–3 weeks minimum. In addition, the information obtained on this large number of species is generally limited to one or a handful of peptides per protein. Therefore, the coverage per protein as well as the entire proteome coverage is limited. The lack of throughput is particularly challenging in proteomics as the proteome is extremely dynamic, and measuring how it changes with respect to time would be a major advance in systems biology.

1.3.4 Metabolomics

The last of the classical "big four" biological molecules are the metabolites. While metabolites are not a direct product of the genome in the way that transcripts and proteins are, they are involved in transcriptional regulation and the regulation of protein activity. Metabolites themselves are acted upon by proteins, sometimes resulting in their conversion to a different metabolite. Analogous to transcriptomics and proteomics, the current major focus of metabolomics is the identification of changes in specific metabolite levels as a result of some perturbation of a cell's metabolome.

Being the youngest of the four major "omics," metabolomics has many unique challenges. Similar to proteomics, its throughput is very slow compared with genomics and transcriptomics. Since it is very dynamic, it is impossible to ascertain what percentage of molecules that make up the metabolome has been interrogated in a given study. The two primary technologies used in metabolomics are NMR spectroscopy and MS. These technologies are quite complementary. For example, NMR has higher throughput but lacks the sensitivity of MS. While fewer metabolites are detectable by NMR spectroscopy, those that are observed can often be readily identified based on their known resonance positions within the spectrum. While MS has the ability to detect a greater number of metabolites, their identification from raw data is difficult as there is no software analogous to that used to interpret tandem MS data of peptides for protein identification. Current "de novo" identification is based on the MS accurate mass and tandem MS data obtained for each metabolite (45). Although the mass accuracy of mass spectrometers has greatly increased in the recent future, it is still not sufficient to unambiguously identify compounds en masse. Also, the tandem MS data obtained for a metabolite is generally not as rich and distinctive as that observed for peptides, making accurate identification based on these data challenging as well.

As mentioned above, the major thrust in metabolomics is to quantify metabolites that are more or less abundant in samples obtained from perturbed systems (i.e., treated, diseased, etc.) compared with controls. This focus is where NMR spectroscopy has a significant advantage over MS. The data obtained from NMR spectroscopy are inherently quantitative as the signal intensity achieved at any specific resonance position is proportional to the concentration of the nucleus giving rise to that signal. Unfortunately, the ability to measure the relative abundances of metabolites by MS is limited to direct comparison of peak intensities. While this approach seems reasonable, signal intensity in a mass spectrum is dramatically influenced by the environment (i.e., other metabolites) of the metabolite. Therefore, a direct comparison of the intensity of two signals in a mass spectrum may not always provide accurate results if the treatment has significantly perturbed the entire metabolome. In addition, there are no available stable isotope-labeling methods available for metabolomics as there are for proteomics.

While metabolomics may be the youngest of the major "omics," studying metabolites in the context of cancer cells has a long history. In the 1920s, Otto Warburg showed that, under aerobic conditions, the metabolism of glucose to lactate is approximately an order of magnitude higher in tumor tissues compared with normal tissues (Figure 1.5) (46). This observation, termed the Warburg effect, was originally misinterpreted as impaired respiration instead of damage to glycolysis regulation. It is now

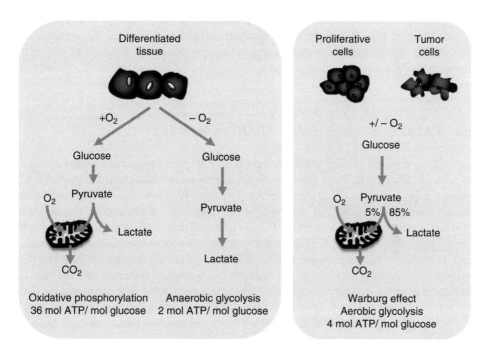

Figure 1.5. Differences between oxidative phosphorylation, anaerobic glycolysis, and aerobic glycolysis (i.e., the Warburg effect). Differentiated tissues in the presence of oxygen first metabolize glucose to pyruvate via glycolysis. Most of the pyruvate is then oxidized to CO_2 during the process of oxidative phosphorylation, which occurs in the mitochondria. In anaerobic glycolysis where oxygen levels are low, pyruvate is metabolized into lactate outside of the mitochondria. Anaerobic glycolysis results in minimal ATP production compared with oxidative phosphorylation (c.f. 2 vs. 36 mol ATP/mol glucose). In the Warburg effect, cancer cells (as well as normal differentiating tissues) convert most of the glucose to lactate (aerobic glycolysis) regardless of the oxygen status and even though mitochondria remain functional and some oxidative phosphorylation continues. While aerobic glycolysis is more efficient than anaerobic glycolysis (c.f. 4 vs. 2 mol ATP/mol glucose), it is still much less efficient at generating ATP than oxidative phosphorylation.

understood that nonproliferating (differentiated) tissues, in the presence of oxygen, initially metabolize glucose to pyruvate via glycolysis. They then completely oxidize most of that pyruvate to carbon dioxide via oxidative phosphorylation, which occurs in the mitochondria. In anaerobic glycolysis where oxygen is limited, pyruvate generated by glycolysis can be redirected away from oxidative phosphorylation by generating lactate. Anaerobic generation of lactate allows glycolysis to continue (through cycling of NADH to NAD^+), however, ATP production is quite low compared with oxidative phosphorylation. Otto Warburg observed that cancer cells, and normal proliferative cells, are inclined to convert most glucose to lactate regardless whether oxygen is present (aerobic glycolysis). Under these conditions, mitochondria respiration remains intact and some oxidative phosphorylation persists in cancer and normal proliferating cells.

The example given above illustrates that studies that demonstrated the importance of metabolites in cancer are still relevant almost a century later. Metabolomics continues this tradition of the importance of metabolites in diseases, just at a more expanded level. While it has been somewhat overshadowed by genomics, transcriptomics, and proteomics, my personal feeling is that metabolomics will have the greatest impact in cancer diagnostics in the future.

1.4 CANCER IS A SYSTEMS BIOLOGY DISEASE

Finding the cure for cancer would be simple if a single characteristic could be used to describe what happens to a cell during the process of malignant transformation. Unfortunately, cancer cells are characterized by a number of aberrant capabilities. These include abnormal growth, the ability to evade apoptosis, hyper-angiogenic activity, and the ability to metastasize (47). These characteristics are often induced through the accumulation of multiple genetic alterations, and approximately 300 genes have been identified as being casually implicated in cancer (48). As anticipated based on the phenotype of cancer cells, many of these genes express proteins that are involved in signal transduction processes that regulate cell cycle progression, apoptosis, angiogenesis, and tissue infiltration.

While genetic abnormalities are known to be prevalent in cancer, they do not act alone. Ultimately, proteins are the effectors that act to produce the phenotypic responses observed in cancer cells. While many basic research laboratories may focus on a single protein's contribution to the cancerous phenotype, it is a universally accepted fact that the effects observed result from a coordinated process of protein interactions. At a fundamental level, protein interactions can be as simple as two identical proteins coming together to form a functional homodimer. As more and more interactions between proteins were discovered, it became apparent that multiple proteins operate within signaling pathways to affect cell function. Adding even more complexity, these signaling pathways often do not occur independently of each other nor do they always function in a linear direction. Pathways function within a complex network of other signaling pathways in which proteins from divergent pathways may interact to influence each other. This influence can be either direct (e.g., phosphorylation of a kinase substrate) or indirect (e.g., transcriptional regulation).

Through years of systematically characterizing genes that are known to be involved in cancer, as well as the proteins that they encode for, signaling pathways important in this disease have been recognized. By bringing together all of the available pieces of information, large "road maps" detailing the protein circuitry involved in cancer cells could be drawn. These diagrams serve to illustrate the complexity that results from a large number of proteins that are involved, the cross talk that occurs between pathways, and the nonlinear relations that occur between the interacting molecules. It is easy to envision why it is extremely difficult, if not impossible, to predict the changes that would occur within the cell if even a small perturbation was introduced. Even if such a prediction could be made for a single type of cancer, it must be realized that different cancers do not behave in the same manner. Therefore, any prediction may only be useful for a single, closed system. While basic, systematic research has provided us with a tremendous knowledge of cancer cell biology, it has also shown us that the complexity of the cancer system is enormous. This complexity leaves us with a deficient understanding of the cancer system and an inability to predict how it responds to treatment.

1.5   MODELING SYSTEMS BIOLOGY

Unfortunately, modeling protein pathways and networks is only scratching the surface of true systems biology. To accurately predict the effect of a perturbation on the cell requires consideration of each type of biomolecule that is present. As a simplification, each molecule can be considered as a data point. Our current computational power and the knowledge of the human cell require us only to consider only a constrained set of data points so that accurate and predictive results can be obtained in a reasonable amount of time. These models must take into accounts both time and space. Models have been proposed that incorporate four spatial scales: atomic, molecular, microscopic, and macroscopic (49).

The atomic scale models dynamic structural events that occur primarily within proteins, peptides, and lipids. These models utilize tools such as molecular dynamics simulations and predict what occurs at the atomic level when molecules interact or their environment changes. These interactions involve nanometer-range distances and nanosecond time scales. Atomic-scale measurements are specific for each specific interaction occurring within the cell.

Molecular-scale modeling is probably what most individuals are thinking about when they contemplate systems biology. This scale models the average property of a specific population and is associated with such events as gene expression or cell signaling. Mathematically theses models utilize differential equations to represent biological reactions and interactions. This type of analysis is the most commonly studied in genomics, transcriptomics, proteomics, and metabolomics today. Spatial lengths on the order of nanometers to micrometers and timescales on the order of microseconds to seconds are modeled at the molecular scale.

The third scale is the microscopic scale. This scale models tissues and multicellular interactions. It is well know that cancer cells interact with their environment and malignant transformation is associated with alterations of cell–cell and cell–matrix interactions. The behavior of a tumor is also related to the heterogeneity of its environment. For example, while many interactions take place within a tumor at various levels (e.g., protein/metabolite, protein/protein, etc.), this model considers the supra-cellular interactions that contribute to a more complex system. Tumor cells interact directly with each other by direct physical contact and indirectly through secreted signaling molecules. Tumor cells also interact with endothelial cells to promote angiogenesis, fibroblasts to ensure stability, and immune-derived cells to escape detection and eventual destruction by the immune response. Microscopic modeling deals with distances in the micrometer to millimeter range and timescales in the range of minutes to hours.

The final modeling scale that is generally considered is the macroscopic scale. These models examine the morphology, shape, extent of vascularization, and invasion of the tumor as a whole under different conditions. This model is critical if we are to develop an accurate predictive capability as to when a primary tumor becomes metastatic. At this scale, the model is an average of all of the cells in the system since the number of cells is too large for them to be considered individually. These macroscopic scale models are associated with millimeter-to-centimeter-range lengths scales and time scales that can range from days to years (even decades).

1.6 DATA INTEGRATION

Conducting a true systems biology study requires integrating all of the data collected on various biomolecules within and outside the cell. This integration is the single greatest barrier facing systems biology. Some of the factors that make this such a difficult challenge are both technical and physiological. On the technical side, it is still impossible to collect and/or interpret all of the available information coded within the genome, transcriptome, proteome, and metabolome. The coverage afforded at the genomic and transcriptomic levels is much greater than that obtainable at the proteome and metabolome levels. Regardless of how complete current proteomic and metabolomic measurements are, there are still "holes" in the datasets. These "holes" refer to the overall coverage afforded by either of these disciplines. Although the exact number of proteins or metabolites that make up a proteome or metabolome is always unknown, it is likely that only a small fraction of each is being recorded in any study. A further complicating factor seen most readily when using MS is that the molecules measured within a study vary significantly between samples. For example, two identical aliquots of the same proteome can be analyzed back-to-back using LC–MS^2 and the overlap in the proteins identified will only be about 60–70%. This uncertainty, combined with inherent irreproducibility inherent with any MS measurement, adds uncertainty to models that use global proteomic or metabolomic measurements. At the physiological level, one of the greatest barriers is the dynamic nature of the cell. The ever-changing state of cellular processes makes it difficult to create a static picture in which biomolecules are behaving in a single defined role. The impact of dividing harvested cells or tissue into aliquots for genomic, transcriptomic, proteomic, or metabolomic analysis is presently an unknown. Considering that even a protein crystal can retain enzymatic activity (50), it is certain that molecules extracted from the cell and kept frozen also retain some semblance of their solution state activity. We presently know very little about the global effects of short- or long-term storage on molecules. Events such as interconversion of metabolites could be occurring during storage and would be very difficult to detect using existing experimental designs.

Beyond these issues, an even greater challenge is our overall lack of knowledge of how molecules within the cell interact on both a spatial and temporal level. While the phenomenon of such biological functions as gene transcription and translation, as well as the concept of protein complexes are easily understood, it is very difficult to put all cellular functions into the same spatial context. Many proteins in the human proteome carry out more than one function. Which function it carries out may be specifically dependent on its cellular location. There is even a lack of temporal relationship between gene transcription, translation, and protein turnover. For instance, how much time elapses between the signaling of a transcriptional complex to transcribe a gene and when the final functional protein product is active? Obviously, the time required for this event is not the same for every gene. What is the final amount of protein produced and when and how is this equilibrium established and maintained? Even these fundamental concepts are challenging when attempting to rationalize the myriad of different events that are occurring dynamically within the cell.

1.6.1   Integrating Transcriptomics and Proteomics

The most popular venture into systems biology has been the integration of transcriptomic and proteomic data. Since quantitative microarray and global proteomic data both provide the relative fold change observed for a particular species, these data sets are easily compared using a simple four quadrant log plot. Unfortunately in most cases, these comparisons have not been as effective as one might anticipate as the correlation between mRNA and protein levels is quite low (51–53). With low correlations, it has been difficult to establish universal rules that predict that guide transcript translation and the stability of both mRNAs and proteins. While this correlative result is disappointing, in retrospect it shouldn't be that unexpected. Most of these comparisons are made on transcripts and proteins that are extracted from cells harvested at the exact same time point. These studies do not take into account any temporal differences between gene transcription and mRNA translation. While the speed at which most genes are transcribed and then translated may be so quick that the temporal difference is minor, this may not be true for all genes.

A recent study examined the relationship between mRNA and protein levels on a global level by measuring the absolute abundance of both biomolecules as well as their turnover in a population of unperturbed mouse fibroblast (NIH 3T3) cells (54). Their results showed that approximately 40% of the variability in protein levels could be explained by the levels of their corresponding mRNAs: a much higher percentage than in previous studies that showed much lower correlation (55). In NIH 3T3 cells, translation efficiency was shown to be the best predictor of protein levels, suggesting that protein abundance is predominantly regulated at the ribosome. Protein stability played a minor role in regulating protein abundance compared with translational control. This result was unexpected as many cellular processes are known to be regulated by protein

(a)

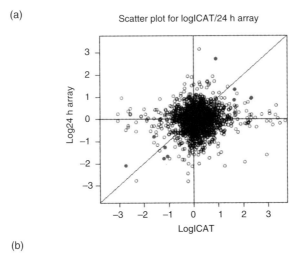

(b)

BioCarta pathway	Pearson correlation	*p* value
Induction of apoptosis	0.963	0.050
Rho signal transduction	0.831	0.049
Integrin- mediated signaling	0.829	0.048
Mitosis	0.825	0.050
G-protein-coupled receptors	0.763	0.046
Cell cycle	0.501	0.047

Figure 1.6. (a) Scatter plot of mRNA and their corresponding protein abundances in osteoblasts treated with inorganic phosphate. (b) Pearson correlation coefficients between mRNA and protein abundances grouped into specific functional pathways.

degradation (56, 57). One common association found between mRNA and proteins was that transcription factors and other genes with cell-cycle-specific function had unstable mRNAs and proteins. Conversely, most mRNAs and proteins that are not required for rapid response to stimuli are stable. This study definitely suggests that linkages between mRNA and protein levels are possible; however, more sophisticated empirical studies are going to be necessary to determine these linkages.

A study in my laboratory conducted a few years ago also showed a striking correlation between mRNA and protein levels if they were grouped based on their functions (Figure 1.6) (58). We compared changes observed in the levels of transcripts with those observed in proteins acquired from murine osteoblasts that were treated with inorganic phosphate. Consistent with most studies, the Pearson correlation between individual mRNA transcripts and proteins was an abysmal 0.09. We then proceeded to look at the mRNA/protein correlation from a slightly different angle. The transcripts and proteins were grouped into Biocarta annotated pathways (i.e., what pathway each protein is known to be associated within). When grouped into pathways, the correlation between mRNA and protein levels was quite high for a variety of cell processes including mitosis, integrin signaling, and apoptosis. This result suggests that comparing transcripts and proteins at the pathway level may provide a more fruitful approach than comparisons done at the individual level.

1.7 CONCLUSIONS

The advent of the omics era has increased our knowledge of the biological molecules that make cells, tissues, and organisms operate. We now have a reasonable estimation of the size of the human genome, whereas just over a decade ago, many thought that the human genome would contain approximately 100,000 genes (59). As the technology evolved, we have been able to gather data at an astounding rate and turn this into useful information. It took about 15 years to generate the first human genome and investigators can compare multiple sequences from many different tumors and delineate patterns of gene mutations associated with specific cancers. These gene mutations can also identify activating mutations in proteins that cluster within specific protein pathways (60). Mass spectrometers can identify over 10,000 proteins and upward of 2000 metabolites in cell populations. Progress is even being made in coordinating changes in the transcriptome and metabolome (61).

While these examples of omic integration are exciting, we are still a long way from realizing a systems biology view that enables us to predict how cells and organisms respond to either internal (e.g., gene mutations) or external (e.g., drug treatment) events. Sometimes it appears that this capability is beyond our reach. As we learn more about known components of the cell, new classes of biological molecules are discovered that have profound effects on how the cell functions. For example, the discovery of miRNAs in 1993 presented a previously undiscovered mechanism of translational repression and gene silencing (62). To further complicate human systems biology, each individual is not simply made up of "human" cells. The number of cells that comprise organisms that reside within humans (i.e., the microbiome) is estimated to outnumber human cells by an order of magnitude (63). These nonhuman organisms need to be considered as their presence is connected to human health, particularly within the gut. While all of these discoveries are exciting, it can make reaching the goal of systems biology frustrating. Just when we think we are making progress in identifying all of the parts that comprise the system, a new "treasure trove" of components is discovered that need to be accounted for in the model. Our "5000-piece flower garden jigsaw puzzle" can begin to seem like a jigsaw puzzle with a vague picture made up of an unlimited number of pieces.

While we do not have a predictive capability as yet, the progress in systems biology itself to this point was probably predictable. At the inception of the Human Genome Project, many of the key scientific leaders at the time predicted it would take about 15 years to sequence the first genome. The first draft of the human genome was actually delivered much quicker than predicted. Many of these same scientific leaders also predicted that it would take about 100 years to understand the code written in our genomes. If this prediction is correct, then a complete systems biological view of human cells will not be available for at least a century. Obviously, small steps forward in isolated systems, such as particular gene and protein pathways, will continue to be generated in the next few years. Predicting the dynamics that occur within these closed systems will be important not only because of their potential importance to disease treatment but they also represent a model for studying additional pathways. A century from now scientists will look upon our era as the time when the tools necessary to identify all of the biological components needed for generating systems biology models were developed and applied. Maybe the job of our generation is not to put the jigsaw puzzle together, but lay the pieces out on the table.

REFERENCES

1. Baker SJ, Fearon ER, Nigro JM, Hamilton SR, Preisinger AC, Jessup JM, vanTuinen P, Ledbetter DH, Barker DF, Nakamura Y, White R, Vogelstein B. Chromosome 17 deletions and p53 gene mutations in colorectal carcinomas. Science 1989;244:217–221.

2. Perutz MF, Rossmann MG, Cullis AF, Muirhead H, Will G, North ACT. Structure of hemoglobin. Nature 1960;185:416–422.

3. Kendrew JC, Bodo G, Dintzis HM, Parrish RG, Wyckoff H, Phillips DC. A three-dimensional model of the myoglobin molecule obtained by X-ray analysis. Nature 1958;181:662–666.

4. Veenstra TD, Gross MD, Hunziker W, Kumar R. Identification of metal-binding sites in rat brain calcium-binding protein. J. Biol. Chem. 1995;270:30353–30358.

5. Veenstra TD, Johnson KL, Tomlinson AJ, Naylor S, Kumar R. Determination of calcium-binding sites in rat brain calbindin D_{28K} by electrospray ionization mass spectrometry. Biochemistry 1997;36:3535–3542.

6. Veenstra TD, Lee L. NMR study of the positions of His-12 and His-119 in the ribonuclease A–uridine vanadate complex. Biophys. J. 1994;67:331–335.

7. Sastre L. New DNA sequencing technologies open a promising era for cancer research and treatment. Clin. Transl. Oncol. 2011;13:301–306.

8. Lockhart DJ, Winzeler EA. Genomics, gene expression and DNA arrays. Nature 2000;405:827–836.

9. Bensimon A, Heck AJ, Aebersold R. Mass spectrometry-based proteomics and network biology. Annu. Rev. Biochem. 2012;81:379–405.

10. Ideker T. Systems Biology 101—What you need to know. Nat. Biotechnol. 2004;22:473–475.

11. http://www.ornl.gov/sci/techresources/Human_Genome/project/info.shtml.

12. Stratton MR, Campbell PJ, Futreal PA. The Cancer Genome. Nature 2009;458:719–724.

13. Fearon ER, Vogelstein B. A genetic model for colorectal tumorigenesis. Cell 1990;61:759–767.

14. Talbot SJ, Crawford DH. Viruses and tumours—An update. Eur. J. Cancer 2004;40:1998–2005.

15. Moore PS, Chang Y. Why do viruses cause cancer? Highlights of the first century of human tumour virology. Nat. Rev. Cancer 2010;10:878–889.

16. Wong KM, Hudson TJ, McPherson JD. Unraveling the genetics of cancer: Genome sequencing and beyond. Annu. Rev. Genomics Hum. Genet. 2011;12:407–430.

17. Shepherd R, Forbes SA, Beare D, Bamford S, Cole CG, Ward S, Bindal N, Gunasekaran P, Jia M, Kok CY, Leung K, Menzies A, Butler AP, Teague JW, Campbell PJ, Stratton MR, Futreal PA. Data mining using the catalogue of somatic mutations in cancer BioMart. Database (Oxford) 2011;2011:bar018.

18. Draht MX, Riedl RR, Niessen H, Carvalho B, Meijer GA, Herman JG, van Engeland M, Melotte V, Smits KM. Promoter CpG island methylation markers in colorectal cancer: The road ahead. Epigenomics 2012;4:179–194.

19. Sandoval J, Esteller M. Cancer epigenomics: Beyond genomics. Curr. Opin. Genet. Dev. 2012;22:50–55.

20. Kasinski AL, Slack FJ. Epigenetics and genetics. MicroRNAs en route to the clinic: Progress in validating and targeting microRNAs for cancer therapy. Nat. Rev. Cancer 2011;11:849–864.

21. Reik W, Dean W, Walter J. Epigenetic reprogramming in mammalian development. Science 2001;293:1089–1093.

22. Rando OJ, Verstrepen, KJ. Timescales of genetic and epigenetic inheritance. Cell 2007; 128:655–668.

23. Peng L, Yuan Z, Ling H, Fukasawa K, Robertson K, Olashaw N, Koomen J, Chen J, Lane WS, Seto E. SIRT1 deacetylates the DNA methyltransferase 1 (DNMT1) protein and alters its activities. Mol. Cell. Biol. 2011;31:4720–4734.

24. Métivier R, Gallais R, Tiffoche C, Le Péron C, Jurkowska RZ, Carmouche RP, Ibberson D, Barath P, Demay F, Reid G, Benes V, Jeltsch A, Gannon F, Salbert G. Cyclical DNA methylation of a transcriptionally active promoter. Nature 2008;452:45–50.

25. Esteller M, Corn PG, Baylin SB, Herman, JG. A gene hypermethylation profile of human cancer. Cancer Res. 2001;61:3225–3229.

26. Aravind L, Abhiman S, Iyer LM. Natural history of the eukaryotic chromatin protein methylation system. Prog. Mol. Biol. Transl. Sci. 2011;101:105–176.

27. Seligson DB, Horvath S, Shi T, Yu H, Tze S, Grunstein M, Kurdistani SK. Global histone modification patterns predict risk of prostate cancer recurrence. Nature 2005;435:1262–1266.

28. Hawkins RD, Hon GC, Ren B. Next-generation genomics: An integrative approach. Nat. Rev. Genet. 2010;11:476–486.

29. Droege M, Hill B The Genome Sequencer FLX System—Longer reads, more applications, straight forward bioinformatics and more complete data sets. J. Biotechnol. 2008;136:3–10.

30. Koboldt DC, Larson DE, Chen K, Ding L and Wilson RK. Massively Parallel Sequencing Approaches for Characterization of Structural Variation. Methods Mol. Biol. 2012;838:369–384.

31. Paszkiewicz K, Studholme DJ. De novo assembly of short sequence reads. Brief. Bioinform. 2010;11:457–472.

32. Orlando L, Ginolhac A, Raghavan M, Vilstrup J, Rasmussen M, Magnussen K, Steinmann KE, Kapranov P, Thompson JF, Zazula G, Froese D, Moltke I, Shapiro B, Hofreiter M, Al-Rasheid KA, Gilbert MT, Willerslev E. True single-molecule DNA sequencing of a pleistocene horse bone. Genome Res. 2011;21:1705–1719.

33. Brakmann S. Single-molecule analysis: A ribosome in action. Nature 2010;464:987–988.

34. Rothberg JM, Hinz W, Rearick TM, Schultz J, Mileski W, Davey M, Leamon JH, Johnson K, Milgrew MJ, Edwards M, Hoon J, Simons JF, Marran D, Myers JW, Davidson JF, Branting A, Nobile JR, Puc BP, Light D, Clark TA, Huber M, Branciforte JT, Stoner IB, Cawley SE, Lyons M, Fu Y, Homer N, Sedova M, Miao X, Reed B, Sabina J, Feierstein E, Schorn M, Alanjary M, Dimalanta E, Dressman D, Kasinskas R, Sokolsky T, Fidanza JA, Namsaraev E,

McKernan KJ, Williams A, Roth GT, Bustillo J. An integrated semiconductor device enabling non-optical genome sequencing. Nature 2011;475:348–352.

35. Schena M, Shalon D, Davis RW, Brown PO. Quantitative monitoring of gene expression patterns with a complementary DNA microarray. Science 1995;270:467–470.

36. Shalon D, Smith SJ, Brown PO. A DNA microarray system for analyzing complex DNA samples using two-color fluorescent probe hybridization. Genome Res. 1996;6:639–645.

37. Yang G, Zhang G, Pittelkow MR, Ramoni M, Tsao H. Expression profiling of UVB response in melanocytes identifies a Set of p53-target genes. J. Invest. Dermatol. 2006;126:2490–2506.

38. Hystad ME, Myklebust JH, Bø TH, Sivertsen EA, Rian E, Forfang L, Munthe E, Rosenwald A, Chiorazzi M, Jonassen I, Staudt LM, Smeland EB. Characterization of early stages of human B cell development by gene expression profiling. J. Immunol. 2007;179:3662–3671.

39. Iwakawa R, Kohno T, Kato M, Shiraishi K, Tsuta K, Noguchi M, Ogawa S, Yokota J. MYC amplification as a prognostic marker of early-stage lung adenocarcinoma identified by whole genome copy number analysis. Clin. Cancer Res. 2011;17:1481–1489.

40. Al-Mulla F. Microarray-based CGH and copy number analysis of FFPE samples. Methods Mol. Biol. 2011;724:131–145.

41. Panzeri E, Conconi D, Antolini L, Redaelli S, Valsecchi MG, Bovo G, Pallotti F, Viganò P, Strada G, Dalprà L, Bentivegna A. Chromosomal aberrations in bladder cancer: Fresh versus formalin fixed paraffin embedded tissue and targeted FISH versus wide microarray-based CGH analysis. PLoS One 2011;6(9):e24237.

42. Chen X, Jorgenson E, Cheung ST. New tools for functional genomic analysis. Drug Discov. Today 2009;14:754–760.

43. Dell'Orso S, Fontemaggi G, Stambolsky P, Goeman F, Voellenkle C, Levrero M, Strano S, Rotter V, Oren M, Blandino G. ChIP-on-Chip analysis of in vivo mutant p53 binding to selected gene promoters. OMICS 2011;15:305–312.

44. Wolters DA, Washburn MP, Yates JR III. An automated multidimensional protein identification technology for shotgun proteomics. Anal. Chem. 2001;73:5683–5690.

45. Clements M, Li L. Strategy of using microsome-based metabolite production to facilitate the identification of endogenous metabolites by liquid chromatography mass spectrometry. Anal. Chim. Acta 2011;685:36–44.

46. Koppenol WH, Bounds PL, Dang CV. Otto Warburg's contributions to current concepts of cancer metabolism. Nat. Rev. Cancer 2011;11:325–337.

47. Martin GS. Cell signaling and cancer. Cancer Cell 2003;4:167–174.

48. http://news.sciencemag.org/scienceinsider/2010/04/a-skeptic-questions-cancer-genom.html.

49. Deisboeck TS, Wang Z, Macklin P, Cristini V. Multiscale cancer modeling. Annu. Rev. Biomed. Eng. 2001;13:127–155.

50. Doscher MS, Richards FM. The activity of an enzyme in the crystalline state: Ribonuclease S. J. Biol. Chem. 1963;7:2399–2406.

51. Evguenieva-Hackenberg E, Klug G. New aspects of RNA processing in prokaryotes. Curr. Opin. Microbiol. 2011;14:587–592.

52. Grønborg M, Kristiansen TZ, Iwahori A, Chang R, Reddy R, Sato N, Molina H, Jensen ON, Hruban RH, Goggins MG, Maitra A, Pandey A. Biomarker discovery from pancreatic cancer secretome using a differential proteomic approach. Mol. Cell. Proteomics 2006;5:157–171.

53. Ghazalpour A, Bennett B, Petyuk VA, Orozco L, Hagopian R, Mungrue IN, Farber CR, Sinsheimer J, Kang HM, Furlotte N, Park CC, Wen PZ, Brewer H, Weitz K, Camp DG 2nd, Pan C, Yordanova R, Neuhaus I, Tilford C, Siemers N, Gargalovic P, Eskin E, Kirchgessner T,

Smith DJ, Smith RD, Lusis AJ. Comparative analysis of proteome and transcriptome variation in mouse. PLoS Genet. 2011;7:e1001393.

54. Schwanhäusser B, Busse D, Li N, Dittmar G, Schuchhardt J, Wolf J, Chen W, Selbach M. Global quantification of mammalian gene expression control. Nature 2011;473:337–342.

55. de Sousa Abreu R, Penalva LO, Marcotte EM, Vogel C. Global signatures of protein and mRNA expression levels. Mol. Biosyst. 2009;5:1512–1526.

56. Mohd-Sarip A, Lagarou A, Doyen CM, van der Knaap JA, Aslan Ü, Bezstarosti K, Yassin Y, Brock HW, Demmers JA, Verrijzer CP. Transcription-independent function of polycomb group protein PSC in cell cycle control. Science 2012;336:744–747.

57. Vogel C, Marcotte EM. Insights into the regulation of protein abundance from proteomic and transcriptomic analyses. Nat. Rev. Genet. 2012;13:227–232.

58. Conrads KA, Yi M, Simpson KA, Lucas DA, Camalier CE, Yu LR, Veenstra TD, Stephens RM, Conrads TP, Beck GR Jr. A combined proteome and microarray investigation of inorganic phosphate-induced pre-osteoblast cells. Mol. Cell. Proteomics 2005;4:1284–1296.

59. Schuler GD, Boguski MS, Stewart EA, Stein LD, Gyapay G, Rice K, White RE, Rodriguez-Tomé P, Aggarwal A, Bajorek E, Bentolila S, Birren BB, Butler A, Castle AB, Chiannilkulchai N, Chu A, Clee C, Cowles S, Day PJ, Dibling T, Drouot N, Dunham I, Duprat S, East C, Edwards C, Fan JB, Fang N, Fizames C, Garrett C, Green L, Hadley D, Harris M, Harrison P, Brady S, Hicks A, Holloway E, Hui L, Hussain S, Louis-Dit-Sully C, Ma J, MacGilvery A, Mader C, Maratukulam A, Matise TC, McKusick KB, Morissette J, Mungall A, Muselet D, Nusbaum HC, Page DC, Peck A, Perkins S, Piercy M, Qin F, Quackenbush J, Ranby S, Reif T, Rozen S, Sanders C, She X, Silva J, Slonim DK, Soderlund C, Sun WL, Tabar P, Thangarajah T, Vega-Czarny N, Vollrath D, Voyticky S, Wilmer T, Wu X, Adams MD, Auffray C, Walter NA, Brandon R, Dehejia A, Goodfellow PN, Houlgatte R, Hudson JR Jr., Ide SE, Iorio KR, Lee WY, Seki N, Nagase T, Ishikawa K, Nomura N, Phillips C, Polymeropoulos MH, Sandusky M, Schmitt K, Berry R, Swanson K, Torres R, Venter JC, Sikela JM, Beckmann JS, Weissenbach J, Myers RM, Cox DR, James MR, Bentley D, Deloukas P, Lander ES, Hudson TJ. A gene map of the human genome. Science 1996;274:540–546.

60. Chapman MA, Lawrence MS, Keats JJ, Cibulskis K, Sougnez C, Schinzel AC, Harview CL, Brunet JP, Ahmann GJ, Adli M, Anderson KC, Ardlie KG, Auclair D, Baker A, Bergsagel PL, Bernstein BE, Drier Y, Fonseca R, Gabriel SB, Hofmeister CC, Jagannath S, Jakubowiak AJ, Krishnan A, Levy J, Liefeld T, Lonial S, Mahan S, Mfuko B, Monti S, Perkins LM, Onofrio R, Pugh TJ, Rajkumar SV, Ramos AH, Siegel DS, Sivachenko A, Stewart AK, Trudel S, Vij R, Voet D, Winckler W, Zimmerman T, Carpten J, Trent J, Hahn WC, Garraway LA, Meyerson M, Lander ES, Getz G, Golub TR. Initial genome sequencing and analysis of multiple myeloma. Nature 2011;471:467–472.

61. Eckel-Mahan KL, Patel VR, Mohney RP, Vignola KS, Baldi P, Sassone-Corsi P. Coordination of the transcriptome and metabolome by the circadian clock. Proc. Natl. Acad. Sci. U.S.A. 2012;109:5541–5546.

62. Lee RC, Feinbaum RL, Ambros V. The C. elegans heterochronic gene Lin-4 encodes small RNAs with antisense complementarity to Lin-14. Cell 1993;75:843–854.

63. Turnbaugh PJ, Ley RE, Hamady M, Fraser-Liggett CM, Knight R, Gordon JI. The Human Microbiome Project. Nature 2007;449:804–810.

64. Geiger T, Wehner A, Schaab C, Cox J, Mann M. Comparative proteomic analysis of eleven common cell lines reveals ubiquitous but varying expression of most proteins. Mol. Cell. Proteomics 2012;11:M111.014050.

2

MASS SPECTROMETRY IN CANCER RESEARCH

2.1 INTRODUCTION

A consistent theme throughout this book is the need to find novel biomarkers that aid in the diagnosis of cancer. As described earlier, expedient diagnosis and treatment of cancer is the number one factor that determines patients' survival rates. One fact that cannot be ignored, however, is that finding cancer biomarkers, whether they be for diagnosis or therapeutic monitoring, is extremely difficult. In particular, discovery of a biomarker for detecting cancer at its earliest stage would have an enormous impact on the survival rates; however, it is probably the most challenging to find this type of biomarker. Why is this statement true? There are a myriad of reasons but the most prevalent are physiological and analytical. From a physiological stand point, many cancers are closely related. This fact is especially true when examining different grades and stages of the same tumor type. While many tumors result in the overexpression of acute-phase proteins, circulatory molecules that are produced in response to many different types of tissue injuries, a specific tumor may not express a protein that is unique to itself. From an analytical stand point, present technology may not have the capabilities necessary to discover a protein that is unique to a specific tumor type. As described in Chapter 8, finding a novel biomarker for a specific tumor may require the ability to detect one protein within a complex mixture that contains anywhere from 10^4 to 10^6 other proteins. This search

Proteomic Applications in Cancer Detection and Discovery, First Edition. Timothy D. Veenstra.
© 2013 John Wiley & Sons, Inc. Published 2013 by John Wiley & Sons, Inc.

is the ultimate "needle-in-a-haystack" experiment. Present technology is incapable of identifying, let alone reliably quantitating, this number of proteins. To compound this challenge, a biomarker may be present at a very low concentration making its detection using even state-of-the-art technologies very difficult.

2.2 MASS SPECTROMETRY: THE TECHNOLOGY DRIVING CANCER PROTEIN BIOMARKER DISCOVERY

Biomarker discovery is a broad field that encompasses a number of technologies. This book is focused primarily on proteins; however, the search for DNA, RNA, and metabolite biomarkers is extremely active. Massively parallel sequencing-by-synthesis methods (i.e., 454 sequencing) for analyzing circulating DNA are now being used to find nucleic acid markers for cancer (1). Genome-wide association studies have identified a large number of genetic loci associated with disease risk, including various cancers (2). The search for mRNA biomarkers for a variety of cancers has been conducted in a number of studies for well over a decade now (3–5). More recently, the search for microRNA molecules that can act as potential biomarkers for diseases has generated a lot of interest (6–8). For the discovery of protein and metabolite biomarkers, mass spectrometry (MS) is far and away the technology of choice. While by no means a recently developed technology, MS has undergone a number of seminal developments that have made it arguably the most powerful analytical tool available today.

It sometimes appears that scientists who are heavily involved with MS technology speak their own language. While terms such as electrospray ionization (ESI), Orbitrap, ion trap, and time-of-flight (TOF) are familiar to most scientists, it is still sometimes difficult for those not involved in MS to keep these terms straight. A mass spectrometer is composed of three basic parts: the ion source, mass analyzer, and detector (Figure 2.1). The role of the ion source is to produce gas-phase ions. These ions are then transported into the mass analyzer where the ions are separated based on their mass-to-charge (m/z) ratio. Finally, the ions move to the detector, which measures the charge induced or current produced when the ions pass by or hit a surface. These three components are all coordinated through a central data system. The aim of the following sections, which may be mundane to mass spectrometrists, will be to provide an understanding of the technology to the casual users of MS so that they will be able to differentiate a "source" from a "mass analyzer" and an "ion trap" from a "triple quadrupole."

2.2.1   Ion Sources

Every so often an individual will contact my laboratory and say "I want to do matrix-assisted laser desorption/ionization (MALDI) on my protein." My reply is always to ask them exactly what the goals of the experiment are. This question illustrates one of the fundamental misunderstandings of MS. Fortunately, questions like these provide an opportunity for discussion on how to design the best experiment to achieve the project's goals and educate our collaborator on how MS works. This question also illustrates the importance that development of novel ionization methods have played in enabling MS

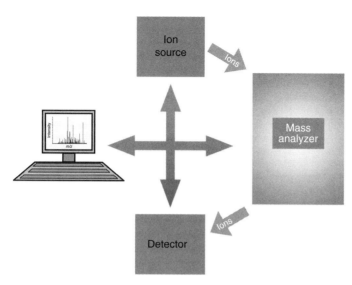

Figure 2.1. A schematic of a typical mass spectrometer. The basic components of a mass spectrometer include an ionization source where molecules are ionized and propelled into the gas phase, a mass analyzer region where the ions are manipulated, and a detector. Many samples are introduced into the mass spectrometer via liquid chromatography. All of these components are controlled and synchronized via software loaded onto a computer workstation.

for biomarker discovery. ESI (9) and MALDI (10) are the two most common methods to ionize proteins and peptides prior to their entering the mass analyzer region of the mass spectrometer. While the seminal work of Dr. John Fenn extended its use to measure biological molecules; the concept of ESI was first conceived by Malcolm Dole, who gained insight into the process through his knowledge of electrospraying automobile paint (11). The importance of Dr. Fenn's work in extending ESI to protein molecules is underscored by his receiving of the Nobel Prize in chemistry in 2002.

2.2.2 Electrospray Ionization

In the ESI process, the sample solution is introduced into the ionization source region of the spectrometer (Figure 2.2) (12). To ionize the sample, a high voltage (typically 3–5 kV) is applied to a stainless steel or other conductively coated needle through which the sample passes. Depending on the polarity applied, the voltage causes the sample to become positively or negatively charged. Positive ionization is used a vast majority of the time for the analysis of proteins and peptides. As it exits the spray tip, submicrometer-sized droplets containing both solute and analyte ions are produced. Beyond ionization, the next key step is to desolvate the ions prior to their entering the mass analyzer region of the spectrometer. Keep in mind that the droplets contain analytes of the same charge. As solvent molecules evaporate from the surface, the size of the droplets shrinks and the distances between the charges decrease. Evaporation of the solvent molecules is

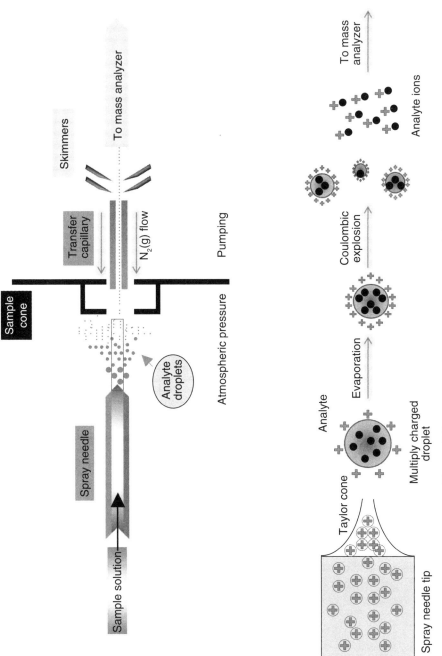

Figure 2.2. A cartoon illustrating the process of electrospray ionization.

aided by heating the capillary which the sample passes through and applying an inert curtain gas (such as nitrogen). As their size continues to decrease, the density of the charges on the surface of the droplet increases until it reaches a critical value (the Rayleigh limit) at which it violently blows apart (known as a Coulombic explosion) to form even smaller droplets. Solvent evaporation, droplet contraction, and Coulombic explosions continue to occur repeatedly until the analyte ions are released from the final droplet.

Electrospray ionization is referred to as a "soft" ionization technique (13). This term refers to the fact that ESI does not generally result in the fragmentation of the molecule. This ionization technique can also result in the production of multiple charged ions. This multiple charging results in larger molecules (e.g., proteins and DNA) having m/z ratios that are often greater than an order of magnitude smaller than their molecular weight. This result allows very high molecular weight molecules to be detectable using most types of mass analyzers. Since ESI is an atmospheric process, it can be easily coupled to separation devices such as high performance liquid chromatography (LC) or capillary electrophoresis (CE). As discussed later, the ability to couple LC with ESI has allowed thousands of proteins to be identified in a single experiment and was arguably one of the major advances that led to the proteomics era using MS.

2.2.3   Matrix-Assisted Laser Desorption/Ionization

While MALDI is a popular ionization technique for proteomics, its impact on cancer research, and in particular biomarker discovery, has not been nearly as great as ESI. In MALDI, ions are generated by irradiating a solid mixture with a pulsed laser beam (14). The mixture contains the proteins or peptides of interest mixed with a saturated solution of a highly conjugated, organic matrix compound (e.g., α-cyano-4-hydroxycinnamic acid, 2,5-dihydroxybenzoic acid, and 3,5-dimethoxy-4-hydroxycinnamic acid). The matrix is prepared in a solvent such as water, acetone, acetonitrile, or tetrahydrofuran. A few microliters of the solid mixture is placed onto a MALDI target plate and allowed to dry causing the proteins or peptides to become a part of the crystal lattice. The target plate is placed into the source region of the mass spectrometer and a short (2–200 Hz) ultraviolet laser pulse is used to ionize and desorb the analyte molecules from the solid mixture (Figure 2.3). The MALDI source region of most spectrometers is maintained at a relatively high pressure (even atmospheric pressure) causing the ions to be drawn into the mass analyzer region of the instrument, which is maintained at a lower pressure. In the MALDI process, peptide and large biomolecular ions are typically singly charged, whereas in ESI, these same molecules can pick up multiple charges. This singly charged character results in proteins having large m/z values, resulting in MALDI interfaces being most commonly combined with TOF mass analyzers that are capable of measuring ions across a broad range.

The developments of ESI and MALDI made the identification of proteins, most often through peptides produced by first tryptically digesting the molecule, routine (15, 16). The ability to identify complex mixtures of proteins separated using two-dimensional

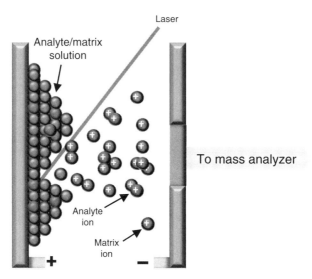

Figure 2.3. A cartoon illustrating the process of matrix-assisted laser desorption/ionization.

polyacrylamide gel electrophoresis (2D-PAGE) (17) became possible because MALDI–MS was able to keep up with the rate at which the protein spots were being robotically processed. Complex mixtures of digested proteins could also be identified as they eluted from reversed-phase (RP) LC columns or CE capillaries directly into a mass spectrometer equipped with an ESI source. This ability to couple LC and CE with ESI–MS has allowed thousands of proteins to be identified in a single LC–MS experiment.

2.3 TYPES OF MASS SPECTROMETERS

For those in basic research laboratories that do not work within proteomic laboratories, MS can be a confusing technology. When people in the field discuss projects, they speak in terms of ion traps, quadrupoles, TOFs, and so forth, not simply MS. While it is not essential for scientists who simply collaborate with a MS laboratory to understand exactly how the technology works, it is important for them to understand what type of mass spectrometer is best suited to meet their research needs. Table 2.1 provides brief relative performance characteristics of some of the instruments describe in this chapter, whereas Table 2.2 provides relative guidelines for what aim each mass spectrometer is most proficient. Why are there so many different types of mass spectrometers? A selection of just a few of the popular types of mass spectrometers in use today is provided in Figure 2.4. While all mass spectrometers measure the masses of ions, they have specialized methods of manipulating ions prior to their detection. It is these manipulations that define the mass spectrometer and give each type a special capability.

TABLE 2.1. Performance Characteristics of Different Types of Mass Spectrometers

Analyzer	Sensitivity	Resolution	Mass Accuracy
Ion trap	Good	Low	Low
Linear ion trap	Excellent	Low	Low
Triple quadrupole	Good	Good	Good
TOF	Good	High	High
FTICR	Excellent	High	High

TABLE 2.2. Comparison of Identification and Quantitation Capabilities of Different Types of Mass Spectrometers

Analyzer	Molecular Identification	Molecular Quantitation
Ion trap	Excellent	Fair
Linear ion trap	Excellent	Average
Triple quadrupole	Good	Excellent
TOF	Good	Average
FTICR	Excellent	Average

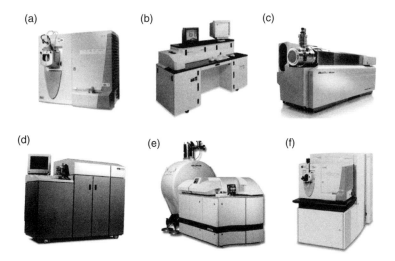

Figure 2.4. Images of six different popular mass spectrometers in use today. (a) LTQ linear ion trap; (b) time of flight (TOF); (c) triple quadrupole; (d) triple quadrupole/TOF; (e) Fourier transform ion cyclotron resonance; and (f) Orbitrap.

2.3.1   Ion-Trap Mass Spectrometer

The workhorse of proteomics and biomarker discovery is the ion-trap mass spectrometer (18). As readers go through this section, they should start to notice that many instruments are named based on a key component within the mass spectrometer or how it manipulates ions. In the case of an ion-trap mass spectrometer, both are relevant. This type of mass spectrometer contains a central trap where, not surprisingly, ions are captured. Fortunately, these instruments do a lot more than just trap the ions. An ion-trap mass spectrometer goes through repetitive cycles of trapping, storing, manipulating, and ejecting ions. In a typical biomarker discovery analysis, peptides are separated using LC directly into the ESI interface of an ion-trap mass spectrometer. At any point in time, a large number of peptide ions will enter the mass analyzer. These ions are guided via a series of quadrupoles, which direct them into the ion trap where they are stored. The instrument then ejects all other ions, except those with a specific m/z value. These isolated peptides are then fragmented by increasing their energy and colliding them with an inert gas such as nitrogen or helium (19). This process is referred to as collisional-induced dissociation (CID). The fragments obtained from the parent peptide ion are then stored in the ion trap and sequentially ejected into a conventional electron multiplier detector and their m/z values are recorded. This process popularly known as tandem MS (abbreviated as MS–MS or MS^2) is the primary method used to identify peptides for global proteomic studies. Beyond conducting MS^2, the ion trap can be used to isolated one of the fragment (or daughter) ions and subject this ion to further rounds of MS^2 (i.e., MS^3 or MS^n), obtaining further structural information. For example, MS^3 is commonly used to identify phosphopeptides when MS^2 results in the loss of the phosphate group but does not fragment the peptide backbone sufficiently to identify the peptide to which it is attached (20). Isolation of the peptide (minus the phosphate group) and fragmenting it using another round of MS will provide structural information describing the sequence of the peptide so that it may be identified.

When analyzing very complex mixtures, the ion trap is typically operated in a data-dependent MS^2 mode (21) in which each full MS scan is followed by a specific number (usually three to five, but sometimes >10 when using a linear ion trap) of MS^2 scans where the most abundant peptide molecular ions are sequentially isolated in the ion trap and subjected to CID (Figure 2.5). State-of-the-art ion traps can perform on the order of 7000 MS^2 experiments per hour affording the identification of over 1000 peptides in a single LC–MS^2 experiment (22, 23).

2.3.2   Fourier Transform Ion Cyclotron Resonance MS

Another commonly used instrument for identifying large numbers of peptides within complex mixtures is the Fourier transform ion cyclotron resonance (FTICR) mass spectrometer. This type of instrument was developed over 35 years ago by Comisarow and Marshall (24). It is only recently, however, that it has been routinely used to generate high quality data within proteomics. A FTICR functions much like an ion trap, having similar ion-trapping and manipulating capabilities. The obvious difference is that the ion trap of an FTICR is housed within a high field (typically on the order of 7–12 Tesla)

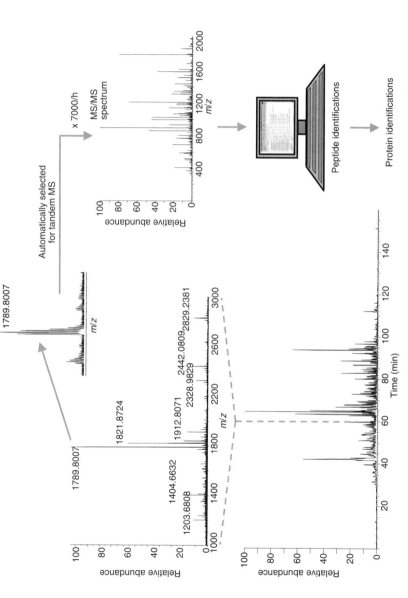

Figure 2.5. Process of protein identification via data-dependent tandem mass spectrometry (MS–MS). The mass spectrometer initially records the mass-to-charge (*m/z*) ratio of the various peptides that are detected at any point during the MS–MS experiment (a and b). (c) Peptides are isolated based on their intensity recorded in the previous MS scan (c). The peptide (in this case *m/z* 1789.8) is isolated (d) within the collision cell and fragmented via collision-induced dissociation (CID) and its MS–MS spectrum is recorded (e). The MS–MS spectrum is analyzed against a database of *in silico* predicted fragmentation patterns of peptides within the appropriate database to determine the peptide's identity (f), which is then correlated to its predicted protein of origin.

35

magnet field (25). The presence of the magnetic field causes the ions within the trap to resonate at their cyclotron frequency. A uniform electric field oscillating at or near the cyclotron frequency of the trapped ions is applied to excite the ions into a larger orbit. The orbit of the ions is measured as they pass by detector plates positioned on opposite sides of the trap. The detector measures the cyclotron frequencies of all of the individual packets of ions stored within the trap and a Fourier transform is used to convert these frequencies into m/z values. As with an ion-trap configuration, a specific peptide ion can be isolated and fragmented and its MS^2 spectrum recorded.

The instrument specifications of a FTICR mass spectrometer are related to strength of the magnet, much like that of a nuclear magnetic resonance spectrometer. Working at higher magnetic fields benefits many of the parameters related to FTICR performance, primarily resolution and mass accuracy (24, 25). While not always the case, FTICR is now widely used in proteomics. Up until the early part of this decade, FTICR instruments were considered too technically challenging and expensive for routine use. This situation changed dramatically when commercial instruments became available. Now FTICR mass spectrometers are ubiquitous in proteomic laboratories around the world.

2.3.3 Orbitrap Mass Spectrometer

While the commercial development of FTICR mass spectrometers brought high mass accuracy, resolving power, sensitivity, and dynamic range to many laboratories, these instruments were still quite large, complex, and costly to maintain owing primarily to the need for cryogens to keep the magnet cool. Fortunately, recent years have seen the development of the Orbitrap mass spectrometer that has comparable performance to the FTICR, yet is less costly and easier to maintain.

The Orbitrap was invented by Alexander Makarov in 1999 (26). In the Orbitrap, trapped ions orbit around a central spindle-like electrode. The frequency of their orbit is dependent on their m/z values. Their harmonic oscillations induce an image current in the outer electrodes, which can be converted into mass spectra by applying a Fourier transform. On the basis of this development, a new hybrid mass spectrometer became commercial available (LTQ-Orbitrap, Thermo Fisher Scientific) (27). This invention was the basis of the commercial instrument provided by Thermo Fisher Scientific, which consists of a LTQ coupled to a C-trap and the Orbitrap. This LTQ-Orbitrap has become an incredibly useful analytical device for proteomics. Drs. Petricoin and Liotta used the LTQ-Orbitrap to identify changes in the phosphoproteome within spleens from mice that were challenged with presymptomatic anthrax (28). After enriching phosphopeptides using titanium dioxide chromatography, the samples were analyzed using LC–MS^2 with an LTQ-Orbitrap. They were able to identify 6248 phosphopeptides (corresponding to 5782 phosphorylation sites). Other studies, such as demonstrated by John Yates' laboratory, have shown that in comparing stable-isotope-labeled proteomes, the improved performance characteristics of the LTQ-Orbitrap provide a fourfold to fivefold improvement in the number and quality of the peptide ratio measurements compared with the LTQ alone (14).

2.3.4 TOF Mass Spectrometer

Time-of-flight mass spectrometers have also been extensively used in proteomic research; however, their use in cancer research is primarily limited to coupling them with 2D-PAGE separations (29, 30). The major traits that make TOF mass spectrometers attractive are their high throughput, sensitivity, mass accuracy, and resolution (31). These spectrometers measure the m/z ratios of ions based on the time it takes for the ions generated in the source to fly the length of the analyzer and strike the detector. The time that it takes for the ions to traverse the length of the analyzer tube is proportional to their m/z value, with larger ions taking longer times to reach the detector.

Time-of-flight mass spectrometers have been coupled primarily with MALDI sources. Their primary application in proteomics has been to identify individual proteins through peptide mass mapping (32, 33). In most experiments, the proteins are separated using 2D- or 1D-PAGE prior to MALDI–TOF–MS analysis. The development of MALDI–TOF–TOF instruments, which incorporates a collision cell between the two TOF tubes, has enabled these instruments to conduct MS^2 so peptides can be identified by sequence-related information (34). Previously, standard TOF instruments did not contain a true collision cell; however, fragmentation could be induced through a process called "post-source decay" (PSD) (35). In this method, the reflectron voltage is adjusted so that fragment ions generated during the ionization and acceleration of the peptide are focused and detected. Unfortunately, PSD analysis is relatively slow and does not satisfy the high throughput demands most proteomic investigators desire.

2.3.5 Triple-Quadrupole Mass Spectrometer

The quadrupole mass spectrometer has been the most commonly used mass analyzer with ESI (36). Quadrupole refers to the geometric arrangement of four rods in parallel (or slightly out of parallel) through which ions are allowed to pass or manipulated. The two most common types of quadrupole mass spectrometers are single-stage and triple quadrupoles. As the names imply, a single quadrupole has one main length of quadrupole, whereas a triple quadrupole has three quadrupoles via which the ions can be manipulated. Although each instrument has other shorter quadrupoles within them to guide ions; for the purpose of this chapter, I will refer only to the main quadrupoles. Single-quadrupole analyzers have limited use in proteomics because they lack true MS^2 abilities; however, triple-quadrupole instruments have recently become extremely popular in proteomics because of their ability to conduct MS^2 and perform accurate quantitation of molecules (37, 38).

To identify peptides using a triple-quadrupole mass spectrometer, the instrument is switched between two different scan modes. In the "full-scan" mode, a broad m/z range of ions is allowed to pass through all of the three quadrupoles onto the detector. In the second scan mode, the first quadrupole acts as a mass filter and allows a specific ion to pass through onto the second quadrupole. The ion is then subjected to fragmentation within the second quadrupole by filling this region with an inert gas. The fragmentation ions then pass through the third quadrupole and are detected.

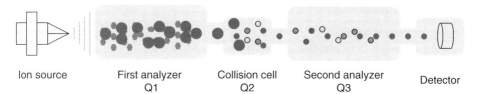

| Ion source | First analyzer
Q1 | Collision cell
Q2 | Second analyzer
Q3 | Detector |

Figure 2.6. A cartoon showing the use of triple-quadrupole mass spectrometer in the detection of an analyte using selected reaction monitoring mass spectrometry (SRM-MS). In Q1, the analyte of interest is isolated and guided into Q2 where it is fragmented by collision-induced dissociation. These fragments pass into Q3, where one specific fragment is allowed to pass onto the detector.

Triple-quadrupole mass spectrometers are extremely versatile, being able to conduct product ion, precursor ion, and neutral loss scanning. Triple quadrupoles have been used to identify proteins extracted from SDS-PAGE gels (39, 40), and the identification of posttranslational modifications such as phosphorylation and glycosylation (41, 42). A recent trend has been to utilize the third quadrupole to allow only specific fragment ions generated in the second quadrupole to pass onto the detector (Figure 2.6). These modes of operation, known as selected reaction monitoring and multiple reaction monitoring, are being used to measure the absolute quantity of peptides in complex biological systems (43).

2.3.6   Triple-Quadrupole TOF Mass Spectrometer

In recent years, different types of mass spectrometers have been developed that combine the best features of two different types of instruments. One of the best examples of these is the quadrupole-TOF mass spectrometer (QqTOF), a versatile instrument that plays a key role in proteomic analysis (44). As its name implies, the QqTOF mass spectrometer combines a triple quadrupole with a TOF analyzer. The quadrupole gives this instrument mass resolving power with a collision cell and the TOF provides high resolution and mass accuracy on both parent and fragment ions (44). This versatility of the instrument extends to the different types of ion sources it can accommodate; ESI, MALDI, as well as Ciphergen sources, are readily interchangeable on this mass spectrometer.

The utility of the QqTOF in proteomics is generated by its ability to perform tandem MS for peptide identification as well as retain high resolution for separating isotopically labeled internal peptide standards for quantitative comparative studies (45). The resolution provided by this instrument is critical for conducting quantitative analysis when isotope labeling strategies such as $^{16}O/^{18}O$ labeling are used. In this strategy, only 4 Da separates the mass of the peptides being compared, resulting in doubly charged ions that are only 2 m/z units apart (46); a difference that is not very well resolved using a conventional ion-trap mass spectrometer.

2.4   PROTEIN FRACTIONATION

While MS has obviously been the technology that has driven the development of proteomics, it would be a grave injustice if the importance of protein fractionation was
not discussed. Simply put, MS-based proteomics would not exist in its present form
without the development of high resolution methods to separate protein samples before
their analysis. Most proteome samples analyzed in cancer research today are much
too complex to be directly infused into a mass spectrometer and expected to produced
meaningful results. This reasoning is obvious when the samples of interest can contain
in excess of 100,000 different protein species and the number of potential isoforms and
modified versions of each protein are considered. Since most studies digest the proteome
into tryptic peptides, there may be $\sim10^6$ different species available for analysis. Infusing
all of these peptides into the instrument at once would result in the identification of only
a handful (i.e., <50) of peptides with the highest abundance and ionization efficiency.
In serum or plasma, two of the most popular types of sample used for cancer biomarker
discovery, a majority of the observable peptides would originate from albumin.

     While mass spectrometers have limits of sensitivity, the need for protein separation is
driven by the instrument's dynamic range. Dynamic range is the ratio between the highest
and lowest signal that an instrument can reliably measure. In most mass spectrometers,
the dynamic range is about two orders of magnitude (10^2). Unfortunately, the dynamic
range of protein concentrations is between 10^5 and 10^{12} in a typical proteomic sample
and current state-of-the-art MS studies can only identify proteins across approximately
seven orders of magnitude concentration difference (28). To identify as many proteins as
possible requires a "divide and conquer" strategy. This strategy means that the number
of peptides that enter the mass analyzer region of a mass spectrometer at any given time
is minimized allowing the instrument time to select and identify as many species as
possible across the entire experiment.

2.4.1   Polyacrylamide Gel Electrophoresis

When I first became acquainted with proteomics, Marc Wilkins in Australia had just
coined the term "proteomics" and the laboratory that he was working in had begun combining 2D-PAGE separation of proteins with identification of the resolved "spots" using
MS and amino acid analysis (47). Two-dimensional PAGE brought two key attributes to
proteomics: (1) the ability to resolve thousands of proteins and (2) the ability to measure
changes in their relative abundances (which will be discussed further in Chapter 5).
Proteomic analysis of a complex sample using 2D-PAGE and MS is quite straightforward. Proteins are separated on the gel and then visualized by staining. The pieces of
gel containing protein are cored from the gel and placed into individual tubes. The gel
pieces are subjected to in-gel tryptic digestion and the peptides extracted from the gel
piece and analyzed by peptide mapping or tandem MS. As one can easily envision,
identifying each protein spot individually is very laborious and time consuming.

     Two-dimensional PAGE separates proteins based on their isoelectric point (pI)
in one dimension followed by molecular mass (M_r) in the second. This orthogonal
separation has been shown to have a routine resolution of 2000, with a recent study

using ultra-zoom gels showing a resolution of almost 5500 proteins extracted from human liver (48). Even this level of resolution, however, results in an undersampling of complex proteome mixtures. Over the years, improvements in 2D-PAGE have included solubilization agents to enable analysis of membrane proteins and narrow pH gradient strips to increase the resolution within a specific pI range. After the proteins have been separated, they are visualized using a staining reagent. While a number of staining reagents are available, the two most commonly used by far are Coomassie blue and silver stain (49). The choice of staining reagent can depend on the amount of material loaded onto the gel or the known abundance of the proteins of interest. While silver stain has greater sensitivity, it must be remembered that the spots are going to be identified using MS. Therefore, the critical parameter is the sensitivity of the mass spectrometer being used. Care must be taken in choosing one that is compatible with MS identification as many silver stains require fixatives such as glutaraldehyde or formaldehyde that result in covalent cross-links within proteins making them virtually impossible to identify using MS. Fortunately, a number of MS-compatible silver stains, including Invitrogen's SilverQuest and Thermo Scientific's Pierce Silver Stain Kit, are commercially available.

Other hurdles that 2D-PAGE has had to overcome is a conception that it is irreproducible and laborious. It is not that 2D-PAGE is horribly irreproducible, it is simply that attempting to locate specific spots between multiple gels (particularly those that are low abundant) requires high precision alignment. With the development of techniques such as differential gel electrophoresis, multiple proteomes can be separated and compared within a single gel, alleviating issues related to irreproducibility between gels. The need for precise alignment has been overcome through software algorithms that "landmark" proteins within a reference gel and then locates them within comparative gels. The perception that 2D-PAGE is laborious needs to be considered beyond simply the time required to run the gels. While it is true that a few hours are required to run a gel, equipment is now available to run them in parallel and this amount of time is not any longer than is required to conduct a typical multidimensional fractionation experiment that is coupled directly to the mass spectrometer. In addition, time can be saved downstream as staining the proteins separated using 2D-PAGE requires only a small percentage of the species to be identified. Typical solution-based studies identify as many proteins as possible, requiring a significant increase in bioinformatic analysis (both in terms of time and complexity) compared with 2D-PAGE-based studies.

Beyond 2D-PAGE, one of the most popular methods of prefractionating proteome samples prior to RPLC–MS analysis is the good "old-fashioned" 1D-PAGE gel (50). The 1D-PAGE gel has been used for a variety of different samples ranging in complexity from simple (e.g., single isolated protein) to extremely complex (e.g., entire cell lysate) (51, 52). While 1D-PAGE does not have the resolution afforded by chromatographic methods such as strong cation exchange (SCX), it is most often coupled with LC–MS, which enhances the resolution both from the RP chromatography separation and mass spectrometer itself. Probably, the greatest use of 1D-PAGE in proteomics is in the characterization of protein complexes. Before the advent of in-gel tryptic digestion and advanced MS, isolated protein complexes were fractionated using 1D-PAGE and the proteins transferred to a polyvinylidene fluoride or nitrocellulose membrane (53). The membrane was then interrogated using antibodies directed against proteins that the investigator believed may bind to the target protein of interest. This hypothesis-driven method, known as Western blotting (54), is very challenging as it relies on not only having the correct hypothesis but also having antibodies that bind to the intended protein with high specificity. Probably, the major drawback is this method inhibits completely novel discoveries from being made. Combining in-gel digestion with MS allows the identity of every band visualized on the 1D-PAGE gel to be identified. In this scheme, individual bands are cut from the gel, digested using trypsin, and the peptides extracted from the gel. These peptides (and hence their protein of origin) are then identified using LC–MS^2. This discovery-driven method does not require any preconceived hypothesis and can identify truly novel proteins. In addition, the MS data can be analyzed to search for modified residues, whereas Western blotting requires antibodies against the specific type of modification being examined. The methods used in hypothesis- and discovery-driven characterization of protein complexes are contrasted in Figure 2.7.

2.4.2 Liquid Chromatography

The combining of LC with MS was a revolutionary step in analytical chemistry (55). While originally designed for metabolite analysis, LC is now the main tool used to fractionate the proteome prior to MS analysis (56–58). RP is the dominant form of LC used in proteomics because of the volatile, acidic nature of the mobile phase employed (e.g., acetonitrile containing ion-pairing agents such as formic acid etc.) and its ability to provide high resolution at low flow rates (which is ideal for ESI). These characteristics have enabled this chromatography to be coupled directly to the mass spectrometer such that peptides eluting from the column directly enter the instrument for analysis. The coupling of RPLC with mass spectrometers has become so ubiquitous that most investigators consider a RPLC–MS to be a single device. In fact, when LC–MS is mentioned, most investigators understand that the type of LC being referred to is RP.

Most protein chemists are familiar with RPLC for protein isolation; however, there are many differences between how this chromatography is conducted when coupled directly to a mass spectrometer. The most obvious difference is column size, as illustrated in Figure 2.8. Preparative RP columns used for protein purification are generally several centimeters (cm) in diameter (e.g., 2–5) and length (15–30). The lengths of the columns coupled directly online with the mass spectrometer are generally similar to those of preparative columns; however, a typical RP column packed with C18 particles for LC–MS analysis is on the order of 50–100 μm in diameter. The particles used in capillary columns are on the order of 3–5 μm in diameter and provide very high resolution separations.

Even with the high resolution separation afforded using RPLC, complex proteome samples are usually prefractionated prior to RPLC–MS analysis. For global proteomic analysis, SCX LC is generally the prefractionation method of choice. In this scheme, the proteome sample is digested into peptides (e.g., using trypsin) and then fractionated into 10–20 SCX fractions. Each of these is then individually analyzed using RPLC–MS. This SCX-RP fractionation scheme developed in the laboratory of Dr. John Yates is commonly known as multidimensional protein identification technology (MudPIT) (59–61). SCX

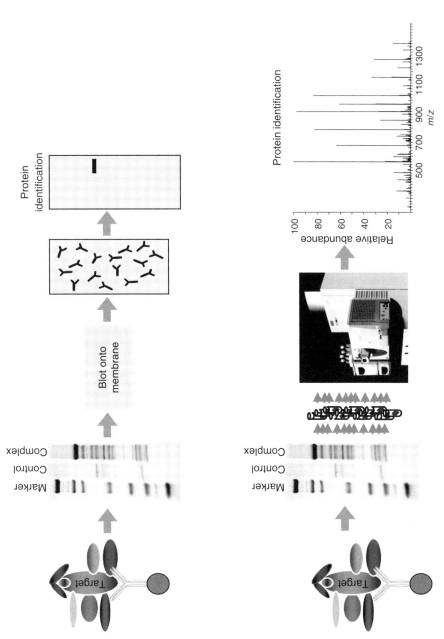

Figure 2.7. A comparison of hypothesis- and discovery-driven methods for identifying protein complexes.

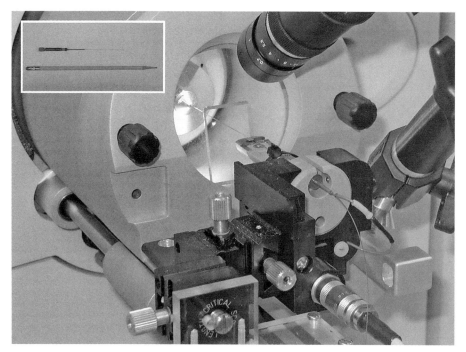

Figure 2.8. Reversed-phase liquid chromatography columns coupled online with mass spectrometers are typically on the order of tens of microns in diameter and a few centimeters in length (inset). The column is positioned within the ion source region allowing peptides being eluted to enter directly into the mass analyzer region of the mass spectrometer.

is a useful fractionation partner with RPLC because its mode of separation is orthogonal to RPLC (charge vs. hydrophobicity), it has lower resolution than RPLC, and peptides eluted from the SCX column can be loaded directly on a RPLC column. The general rule in combining separate fractionation strategies for separating complex proteome mixtures is that the first dimension of fractionation should have the lowest resolution and subsequent techniques should have progressively higher resolution. This rule is one of the reasons that SCX is performed prior to RPLC (Figure 2.9). A combined SCX–RPLC–MS strategy is used for protein identification, quantitating abundance differences, and identifying modified residues across complex proteome samples such as tissue, blood, and cell culture lysates (62–65).

Even with the high speed at which modern mass spectrometers can select and subject peptides to CID for obtaining sequence information, there is still no possible way to obtain complete coverage of a complex proteome sample. This lack of coverage is not solely due to the speed of the mass spectrometer but is also related to peptide ionization efficiency, abundance, and so forth. However, even using a MudPIT separation scheme does not allow every peptide observed by the mass spectrometer to be subjected to CID. To maximize the number of peptides chosen for CID (and potential identification),

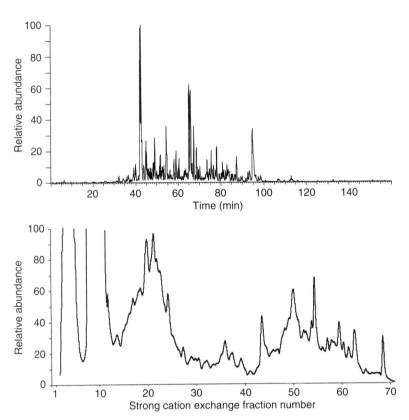

Figure 2.9. A comparison of peptide separation using strong cation exchange (SCX; top) and reversed-phase liquid chromatography (bottom). Since SCX has lower resolution, it is used prior to reversed-phase separation in the multidimensional protein identification scheme designed by John Yates' laboratory (59).

investigators have resorted to three- and four-dimensions of fractionation prior to MS analysis (66–68).

Probably, no sample type has been subjected to the degree of prefractionation as human plasma and serum. The primary reason is that these two samples more than any others have been analyzed in the search for biomarkers to serve as indicators of disease (69). Both serum and plasma, however, are very complex and possess extremely large dynamic ranges of protein concentrations. Therefore, to detect low abundance proteins, reduction in sample complexity through multidimensional prefractionation strategies is critical. As will be discussed later, over 90% of the protein content of serum and plasma is made up of less than 20 proteins (70). Albumin itself can comprise 40–60% of the protein content of serum and plasma. If these blood samples are digested with trypsin and directly analyzed using LC–MS^2, approximately 80% of the identified peptides will be from albumin. Dr. David Speicher at the Wistar Institute in Philadelphia, PA, has developed a four-dimensional separation strategy for profiling blood samples (Figure 2.10) (71).

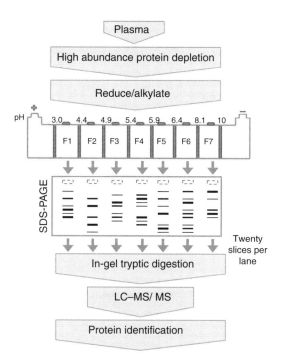

Figure 2.10. Diagram of the four-dimensional separation strategy designed by Dr. David Speicher for in-depth plasma proteome characterization (71). The strategy consists of four separation steps (major protein depletion, MicroSol-isoelectric focusing fractionation, 1D gel separation, and reversed-phase liquid chromatography coupled to tandem mass spectrometry analysis).

It includes immunoaffinity-based abundant protein (e.g., albumin, transthyretin, etc.) depletion, followed sequentially by microscale solution isoelectrofocusing and 1D SDS-PAGE, and finally RP separation of tryptic peptides prior to MS^2. Using this strategy with only 300 μL of plasma, his laboratory routinely identifies proteins over nine orders of magnitude in concentration, including a large number of proteins at the low ng/mL or lower levels.

While multidimensional fractionation of complex proteome samples has been instrumental in providing greater coverage of proteome samples, the number of dimensions used has to be carefully weighed against the time required to complete the study. While investigators often use the phrase "high throughput" to describe MS-based proteomics, its speed pales in comparison with techniques such as microarray analysis. While running a single fractionation dimension may not take a long time, it multiplies the number of samples that require separation in the next dimension. In addition, the analysis can quickly reach a point of diminishing returns where the number of new proteins identified does not linearly increase with the number of separation dimensions added.

2.5   IMPACT OF MS IN CANCER

It is impossible to detail all of the impacts that MS has had on cancer research in this chapter. The various impacts are better detailed based on their application in relevant chapters (e.g., Chapter 7 on posttranslational modifications, etc.). Generally, however, the main impact that MS has had on cancer research is related to its data gathering capability. Throughout the chapters in this book, examples that contrast hypothesis-versus discovery-driven research are provided. Many of the results generating using MS in a discovery-driven experiment would simply not have been discovered using conventional hypothesis-driven approaches. Below, I describe two recent studies in which the ability to gather reams of proteomic data and compare it with other large datasets identified a drug target and circulating biomarkers for lung adenocarcinoma.

2.5.1   Identification of a Drug Target

A recent study screened almost 190,000 compounds to identify small molecule inhibitors of human lung adenocarcinoma cell lines that contained epidermal growth factor receptor (EGFR) or KRAS mutations. Several compounds that inhibited the growth of one or more of the four human lung adenocarcinoma cell lines tested more potently than nontumor cells were found (72). Four compounds that possessed strong structure–activity relationships and had chemical structures that differed from traditional kinase inhibitors were selected for further study. One of these four compounds, LCS-1, was tested on a 27 additional human lung adenocarcinoma cell lines, including some with *EGFR*, *ERBB2*, *KRAS*, *NRAS*, or *MAP2K4* mutations or amplification of *MET* or *TITF* (*NKX2-1*) (72). A total of 10 of the cell lines were ~10-fold more sensitive to the growth inhibitory effect of LCS-1 than normal human bronchial epithelial cells (a negative control). Finding a compound that inhibits cancer cell growth in itself is exciting; the next logical step is to find which biomolecule it interacts with to cause the desired effect.

To identify molecular targets of LCS-1, the data from two different approaches were integrated: affinity chromatography combined with MS and gene expression arrays to find pathways that were perturbed by compound treatment (73). The overall schema and abbreviated results are shown in Figure 2.11. The LCS-1 compound was covalently linked to Sepharose 6B beads, and cell lysates from LCS-1 responsive cells (H358 and H1975) were passed over different affinity columns. After washing, proteins bound to the columns were eluted and identified using LC–MS^2. The list of proteins bound to the column exceeded well over 100; an almost impossible number to attempt to validate. To narrow down the possibilities of LCS-1 targets, a second-affinity column was prepared using an LCS-1 (i.e., LCS-1.28) analog that possessed a very similar activity profile against human lung adenocarcinoma cell lines. Extracts from H358 and H1975 were passed over this affinity column and bound proteins identified using LC–MS^2. Cross validation found that 98 proteins bound to both the LCS-1 and LCS-1.28 affinity columns. As a negative control, proteins that bound to a gefitinib affinity column were subtracted from the LCS-1 and LCS-1.28 datasets. This analysis left 34 proteins as potential molecular targets of LCS-1.

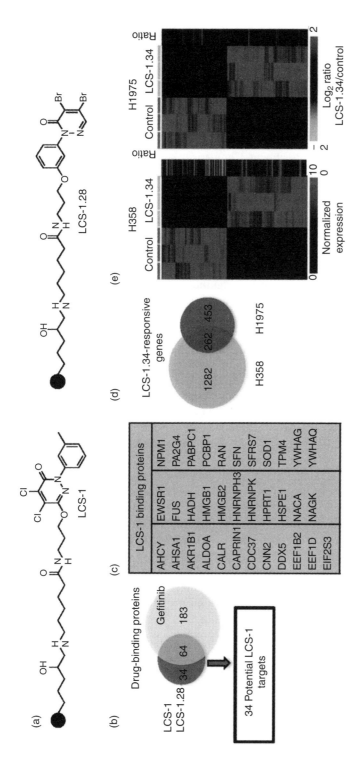

Figure 2.11. Two 2-phenylpyridazin-3(2H)-one analogs (LCS-1 and LCS-1.28) were found among a screening of almost 190,000 compounds to inhibit human lung adenocarcinoma cell line growth. (a) These compounds were immobilized to beads and used to isolate binding partners in cell extracts of lung adenocarcinoma cells. (b) Thirty four potential protein targets were found to bind to these compounds that did not bind to the negative control (gefitinib column). (c) These 34 proteins that were considered candidate targets of LCS-1. (d) Two hundred and sixty two genes were found to be regulated in two cell lines (H358 and H1975) treated with LCS-1. (e) The gene expression data were used to generate heatmaps that show the 262 LCS-1 genes within the signature from H358 and H1975 cells. Only SOD-1 was found both within the gene signature and among the 34 proteins that bound to the LCS-1 and LCS-1.28 affinity columns. (Reproduced with permission from Ref. 73.)

47

While 34 proteins is a significant reduction in the number of existing possibilities, it is still a challenging number to validate through individual experiments. Therefore, the study's investigators used an ingenious approach combining proteomics and genomics data (something that is too rarely done in current translational science). The H358 and H1975 cells were treated with LCS-1.34, an active derivative of LCS-1 that has greater stability. A total of 1,544 and 715 genes were altered in the H358 and H1935 cells, respectively, by LCS-1.34 treatment. Comparison of gene expression and affinity proteomics data showed only one protein, superoxide dismutase 1 (SOD1) was common to both datasets.

Now that the results from the discovery studies point toward a single molecular target, it was time to move into a hypothesis-driven mode to validate the finding. The investigators tested to determine whether there was any relationship between sensitivity to LCS-1 and *SOD1* expression by analyzing 18 lung adenocarcinoma cell lines for which *SOD1* expression and sensitivity to LCS-1 had been acquired. Tumor cell lines that were sensitive to LCS-1 treatment had significantly lower levels of *SOD1* RNA compared with those considered insensitive. The mean SOD1 RNA levels in sensitive cell lines were approximately half of that found in insensitive lines. SOD1 protein levels also showed a similar inverse correlation to LCS-1 sensitivity. Overall, the results of this study suggest that LCS-1 sensitivity is determined by SOD1 levels in that cells with higher SOD1 levels show less response to LCS-1 treatment.

This study exhibits many of the hallmarks of how MS can impact cancer research. It is rare that a single proteomics study that utilizes MS in a discovery-driven mode will provide an obvious, high impact, result. Most often, results from additional studies (and replicates) need to be used as a comparison to find the most interesting result. The above-mentioned example illustrates this principle very well as finding a drug's molecular target within a cell that contains at least 10,000 unique proteins is challenging as many proteins will have some affinity for any small molecule being tested. Testing each individual protein's affinity to LSC-1 would have been impossible due to a lack of time, money, and reagents. Therefore, the most efficient way to conduct these studies was to collect lots of data using different experimental variants and look for commonalities. The results of the individual affinity experiments provided an expected result: hundreds of proteins bound to the small molecule bait. This result is illustrative of one of the biggest challenges in proteomics research; finding the positive result within a background of negative results. I liken this exercise as "finding a needle in a haystack" or "separating the wheat from the chaff". By looking at the complex results from several different affinity experiments and gene expression data, the preponderance of the evidence suggested that SOD1 was the target of LSC-1.

2.5.2   Global Profiling for Identifying Cancer Biomarkers

One of the biggest impacts that MS and separations can have on cancer is the discovery of cancer-related proteins, whether they be druggable targets or biomarkers. While MS has tremendous data-gathering capabilities, in most discovery studies, it is unbiased in what it collects and lacks the quantitative accuracy to make definitive conclusions based on data acquired in a single experiment. For these reasons, data from multiple experiments and models are often compared to discriminate important results from background noise. If multiple models are going to be required, the sample source needs to be critically evaluated. For example, obtaining highly reproducible samples from humans with cancer that are not "contaminated" by population variation is challenging. Therefore, as illustrated in the example below, conducting discovery studies using mouse models is a popular choice.

To identify plasma protein signatures that reflect cell lineages of lung cancer, or protein pathways that direct tumor development, plasma proteins from tumor-bearing mice and age-matched littermate controls were subjected to quantitative profiling using stable isotope labeling, fractionation, and MS (74). Plasma was collected at various time points after doxycycline treatment of *TetO-EGFR*^{L858R}*/CCSP-rtTA* (Lung-EGFR) (75) and *TetO-Kras4b*^{G12D}*/CCSP-rtTA* (Lung-Kras) (76) lung adenocarcinoma mouse models as well as a urethane-treated mice that had multiple adenomas and adenocarcinomas containing frequent Kras and Trp53 mutations (77). Plasma was also collected from a small-cell-lung-cancer (Lung-SCLC) mouse model that develops neuroendocrine tumors (78).

The entire global proteomics effort entailed 14 profiling experiments with the collection of ~13 million spectra (74). A total of 5361 unique proteins were identified, of which concentration ratios could be measured for 2261 of these. The resulting proteomic data from the four lung cancer mouse models and controls were compared with data previously obtained of plasma from other mouse models of cancer, including breast, prostate, colon, ovarian, and pancreatic. The plasma proteomes of inflammatory disease models were analyzed to aid in the recognition of proteins found within the tumor models that were related to general inflammation.

Again, it is very challenging to gain specific insight into any cancer trying to make 2261 pieces of data coalesce into a clear picture. The investigators initially filtered the data by only considering proteins that were significantly elevated in mice bearing lung adenocarcinoma tumors or only identified in tumor-bearing mice from at least two of the three lung adenocarcinoma models. Proteins found in the other mouse models were then removed from this list. This strategy yielded 13 proteins (Table 2.3). An additional 16 proteins were found to be differentially abundant in multiple cancer, but not confounder mouse models, and may be potentially useful as markers of broad epithelial tumors. One of these proteins, Wfdc2, had previously been associated with ovarian cancer (79), but was not known to be elevated in lung cancer plasmas (80).

The progression in this study was to quantitatively measure the proteomes of 21 human lung adenocarcinoma cell lines using LC–MS^2 to identify intracellular, cell surface, and secreted proteins. Twenty-five and 26 proteins found at increased levels in plasmas from mice with lung adenocarcinoma were identified in the conditioned media and cell surface compartment of the adenocarcinoma cell lines, respectively. Twenty-one proteins were more than fivefold enriched in the conditioned media compared with the intracellular compartment. These data suggested that lung cancer cells can secrete proteins related to adenocarcinoma into the plasma.

Two of the secreted proteins that were elevated in lung adenocarcinoma plasma profiles, Sftpb and Sftpd, are under the transcriptional control of Titf1/Nkx2-1 (81, 82). Using this piece of evidence, the investigators sought other targets of this transcriptional

TABLE 2.3. List of Plasma Proteins Found in Higher Abundance in Tumor-Bearing Mice from the Lung Adenocarcinoma Models

	Lung Mouse Models			
Gene	Kras	EGFR	Urethane	SCLC
Sftpb	3.41	4.83	1.99	–
Morc3	10	*a*	10	–
Fgfr2	*a*	9.8	–	–
Adam10	1.5	7.85	–	–
Man2b2	1.46	4.22	–	–
Tfpi2	*a*	10	–	–
Son	10	*a*	–	–
Eif2ak3	9.03	*a*	–	–
Hamp2	3.07	*a*	–	–
Ncan	1.45	*a*	–	–
Sftpd	–	10	10	–
Mocs1	–	1.71	1.3	–
Nup188	*a*	–	10	–
Igsf4a	1.48	3.55	1.55	3.32
Ppbp	1.7	1.63	5.12	1.39
Prtg	10	10	–	10

The measurements are based on the case–control ratio determined from the peak area obtained from labeling cysteine-containing peptides with heavy and light acrylamide isotopes.

[a]Protein only identified in plasma from tumor-bearing mice.

(–) Quantitative ratio between case and control was not calculated.

regulator in the plasma proteome of lung adenocarcinoma models. To further delineate a Titf1/Nkx2-1 plasma proteome signature, mRNA microarray data from a set of 111 human non-small cell lung carcinoma tumors, including both adenocarcinomas and squamous cell carcinomas, was examined (83). To do this, they examined the mRNA microarray data from 111 human tumors and identified four orthologous genes that encoded proteins that exhibited increased levels in the murine lung adenocarcinoma plasmas (*SFTPB, SFTPD, NPC2,* and *WFDC2*) and showed a strong positive correlation with *TITF1/NKX2-1* mRNA levels. The cholesterol-transfer protein, *NPC2*, also showed a strong correlation with *TITF1/NKX2-1* mRNA levels. This protein has been implicated in Niemann–Pick disease (84), but never shown as elevated in lung cancer plasma. Immunohistochemistry analysis of mouse lung cancer tissue showed positive expression for Npc2, Sftpb, and Titf1/Nkx2-1. Reexamination of the proteomics data collected from the media of the 21 lung adenocarcinoma cell lines also showed identification of NPC2.

To elucidate the role of Titf1/Nkx2-1 in regulating protein expression in lung cancer models, the study also examined the effects of inhibiting Titf1/Nkx2-1 in two lung adenocarcinoma cell lines that expressed high levels of TITF1/NKX2-1 via gene expression analysis. Of the 964 potential TITF1/NKX2-1-regulated genes, 34 of the protein products were identified in the previously acquired plasma data, including Sftpa1, Sftpb,

Sftpd, and Npc2. The average plasma abundance ratios of these proteins in lung adeno-carcinoma tumor-bearing mice compared with controls were significantly higher than plasma ratios for other mouse models (mean ± standard deviation: 2.28 ± 0.07 vs. 1.20 ± 0.37, respectively). This evidence further supported the existence of a group of Titf1/Nkx2-1-regulated proteins in plasmas of mice with lung adenocarcinoma.

While finding disease-related biomarkers in mouse models is quite an accomplish-ment, the ultimate prize is discovering them in human samples. After distilling the overall results to a group of interesting proteins, the scientists used ELISAs to quantitate the amount of several proteins, including SFTPB and WFDC2, in plasma samples acquired from 28 newly diagnosed human smokers with operable NSCLC and matched control subjects. For the group of newly diagnosed subjects, SFTPB and WFDC2 protein levels were elevated with statistical significance in subjects with lung cancer in agreement with the mouse model results. These proteins, along with two other proteins (EGFR and ANGPTL3), were measured in 26 samples obtained from individuals 0–11 months prior to their diagnosis of NSCLC. A receiver operating characteristic analysis of the panel resulted in an area under the curve of 0.808, suggesting this panel was useful at diagnosing NSCLC prior to current clinical detection.

This study provides an excellent example of what it is going to take to find novel biomarkers for diseases. MS was a major part of the entire study and reams of proteomic data was collected. This ability is becoming standard in most laboratories as MS tech-nology has become more powerful and user friendly. The proverbial "smoking gun" was not found in the mouse plasma data even after the acquisition of $\sim$13 million spectra. The "smoking gun" was not found by simply applying sophisticated bioinformatic tools to the MS data. Much like trying to solve a crime, the MS data provide leads that require additional pieces of evidence to clarify exactly which proteins are critical to the disease being studied. In the previous study, the MS data led to a series of cell line experiments, mRNA microarray measurements, and finally targeted ELISAs that tested only a few proteins that continued to be supported by evidence through the various experiments. The progression of results through cell lines, mouse models, and finally human samples illustrates a sound and reasonable approach in accumulating enough circumstantial evi-dence to make solid conclusions on the value of specific biomarkers discovered initially using MS.

2.6 CONCLUSIONS

While proteomics is not limited to one instrumental platform, there is no denying that the growth of proteomics is directly tied to the growth of faster, more sensitive mass spectrometers. While they have been around for a century, the development of mass spectrometers has skyrocketed over the past 20 years. Mass spectrometers have quickly risen from that "large box in the basement of the chemistry department that only one person knew how to operate" to an analytical device that is in constant demand by chemists and biochemists alike. While mass spectrometers have been a driving force, these instruments could not have taken proteomics to where it is today without chromatography. Analyzing complex samples such as cell lysates, tissues, blood, and urine could not be achieved without first fractionating their proteins prior to introducing them into the mass spectrometer. Arguing the relative importance of either to the field of proteomics is fruitless as it is the combination of MS and chromatography that has allowed investigators to even consider experiments such as discovering biomarkers for human diseases. The two examples provided at the end of the chapter illustrate two very important concepts: (1) finding an important result within the complex mixtures that comprise proteomes requires accumulating a lot of data using various different experimental models and (2) analyzing data acquired at the various "omic" levels (i.e., genomic, transcriptomic, proteomic, and metabolomic) is critical for solidifying the evidence of any possible drug target or biomarker that is discovered.

REFERENCES

1. van der Vaart M, Semenov DV, Kuligina EV, Richter VA, Pretorius PJ. Characterisation of circulating DNA by parallel tagged sequencing on the 454 platform. Clin. Chem. Acta 2009;409:21–27.

2. Frullanti E, Galvan A, Falvella FS, Manenti G, Colombo F, Vannelli A, Incarbone M, Alloisio M, Nosotti M, Santambrogio L, Gonzalez-Neira A, Pastorino U, Dragani TA. Multiple genetic loci modulate lung adenocarcinoma clinical staging. Clin. Cancer Res. 2011;17:2410–2416.

3. Boffetta P. Biomarkers in cancer epidemiology: An integrative approach. Carcinogenesis 2010;31:121–126.

4. Jotwani AC, Gralow JR. Early detection of breast cancer: New biomarker tests on the horizon? Mol. Diagn. Ther. 2009;13:349–357.

5. Roepman P, Horlings HM, Krijgsman O, Kok M, Bueno-de-Mesquita JM, Bender R, Linn SC, Glas AM, van de Vijver MJ. Microarray-based determination of estrogen receptor, progesterone receptor, and HER2 receptor status in breast cancer. Clin. Cancer Res. 2009;15:7003–7011.

6. Trang P, Weidhaas JB, Slack FJ. MicroRNAs as potential cancer therapeutics. Oncogene 2008;27:S52–S57.

7. Cortez MA, Bueso-Ramos C, Ferdin J, Lopez-Berestein G, Sood AK, Calin GA. MicroRNAs in body fluids—The mix of hormones and biomarkers. Nat. Rev. Clin. Oncol. 2011;8:467–477.

8. Yendamuri S, Kratzke R. MicroRNA biomarkers in lung cancer: MiRacle or QuagMiRe? Transl. Res. 2011;157:209–215.

9. Fenn JB, Mann M, Meng CK, Wong SF. Electrospray ionization for mass spectrometry of large biomolecules. Science 1989;246:64–71.

10. Karas M, Bachmann D, Bahr U, Hillenkamp F. Matrix-assisted ultraviolet laser desorption of non-volatile compounds. Int. J. Mass Spectrom. Ion Process. 1987;78:53–68.

11. Dole M, Ferguson LD, Hines RL, Mobley RC, Ferguson LD, Alice MB. Molecular beams of macroions. J. Chem. Phys. 1968;49:2240–2249.

12. Kebarle P, Verkerk UH. Electrospray: From ions in solution to ions in the gas phase, what we know now. Mass Spectrom. Rev. 2009;28:898–917.

13. Cañas B, López-Ferrer D, Ramos-Fernández A, Camafeita E, Calvo E. Mass spectrometry technologies for proteomics. Brief Funct. Genomic Proteomic 2006;4:295–320.

14. Hillenkamp F, Karas M, Beavis RC, Chait BT. Matrix-assisted laser desorption/ionization mass spectrometry of biopolymers. Anal. Chem. 1991;63:1193A–1203A.

15. Aebersold RA, Mann M. Mass spectrometry-based proteomics. Nature 2003;422:198–207.

16. Schuchardt S, Sickmann A. Protein identification using mass spectrometry: A method overview. EXS 2007;97:141–170.

17. O'Farrell PH. High resolution two-dimensional electrophoresis of proteins. J. Biol. Chem. 1975;250:4007–4021.

18. Stafford GC, Kelley PE, Syka JEP, Reynolds WE, Todd JFJ. Recent improvements in and analytical applications of advanced ion-trap technology. Int. J. Mass Spectrom. Ion Process. 1984;60:85–98.

19. Martin SA, Rosenthal RS, Biemann K. Fast atom bombardment mass spectrometry and tandem mass spectrometry of biologically active peptidoglycan monomers from *Neisseria gonorrhoeae*. J. Biol. Chem. 1987;262:7514–7522.

20. Villen J, Beausoleil SA, Gygi SP. Evaluation of the utility of neutral-loss-dependent MS^3 strategies in large-scale phosphorylation analysis. Proteomics 2008;8:4444–4452.

21. Gatlin CL, Eng JK, Cross ST, Detter JC, Yates JR 3rd. Automated identification of amino acid sequence variations in proteins by HPLC/microspray tandem mass spectrometry. Anal. Chem. 2000;72:757–763.

22. Michalski A, Cox J, Mann M. More than 100,000 detectable peptide species elute in single shotgun proteomics run but the majority is inaccessible to data-dependent LC–MS/MS. J. Proteome Res. 2011;10:1785–1793.

23. Nagaraj N, D'Souza RC, Cox J, Olsen JV, Mann M. Feasibility of large-scale phospho-proteomics with higher energy collisional dissociation fragmentation. J. Proteome Res. 2010;9:6786–6794.

24. Comisarow MB, Marshall AG. Fourier transform ion cyclotron resonance spectroscopy. Chem. Phys. Lett. 1974;25:282–283.

25. Marshall AG, Hendrickson CL, Jackson GS. Fourier transform ion cyclotron resonance mass spectrometry: a primer. Mass Spectrom. Rev. 1998;17:1–35.

26. Makarov A. Electrostatic axially harmonic orbital trapping: a high-performance technique of mass analysis. Anal. Chem. 2000;72:1156–1162.

27. Makarov A, Denisov E, Kholomeev A, Balschun W, Lange O, Strupat K, Horning S. Performance evaluation of a hybrid linear ion trap/orbitrap mass spectrometer. Anal. Chem. 2006;78:2113–2120.

28. Manes NP, Dong L, Zhou W, Du X, Reghu N, Kool AC, Choi D, Bailey CL, Petricoin EF 3rd, Liotta LA, Popov SG. Discovery of mouse spleen signaling responses to anthrax using label-free quantitative phosphoproteomics via mass spectrometry. Mol. Cell. Proteomics 2011;10:M110.000927.

29. Olthoff JK, Lys IA, Cotter RJ. A pulsed time-of-flight mass spectrometer for liquid secondary ion mass spectrometry. Rapid Commun. Mass Spectrom. 1988;2:171–175.

30. Ahmed FE. Utility of mass spectrometry for proteome analysis: Part I. Conceptual and experimental approaches. Expert Rev. Proteomics 2008;5:841–864.

31. Kammerer B, Frickenschmidt A, Gleiter CH, Laufer S, Liebich H. MALDI–TOF MS analysis of urinary nucleosides. J. Am. Soc. Mass Spectrom. 2005;16:940–947.

32. Yates JR 3rd. Mass spectrometry and the age of the proteome. J. Mass Spectrom. 1998;33: 1–19.

33. Kim H, Eliuk S, Deshane J, Meleth S, Sanderson T, Pinner A, Robinson G, Wilson L, Kirk M, Barnes S. 2D gel proteomics: An approach to study age-related differences in protein abundance or isoform complexity in biological samples. Methods Mol. Biol. 2007;371:349–391.

34. Medzihradszky KF, Campbell JM, Baldwin MA, Falick AM, Juhasz P, Vestal ML, Burlingame AL. The characteristics of peptide collision-induced dissociation using a high-performance MALDI–TOF/TOF tandem mass spectrometer. Anal. Chem. 2000;72:552–558.

35. Kaufmann R, Chaurand P, Kirsch D, Spengler B. Post-source decay and delayed extraction in matrix-assisted laser desorption/ionization–reflectron time-of-flight mass spectrometry. Are there trade-offs? Rapid Commun. Mass Spectrom. 1996;10:1199–1208.

36. Yost RA, Boyd RK. Tandem mass spectrometry: Quadrupole and hybrid instruments. Methods Enzymol. 1990;193:154–200.

37. Prakash A, Tomazela D, Frewen B, Merrihew G, Maclean B, Peterman S, MacCoss M. Expediting the development of targeted SRM assays: Using data from shotgun proteomics to automate method development. J. Proteome Res. 2009;8:2733–2739.

38. Gallien S, Duriez E, Domon B. Selected reaction monitoring applied to proteomics. J. Mass Spectrom. 2011;46:298–312.

39. Shevchenko A, Chernushevic I, Shevchenko A, Wilm M, Mann M. "De novo" sequencing of peptides recovered from in-gel digested proteins by nanoelectrospray tandem mass spectrometry. Mol. Biotechnol. 2002;20:107–118.

40. Kuhn E, Wu J, Karl J, Liao H, Zolg W, Guild B. Quantification of C-reactive protein in the serum of patients with rheumatoid arthritis using multiple reaction monitoring mass spectrometry and ^{13}C-labeled peptide standards. Proteomics 2004;4:1175–1186.

41. Kocher T, Allmaier G, Wilm M. Nanoelectrospray-based detection and sequencing of substoichiometric amounts of phosphopeptides in complex mixtures. J. Mass Spectrom. 2003;38:131–137.

42. Jiang H, Desaire H, Butnev VY, Bousfield GR. Glycoprotein profiling by electrospray mass spectrometry. J. Am. Soc. Mass Spectrom. 2004;15:750–758.

43. Issaq HJ, Veenstra TD. Would you prefer multiple reaction monitoring or antibodies with your biomarker validation? Expert Rev. Proteomics 2008;5:761–763.

44. Chernushevich IV, Loboda AV, Thomson BA. An introduction to quadrupole-time-of-flight mass spectrometry. J. Mass Spectrom. 2001;36:849–865.

45. Chu F, Mahrus S, Craik CS, Burlingame AL. Isotope-coded and affinity-tagged cross-linking (ICATXL): An efficient strategy to probe protein interaction surfaces. J. Am. Chem. Soc. 2006;128:10362–10363.

46. Miyagi M, Rao KC. Proteolytic ^{18}O-labeling strategies for quantitative proteomics. Mass Spectrom. Rev. 2007;26:121–136.

47. Wheeler CH, Berry SL, Wilkins MR, Corbett JM, Ou K, Gooley AA, Humphery-Smith I, Williams KL, Dunn MJ. Characterization of proteins from two-dimensional electrophoresis gels by matrix-assisted laser desorption mass spectrometry and amino acid composition analysis. Electrophoresis 1996;17:580–587.

48. Mi W, Liu X, Jia W, Cai Y, Ying W, Qian X. Toward a high resolution 2-DE profile of the normal human liver proteome using ultra-zoom gels. Sci. China Life Sci. 2011;54:25–33.

49. Steinberg TH. Protein gel staining methods: An introduction and overview. Methods Enzymol. 2009;463:541–563.

50. Laemmli UK. Cleavage of structural proteins during the assembly of the head of bacteriophage T4. Nature 1970;227:680–685.

51. Zhang WB, Wang L. Label-free quantitative proteome analysis of skeletal tissues under mechanical load. J. Cell Biochem. 2009;108:600–611.

52. Zhang K, Fischer T, Porter RL, Dhakshnamoorthy J, Zofall M, Zhou M, Veenstra T, Grewal SI. Clr4/Suv39 and RNA quality control factors cooperate to trigger RNAi and suppress antisense RNA. Science 2011;331:1624–1627.

53. Renart J, Reiser J, Stark GR. Transfer of proteins from gels to diazobenzyloxymethyl-paper and detection with antisera: A method for studying antibody specificity and antigen structure. Proc. Natl. Acad. Sci. U.S.A. 1979;76:3116–3120.

54. Burnette WN. "Western blotting": Electrophoretic transfer of proteins from sodium dodecyl sulfate–polyacrylamide gels to unmodified nitrocellulose and radiographic detection with antibody and radioiodinated protein A. Anal. Biochem. 1981;112:195–203.

55. Eckers C, Skrabalak DS, Henion J. On-line direct liquid introduction interface for micro-liquid chromatography/mass spectrometry: Application to drug analysis. Clin. Chem. 1982;28:1882–1886.

56. Chen G, Pramanik BN. Application of LC/MS to proteomics studies: Current status and future prospects. Drug Discov. Today 2009;14:465–471.

57. Johann DJ Jr, Wei BR, Prieto DA, Chan KC, Ye X, Valera VA, Simpson RM, Rudnick PA, Xiao Z, Issaq HJ, Linehan WM, Stein SE, Veenstra TD, Blonder J. Combined blood/tissue analysis for cancer biomarker discovery: Application to renal cell carcinoma. Anal. Chem. 2010;82:1584–1588.

58. Xie F, Liu T, Qian WJ, Petyuk VA, Smith RD. Liquid chromatography–mass spectrometry-based quantitative proteomics. J. Biol. Chem. 2011;286:25443–25449.

59. Washburn MP, Wolters D, Yates JR 3rd. Large-scale analysis of the yeast proteome by multidimensional protein identification technology. Nat. Biotechnol. 2001;19:242–247.

60. Liu H, Lin D, Yates JR 3rd. Multidimensional separations for protein/peptide analysis in the post-genomic era. Biotechniques 2002;32:898–902.

61. Yates JR, Ruse CI, Nakorchevsky A. Proteomics by mass spectrometry: Approaches, advances, and applications. Annu. Rev. Biomed. Eng. 2009;11:49–79.

62. Gonzalez-Begne M, Lu B, Liao L, Xu T, Bedi G, Melvin JE, Yates JR. Characterization of the human submandibular/sublingual saliva glycoproteome using lectin affinity chromatography coupled to multidimensional protein identification technology. J. Proteome Res. 2011;10:5031–5046.

63. Bonner MK, Poole DS, Xu T, Sarkeshik A, Yates JR 3rd, Skop AR. Mitotic spindle proteomics in Chinese hamster ovary cells. PLoS One 2011;6:e20489.

64. Zhu W, Fang C, Gramatikoff K, Niemeyer CC, Smith JW. Proteins and an inflammatory network expressed in colon tumors. J. Proteome Res. 2011;10:2129–2139.

65. Musselman I, Speicher DW. Human serum and plasma proteomics. Curr. Protoc. Protein Sci. 2005;Chapter 24:Unit 24.1

66. Lau E, Lam MP, Siu SO, Kong RP, Chan WL, Zhou Z, Huang J, Lo C, Chu IK. Combinatorial use of offline SCX and online RP–RP liquid chromatography for iTRAQ quantitative proteomics applications. Mol. Biosyst. 2011;7:1399–1408.

67. Xie H, Onsongo G, Popko J, de Jong EP, Cao J, Carlis JV, Griffin RJ, Rhodus NL, Griffin TJ. Proteomics analysis of cells in whole saliva from oral cancer patients via value-added three-dimensional peptide fractionation and tandem mass spectrometry. Mol. Cell. Proteomics 2008;7:486–498.

68. Selvaraju S, El Rassi Z. Reduction of protein concentration range difference followed by multi-column fractionation prior to 2-DE and LC–MS/MS profiling of serum proteins. Electrophoresis 2001;32:674–685.

69. Issaq HJ, Xiao Z, Veenstra TD. Serum and plasma proteomics. Chem. Rev. 2007;107:3601–3620.

70. Anderson NL, Anderson NG. The human plasma proteome: History, character, and diagnostic prospects. Mol Cell Proteomics 2002;11:845–867.

71. Tang HY, Beer LA, Speicher DW. In-depth analysis of a plasma or serum proteome using a 4D protein profiling method. Methods Mol. Biol. 2011;728:47–67.

72. Somwar R, Shum D, Djaballah H, Varmus H. Identification and preliminary characterization of novel small molecules that inhibit growth of human lung adenocarcinoma cells. J. Biomol. Screen. 2009;14:1176–1184.

73. Somwar R, Erdjument-Bromage H, Larsson E, Shum D, Lockwood WW, Yang G, Sander C, Ouerfelli O, Tempst PJ, Djaballah H, Varmus HE. Superoxide dismutase 1 (SOD1) is a target for a small molecule identified in a screen for inhibitors of the growth of lung adenocarcinoma cell lines. Proc. Natl. Acad. Sci. U.S.A. 2011;108:16375–16380.

74. Faca VM, Song KS, Wang H, Zhang Q, Krasnoselsky AL, Newcomb LF, Plentz RR, Gurumurthy S, Redston MS, Pitteri SJ, Pereira-Faca SR, Ireton RC, Katayama H, Glukhova V, Phanstiel D, Brenner DE, Anderson MA, Misek D, Scholler N, Urban ND, Barnett MJ, Edelstein C, Goodman GE, Thornquist MD, McIntosh MW, DePinho RA, Bardeesy N, Hanash SM. A mouse to human search for plasma proteome changes associated with pancreatic tumor development. PLoS Med. 2008;5:e123.

75. Politi K, Zakowski MF, Fan PD, Schonfeld EA, Pao W, Varmus HE. Lung adenocarcinomas induced in mice by mutant EGF receptors found in human lung cancers respond to a tyrosine kinase inhibitor or to down-regulation of the receptors. Genes Dev. 2006;20:1496–1510.

76. Fisher GH, Wellen SL, Klimstra D, Lenczowski JM, Tichelaar JW, Lizak MJ, Whitsett JA, Koretsky A, Varmus HE. Induction and apoptotic regression of lung adenocarcinomas by regulation of a K-ras transgene in the presence and absence of tumor suppressor genes. Genes Dev. 2001;15:3249–3262.

77. Horio Y, Chen A, Rice P, Roth JA, Malkinson AM, Schrump DS. Ki-ras and p53 mutations are early and late events, respectively, in urethane-induced pulmonary carcinogenesis in A/J mice. Mol. Carcinog. 1996;17:217–223.

78. Meuwissen R, Linn SC, Linnoila RI, Zevenhoven J, Mooi WJ, Berns A. Induction of small cell lung cancer by somatic inactivation of both Trp53 and Rb1 in a conditional mouse model. Cancer Cell 2003;4:181–189.

79. Hellström I, Raycraft J, Hayden-Ledbetter M, Ledbetter JA, Schummer M, McIntosh M, Drescher C, Urban N, Hellström KE. The HE4 (WFDC2) protein is a biomarker for ovarian carcinoma. Cancer Res. 2003;63:3695–700.

80. Bingle L, Cross SS, High AS, Wallace WA, Rassl D, Yuan G, Hellström I, Campos MA, Bingle CD. WFDC2 (HE4): A potential role in the innate immunity of the oral cavity and respiratory tract and the development of adenocarcinomas of the lung. Respir. Res. 2006;7:61–70.

81. Tanaka H, Yanagisawa K, Shinjo K, Taguchi A, Maeno K, Tomida S, Shimada Y, Osada H, Kosaka T, Matsubara H, Mitsudomi T, Sekido Y, Tanimoto M, Yatabe Y, Takahashi T. Lineage-specific dependency of lung adenocarcinomas on the lung development regulator TTF-1. Cancer Res. 2007;67:6007–6011.

82. Weir BA, Woo MS, Getz G, Perner S, Ding L, Beroukhim R, Lin WM, Province MA, Kraja A, Johnson LA, Shah K, Sato M, Thomas RK, Barletta JA, Borecki IB, Broderick S, Chang AC, Chiang DY, Chirieac LR, Cho J, Fujii Y, Gazdar AF, Giordano T, Greulich H, Hanna M, Johnson BE, Kris MG, Lash A, Lin L, Lindeman N, Mardis ER, McPherson JD, Minna JD, Morgan MB, Nadel M, Orringer MB, Osborne JR, Ozenberger B, Ramos AH, Robinson J, Roth JA, Rusch V, Sasaki H, Shepherd F, Sougnez C, Spitz MR, Tsao MS, Twomey D, Verhaak RG, Weinstock GM, Wheeler DA, Winckler W, Yoshizawa A, Yu S, Zakowski MF, Zhang Q, Beer DG, Wistuba II, Watson MA, Garraway LA, Ladanyi M, Travis WD, Pao W, Rubin MA, Gabriel SB, Gibbs RA, Varmus HE, Wilson RK, Lander ES, Meyerson M. Characterizing the cancer genome in lung adenocarcinoma. Nature 2007;450:893–898.

83. Bild AH, Yao G, Chang JT, Wang Q, Potti A, Chasse D, Joshi MB, Harpole D, Lancaster JM, Berchuck A, Olson JA Jr, Marks JR, Dressman HK, West M, Nevins JR. Oncogenic pathway signatures in human cancers as a guide to targeted therapies. Nature 2006;439:353–357.

84. Vanier MT, Millat G. Structure and function of the NPC2 protein. Biochim. Biophys. Acta 2004;1685:14–21.

3

QUANTITATIVE PROTEOMICS

3.1 INTRODUCTION

There are generally two parameters to any measurement: magnitude and dimension. Both are critical to understand just about every measurement imaginable. Imagine if you had a recipe book that told you what ingredients to put in a dish, but not the amount of each. What would you do if you were driving along a road that had speed limit signs that only displayed the units of speed? How would you feel about your doctor if he/she measured your heart rate and told you that your heart was beating but did not tell you the number of times per minute? Proteomics is very similar. The initial goal of proteomics was to measure all of the proteins within a cell, tissue, or organism. But what necessitates measure? Is simply identifying the presence of a protein enough? Sometimes it is. However, since proteins interact with other proteins and respond to stimuli, simply knowing a protein is present does not shed much light on what it is doing.

Beyond knowing the identities of the proteins present in a proteome, the next level of information most commonly measured is their abundance. Since the proteome is dynamic, measuring a protein's abundance can indicate whether it is affected by external stimuli or by the presence of a disease state. Almost every aspect of proteomics measurement relies at least in part on measuring protein abundances. The attempt to find

Proteomic Applications in Cancer Detection and Discovery, First Edition. Timothy D. Veenstra.
© 2013 John Wiley & Sons, Inc. Published 2013 by John Wiley & Sons, Inc.

protein biomarkers is primarily predicated on measuring differences in the abundance of proteins in clinical samples taken from control individuals (i.e., healthy) and disease-affected patients. Identifying cellular proteins that respond to a specific treatment is based on measuring abundance differences between untreated and treated cell lines. Even recognizing specifically and nonspecifically bound proteins within a complex relies on quantitating protein abundances. The problem, however, is that the primary technology for proteomics (mass spectrometry; MS) is not inherently quantitative. If it were, every peak in a tryptic digest from a single protein would have the same height or peak area in a MS spectrum, whether it be recorded using electrospray ionization or matrix-assisted laser desorption/ionization (MALDI). In reality, the various peptides produce peaks of various intensities and many do not even produce a signal. The harsh reality is that the ionization efficiency of proteins and peptides widely varies and is unpredictable. Beyond this parameter, the matrix surrounding the protein or peptide of interest has a major impact on the signal it produces. With these inherent limitations, investigators have had to use creative means to exhume quantitative information from MS data.

3.2   WHAT IS BEING MEASURED IN QUANTITATIVE PROTEOMICS?

Prior to discussing the techniques and applications of quantitative proteomics, it needs to be clarified exactly what is being measured in a quantitative proteomic experiment. The quantitative measurement provided in a large majority of experiments is relative, not absolute. In a relative measurement, the abundance of a protein is reported in reference to that of the same protein in a different sample. In an absolute measurement, the exact concentration of a specific protein is reported and is not dependent on the result obtained from another sample. Most quantitative proteomics data, whether these are acquired using MS or protein arrays, are reported as relative amounts similar to gene expression profiling. To measure the absolute abundance of a protein, known amounts of internal standards must be added to the sample or external calibration curves need to be generated. The following section will discuss strategies used to quantitate proteomes and illustrate how these data are, or has been, used in cancer research.

3.3   TWO-DIMENSIONAL POLYACRYLAMIDE GEL ELECTROPHORESIS

Developed in the early 1970s, two-dimensional polyacrylamide gel electrophoresis (2D-PAGE) (1) was the first quantitative method used in modern proteomics (2). Once the two technologies had matured to a necessary level, combining 2D-PAGE with MS was an obvious match. 2D-PAGE brought two key attributes to proteomics: the ability to (1) fractionate thousands of proteins and (2) quantitate the relative abundances of proteins between proteomes by comparing their stained intensities on different gels. MS complemented the relationship through its ability to identify the proteins. While it may not be used to the extent to what it once was, it is difficult to perceive proteomics being where it is today without 2D-PAGE. The complexity of even the simplest proteome

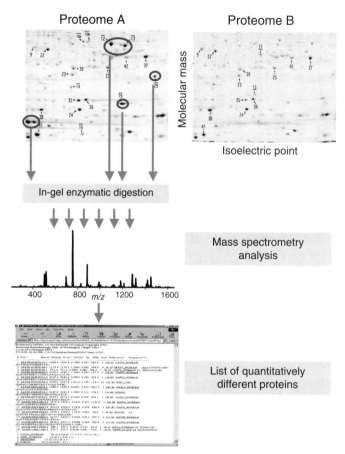

Figure 3.1. Quantitative proteomics using two-dimensional polyacrylamide gel electrophoresis.

requires fractionation to enable the mass spectrometer to confidently identify proteins within the sample.

The concept of using 2D-PAGE for quantitative proteomics is very straightforward (Figure 3.1). The proteome sample extracted from a biological sample is fractionated based on isoelectric point (pI) and molecular mass (M_r) in the first and second dimensions, respectively. Each sample is separated on its own gel, although as described later, newer technologies enable running different proteomes on a single gel. After the proteins are resolved, the gels are stained typically using Coomassie blue or silver stain. The positions of the protein spots are aligned enabling their staining intensity to be synchronized across the gels. Protein spots that show a differential staining on one or more gels compared to a reference (i.e., control) gel are cored from the gel. The protein is in-gel digested, typically using trypsin, and identified using MALDI–MS or LC–MS[2]. Although high resolution 2D-PAGE gels can contain thousands of visualized spots, only those that show a quantitative difference between proteomes need to be selected for MS analysis.

Not unlike most areas of proteomics (and genomics for that matter), the trend in 2D-PAGE over the years has been to increase the number of proteins that can be resolved providing higher proteome coverage. This trend has included the use of solubilization agents and extraction methods enabling the analysis of membrane proteins (3, 4) and overlapping narrow pH gradients that resolve proteins within specific pI ranges with greater resolution (5,6). Investigators have also prefractionated proteome samples using chromatographic methods that are not based on protein pI or M_r and applied each fraction to separate gels (7, 8). While this increases the number of protein spots that can be detected, it also increases the overall complexity of the study by increasing the number of gels that need to be run and analyzed.

While 2D-PAGE has been a foundational technology, it started to gain a reputation of irreproducibility (9) and too labor intensive (10), as newer fractionation methods entered the proteomics scene. It is not that 2D-PAGE is horribly irreproducible; however, attempting to locate a specific spot between multiple gels showing a complex pattern of proteins requires high precision. Add to the fact that some of the proteins spots will vary in intensity and it is easy to see why gel alignment was such an issue. Much of the alignment issue was overcome via software solutions that could generate "landmark" proteins within a reference gel and align the corresponding spots on comparative gels (11, 12). The laboriousness issue was related to the need to run multiple gels for each study, however, equipment for running multiple gels simultaneously have been developed. Frankly, the laboriousness of 2D-PAGE was never that much more considerable than other fractionation techniques, which could only fractionate a single sample per run.

For proteins to be visualized on a 2D-PAGE gel, they must be stained or labeled with a fluorophore (see below). The most commonly used stains are Coomassie blue and silver stain (13). The choice of either depends on the amount of material loaded onto the gel and the desire to visualize lower abundance proteins. The most important issue that nonproteomic scientists need to be aware of is that using silver stains that require fixatives such as glutaraldehyde or formaldehyde are not compatible with MS identification. There are a large number of commercial available Coomassie and silver stains that are MS compatible.

In the field of quantitative proteomics, 2D-PAGE acts as a discovery tool to identify proteins whose abundance is associated with some difference between comparative samples. In cancer research, the major application of 2D-PAGE has been the discovery of biomarkers. In a recent study, a search for prognostic markers for colorectal cancer was conducted using 2D-PAGE analysis of fresh frozen tissue sections of 28 paired Dukes B colorectal cancer and normal colorectal mucosa (14). Approximately 1200 unique spots were resolved on the gels. Comparison of the protein expression profiles of the gels using hierarchical cluster analysis and principal component analysis showed that the colorectal cancer and normal colonic mucosa clustered into distinct patterns of protein expression. Forty-five proteins that showed a minimum 1.5 times increased expression in colorectal cancer were identified using LC–MS^2. Now that 2D-PAGE and MS had identified possible prognostic biomarker candidates, the next step is to validate these findings. Fifteen of potential biomarkers were validated by

immunohistochemistry (IHC) of 515 primary colorectal cancer, 224 lymph node metastasis, and 50 normal colonic mucosal samples using a tissue microarray. Heat shock
protein 60 ($p < 0.001$), S100A9 ($p < 0.001$), and translationally controlled tumor
protein ($p < 0.001$) showed the highest overexpression in primary colorectal cancer
compared with normal colonic mucosa. Analysis of proteins individually identified 14-
3-3β as a prognostic biomarker ($\chi^2 = 6.218$, $p = 0.013$). Hierarchical cluster analysis
identified distinct phenotypes associated with survival, and a two-protein signature consisting of 14-3-3β and aldehyde dehydrogenase 1 was identified as showing prognostic
significance ($\chi^2 = 7.306$, $p = 0.007$, HR $= 0.504$, 95% CI 0.303–0.838) associated
with survival.

In another recent 2D-PAGE study, a search for prognostic and therapeutic markers for lung cancer was conducted by comparing the proteomes of human pulmonary
adenocarcinoma tissue and paired surrounding normal tissue (15). Thirty-two proteins
that showed a minimum twofold abundance difference between pulmonary adenocarcinoma and normal tissues were identified using LC–MS^2. Two of these proteins (PKM2
and cofilin-1) were further validated in a series of lung adenocarcinoma tissues using
IHC, and their abundance was found to be correlated with the severity of epithelial
dysplasia and relatively poor prognosis. To further validate their association, PKM2
expression was knocked down in SPC-A1 cells *in vitro* and in a xenograft model
in vivo. A significant decrease in cell proliferation and increase in apoptosis were
observed in SPC-A1 cells, while the xenograft showed a significant decrease in tumor
growth. Knocking down cofilin-1 expression in a LL/2 metastatic mouse model significantly inhibited tumor metastases and prolonged survival. This study, along with
the colon cancer study, illustrates the effectiveness of 2D-PAGE in finding potential
leads that increase our understanding of cancer and may lead to useful biomarkers in
the future.

3.4 TWO-DIMENSIONAL DIFFERENCE GEL ELECTROPHORESIS

As mentioned above, variability between gels has been one of the biggest criticisms of 2D-PAGE. To minimize variability, 2D-difference gel electrophoresis (DIGE)
was developed to allow multiple proteomes to be fractionated and visualized on a
single gel (16–18). Since the first report in 1997 (16), there have been approximately 1000 manuscripts published using 2D-DIGE, with the number growing every
year. In 2D-DIGE, three different proteome samples are labeled with three distinct fluorophores: 3-[(4-carboxymethyl)phenylmethyl]-3′-ethyloxacarbocyanine halide
N-hydroxysuccinimidyl ester (Cy2), 1-(5-carboxypentyl)-1′-propylindocarbocyanine
halide *N*-hydroxysuccinimidyl ester (Cy3), and 1-(5-carboxypentyl)-1′-methylindodi-
carbocyanine halide *N*-hydroxysuccinimidyl ester (Cy5) (19). The fluorophore dyes are
structurally similar and react with primary amine groups of Lys residues. The dyes are
positively charged to offset the loss of the charge on the Lys residue, which is modified
during the labeling reaction. In a typical experiment, two proteomes to be compared
are labeled with Cy3 and Cy5 dyes, while the control (which can be made up by combining equal amounts of the two proteomes being compared) is labeled with the Cy2

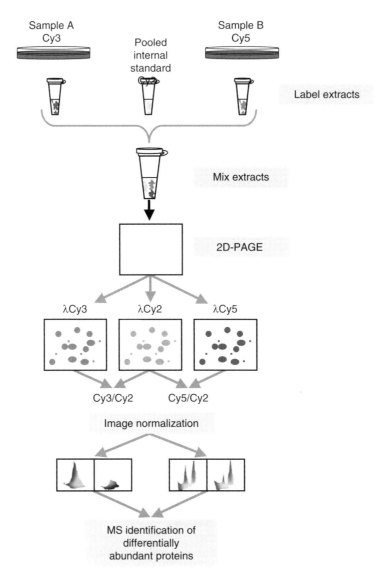

Figure 3.2. A schematic illustrating the use of two-dimensional difference gel electro-phoresis.

dye (Figure 3.2). After covalent labeling, the proteomes are combined and fractionated on a single 2D-PAGE gel. A laser scanning device or xenon-arc-based instrument with different excitation/emission filters are used to detect and quantitate the proteins within the separate proteomes. The fluorescent images are aligned using software and the protein spots quantified. A pseudo-colored image is used to recognize differences in protein abundances. As with normal 2D-PAGE, protein spots can be selected from the gel, in-gel digested, and identified using MS.

A recent study used 2D-DIGE to examine saliva for biomarkers of lung cancer. Early detection of cancers is critical in improving survival rates, and saliva is an attractive fluid owing to its noninvasive collection (20). In this study, equal amounts of proteins were pooled from saliva samples that were prepared from 10 lung cancer patients and 10 healthy controls. Approximately 200 μg of total protein in each saliva sample was labeled by Cy3 (lung cancer) and Cy5 (healthy controls) dyes. After combining the two labeled samples, they were analyzed using 2D-DIGE. Sixteen candidate protein biomarkers were identified by 2D-DIGE and MS to show an abundance difference between lung cancer patients and healthy controls. A series of immunoassays was performed using an additional set of saliva samples, and it was shown that three of the salivary proteins (HP, AZGP1, and human calprotectin) still showed a significant abundance difference in lung cancer patients and healthy controls, exhibiting p values of 1.48E-4, 1.05E-4, and 4.56E-5, respectively. An additional piece of evidence showed that the concentration of these three proteins was higher in two lung cancer cell lines (NCI-H460 and NCI-H1299) than healthy lung cells (MRC-5).

3.5 SOLUTION-BASED QUANTITATIVE METHODS

As scientists tend to do, they began looking for other methods to quantitate changes in proteomes. After John Yates published MudPIT (21), scientists sought to find ways to circumvent gels entirely but still obtain quantitative information. The quandary was that MS did not provide signals that could be directly correlated to the abundance of the molecules it was measuring. As scientists also tend to do, they found novel ways around this problem.

3.5.1 Stable Isotope Labeling

The premise of using isotope labeling for quantitative proteomics relies on the ability of a mass spectrometer to resolve species based on their m/z values. It is important to state here that only stable isotopes (i.e., ^{13}C, ^{15}N, etc.) are used in combination with MS; injecting radioactive isotopes into your mass spectrometer is not advisable. Similar to 2D-DIGE wherein differential fluorescence is used to measure protein relative abundances, stable isotope labeling labels two (or more) proteomes with reagents that are chemically and structurally identical but differ in their mass (19, 22). The different methods of isotope tagging can be crudely separated into two categories: chemical and metabolic labeling. A schematic illustrating the general process of conducting quantitative comparison of entire proteomes and where the various metabolic or chemical isotope labeling and fractionation steps are incorporated is shown in Figure 3.3. In chemical labeling, the isotope tags are reacted with the proteome sample after it has been extracted from a cell, tissue, or organism. The isotope tags are part of a synthetic molecule that covalently modifies a specific reactive group on the protein (i.e., amines, thiols, etc.). In metabolic labeling, the isotope tags are incorporated by the cell's own machinery during protein translation. This incorporation is usually accomplished by culturing cells in medium or feeding animals feed containing isotope-labeled amino acids (e.g., ^{13}C-labeled lysine).

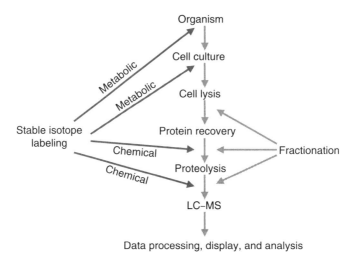

Figure 3.3. A schematic illustrating the quantitative comparison of entire proteomes and where metabolic and chemical isotope labeling and fractionation steps are incorporated.

While chemical labeling was initially developed to conduct quantitative proteomics using stable isotopes, recent developments in metabolic labeling have made it the dominant technology.

3.5.2 Isotope-Coded Affinity Tags

One of the earliest isotopic tagging methods for quantitative proteomics was isotope-coded affinity tags (ICAT) (23). The ICAT reagent was composed of three main units: an iodoacetamide group, a biotin moiety, and a linker region. The iodoacetamide group was incorporated for the modification of cysteinyl (Cys) residues, while the biotin group enabled ICAT-modified peptides to be extracted using streptavidin chromatography. The mass difference between the heavy and light forms of the ICAT reagents was provided by the linker region. This hydrocarbon change was composed of either exclusively hydrogen atoms (ICAT-d_0) or contained eight deuterons in place of hydrogen atoms (ICAT-d_8). This isotopic substitution created two reagents that differed in mass by 8.04 Da, which is resolvable using an ion-trap mass spectrometer. While this first generation of ICAT reagents proved to be quite successful, it did suffer from a couple of minor issues. The ICAT-d_8-tagged peptides elute from the reversed-phase column slightly earlier than their ICAT-d_0-tagged counterparts. This differential elution affected the ability to quantitatively compare the peptides' abundances. Second, the ICAT tag was quite a large modification (i.e., >500 Da) compared to the size of the average tryptic peptide (i.e., 600–3500 Da). This large addition negatively impacted the quality of the MS^2 decreasing the number of peptides that were confidently identified.

Fortunately, a second generation of ICAT reagents has been developed. Instead of using deuterium, the new reagents replaced nine carbon atoms with carbon-13 (^{13}C) atoms (ICAT–^{13}C$_9$) (24,25). This substitution eliminated the differential elution problem.

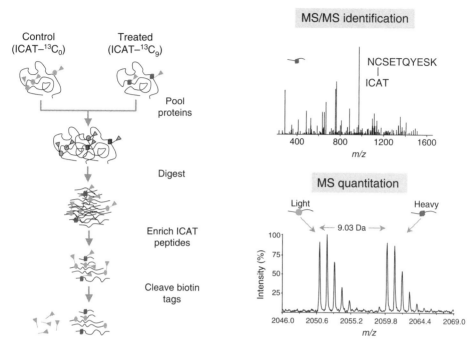

Figure 3.4. The isotope-coded affinity tags (ICAT) strategy for quantifying differential protein abundance in two proteome samples.

An acid-labile bond was incorporated into the linker region allowing modified peptides to be released from the monomeric avidin column using low pH. This labile bond reduced the size of the modification, resulting in higher quality MS^2 spectra. This bond also allowed more stringent washing of the monomeric avidin column, resulting in fewer nonspecific (i.e., unmodified peptides) peptides being extracted.

The use of the ICAT reagents, as well as other stable isotopes for quantitative proteomics, is quite simple (Figure 3.4) (23). The proteomes from two different cell (or organism or tissue) populations are extracted. One is modified with the light ICAT reagent and the other with the heavy ICAT reagent. After labeling, the proteome samples are combined and digested into tryptic peptides. The ICAT-modified peptides (i.e., Cys containing) are then extracted using an avidin column. These combined steps obviously decrease the complexity of the peptide mixture; however, strong cation exchange (SCX) chromatography is typically used to separate the mixture into fractions (i.e., 10–20 fractions), which are then individually analyzed by LC–MS^2 using an ion-trap mass spectrometer.

Relative quantitation and identification of the peptides are provided by the mass spectrometer. Identification is obtained through collision induced dissociation (CID) of either one of the labeled pair of peptides. Quantitation is obtained by integrating the areas under the peaks corresponding to reconstructed ion chromatograms of the individual ICAT-labeled peptides. The areas of the respective peptides are then compared to obtain a ratio of the relative abundance of the peptides from the different samples. The ratio determined from an ICAT experiment and a Western blot of a study analyzing the effect of inorganic phosphate on osteoblasts is shown in Figure 3.5. While the ICAT and densitometry ratios do not exactly match, they are in close agreement. Unlike 2D-PAGE, the ICAT strategy does not allow specific peaks to be selected for identification and quantitation. Peptides are selected in a data-dependent mode for MS^2 based on their intensity. Using this type of nonbiased approach means there is always a good chance that a protein of particular interest will not be analyzed. This fact is one of the features of discovery-driven science, where the data can be crudely described as "you get what you get."

In our laboratory, we used ICAT labeling to investigate the effects of microRNA 34a (miR-34a) on neuroblastoma cells (26). miR-34a is a potential tumor suppressor gene that is a miRNA component of the p53 network (27). To better understand the biological pathways involved in miR-34a action, a global quantitative proteomics study was performed on miR-34a-treated neuroblastoma cells (IMR32) to understand the protein pathways involved in miR34a's action. A total of 1495 proteins were identified from the ICAT analysis by at least two peptides. A total of 143 and 192 proteins were significantly up- or downregulated by miR-34a, respectively. To sort out the major functions represented by these proteins, the differentially expressed proteins were grouped into protein pathways using Ingenuity Pathway software. This grouping showed that proteins related to apoptosis and cell death were upregulated, while those involved in cell cycle regulation (especially ribosomal proteins) were downregulated proteins by miR-34a. Biological network analysis of the data identified the ubiquitous transcription factor YY1, as well as its downstream proteins, as being significantly less abundant in cells treated with miR-34a. While this satisfies the first goal of quantitative proteomics used in a discovery fashion, the next step is to validate whether this finding can be substantiated using a more targeted, orthogonal approach. Western blotting using an anti-YY1 antibody on IMR32 neuroblastoma cells treated with miR-34a confirmed the ICAT results. To further investigate the mechanism of miR-34a on YY1, the $3'$ untranslated region (UTR) of YY1, which contains a miR-34a binding site, was incorporated into a luciferase reporting system. A similar luciferase reporter vector was constructed; however, it did not contain the miR-34a binding site of the YY1 $3'$ UTR. These reporter constructs were transfected into SKNAS cells, a neuroblastoma cell line that does not express miR-34a. These transfected cells were then treated with miR-34a or mimic control miRNA. The reporter vector containing the YY1 $3'$ UTR with a wild-type miR-34a binding site showed decreased activity when treated with miR-34a but not the control miRNA. The mutated YY1 $3'$ UTR vector did not respond to either miR-34a or the control miRNA. Overall, these results not only showed that miR-34a targets the *YY1* gene, but does it through binding to its $3'$ UTR. As YY1 is known to be a negative regulator of p53, the results of this study aid in elucidating the mechanisms by which this ubiquitous transcription factor plays an important role in neuroblastomas.

The use of ICAT peaked in the early part of the previous decade. Unfortunately, a number of laboratories did not seem to be able to achieve good results using this technology. I can remember at least three instances in which I gave a presentation showing results obtained using ICAT immediately after a speaker who derided the

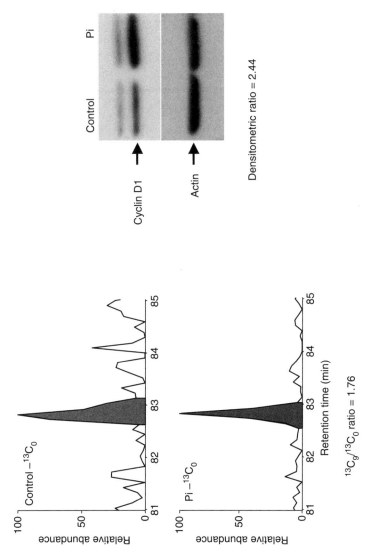

Figure 3.5. Comparison of ICAT and Western blot densitometry ratios for the protein cyclin D1, which was quantitated in a study analyzing the effects of treating osteoblasts with inorganic phosphate (Pi).

69

technology as not working very well. Some of the criticisms against ICAT were its laboriousness, not all proteins contain Cys residues (although at least 95% of human proteins do), and in most cases, proteins were only represented by a single peptide. While it is still being used, ICAT is not nearly as popular as it was several years ago.

With the popularity of ICAT, a number of different methods to quantitative changes in protein abundances between proteomes by chemical modification using stable isotopes were developed. These included using stable-isotope-labeled versions of molecules such as acrylamide and succinic anhydride to label Cys and Lys residues, respectively (28,29). Most of these strategies never gained the momentum required to make them widely used. The next big development in using stable isotopes to chemically modify proteins was isobaric tag for relative and absolute quantitation (iTRAQ) (30,31).

3.5.3 Isobaric Tag for Relative and Absolute Quantitation

The iTRAQ methods label primary amines on peptides using N-hydroxy-succinimide groups on the reagents that also incorporate different combinations of stable isotopes (32). iTRAQ, which is trademarked by Applied Biosystems, permits multiplexed quantitation through the use of isobaric stable isotope labeling reagents that allow samples to be pooled and analyzed in a single LC–MS^2 analysis. While the first version of the reagents permitted four samples to be concurrently analyzed, the newer versions permit up to eight. The four-plex reagents consist of four reporter ions (isobaric tags) designated 114, 115, 116, and 117, while the eight-plex includes these reagents and adds four additional tags, designated 113, 118, 119, and 121 (Figure 3.6) (33). The use of the iTRAQ reagents is similar to that of the ICAT reagents; however, it can be done for up to eight different proteome samples (Figure 3.7). Proteome samples are enzymatically digested (usually with trypsin) and labeled at primary amines with the isobaric tags. The labeling is facilitated by the N-hydroxy-succinimide group on the iTRAQ reagents and result in covalent modification of N-terminus and Lys side chains. The labeled proteomes are then pooled, and analyzed using LC–MS^2. Prefractionation methods such as SCX are also compatible with iTRAQ. The isobaric tags have the same charge and overall mass resulting in peptides from the different samples producing a single peak. After CID, however, the tags produce different low mass peaks (i.e., 113–121 Da). Quantitation is achieved by comparing the MS^2 spectra peak areas associated with the eight reporter ions.

3.5.4 Stable Isotope Labeling of Amino Acids in Culture

Stable isotope labeling of amino acids in culture, or SILAC (34), is a metabolic labeling approach in which stable-isotope-labeled amino acids are added to the cell culture medium in place of their normal isotopic abundance counterparts. The origins of using SILAC for proteomics go back to the year 2000, where Richard Smith's laboratory cultured auxotrophic strains of *Escherichia coli* in medium containing a heavy version of the amino acid leucine (Leu-d$_{10}$) and analyzed these cultures combined with those cultured in normal medium (35). A follow-up study expanded this concept to *S. cerevisiae* and additional stable-isotope-labeled amino acids (i.e., His, Arg, Ile, Phe, and Lys) (36).

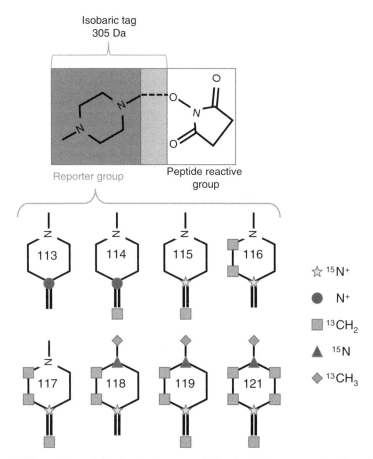

Figure 3.6. Illustration of the basic structure of the iTRAQ reagents showing the amine reactive group and reporter ion (top). Chemical composition of the various reporter groups showing how ions of various *m/z* values can be measured simultaneously for proteomes labeled with the eight different iTRAQ reagents (bottom).

While SILAC has become highly popular in MS-based proteomics in the past 5 years, metabolic labeling of proteins with stable isotopes (i.e., ^{2}H, ^{13}C, or ^{15}N) had previously been used for many years in the determination of protein structures using NMR (37,38). The earliest example of a SILAC-like approach for proteomics was demonstrated by the culturing of mammalian cells in nitrogen-15 (i.e., ^{15}N)-enriched medium (39). Cells were cultured in media containing either natural abundance of the isotopes of nitrogen or in ^{15}N-enriched media. The two cell cultures were harvested and combined. The extracted proteins were covalently modified with the light ICAT tagged and digested into tryptic peptides. The ICAT tagged peptides were extracted using avidin chromatography and then analyzed using LC–MS^2. The origin of the peptides can be distinguished by MS since one set contains the natural abundance of nitrogen, whereas proteins from the

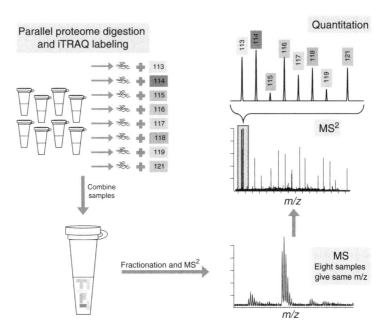

Figure 3.7. The isobaric tag for relative and absolute quantitation (iTRAQ) strategy for quantifying differential protein expression.

other set will have ^{15}N incorporated. The ratio of this pair of peaks is used to determine the relative abundance of the protein between the two samples. Importantly, the MS^2 spectra of ^{14}N and ^{15}N peptides are qualitatively similar; therefore, the peptide can be identified if either is selected in a data-dependent experiment.

While differential labeling of proteins with ^{14}N/^{15}N isotopes was useful, it had one fundamental difficulty. The mass difference between the pairs of peptides was inconsistent since it depended on the number of N atoms within each. This inconsistency provided a challenge to the software trying to determine which two peaks in the MS spectrum constituted a pair. Fortunately SILAC labeling eliminated this challenge as the isotope labels were furnished by the addition of specifically labeled amino acids added to the cell culture medium. The original stable amino acid used in SILAC was Leu-d$_3$ (34); however, stably labeled versions of Lys and Arg residues are presently the most popular (40). The use of Lys and Arg is well suited to current proteomic analysis, as both are essential amino acids to mammals and their use also guarantees at least one labeled amino acid in every peptide (except possibly the C-terminal peptide) generated via the standard tryptic digest of a proteome sample.

Conducting quantitative proteomic studies using SILAC is very straightforward as shown in Figure 3.8. In the comparison of two different cell cultures, one is cultured in normal isotopic abundance medium (light) and the other in medium containing the heavy versions of Lys or Arg (heavy). It is recommended that the cells be cultured for at least three cell passages to make sure that the proteins are fully incorporated with the heavy

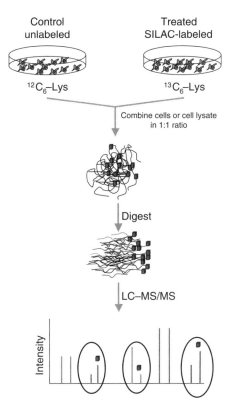

Figure 3.8. Illustration of the use of stable-isotope labeling of amino acids in culture (SILAC) for conducting global quantitative proteomics.

isotope (41). Incomplete incorporation of the isotope will provide incorrect quantitative results. After being cultured and treated as per the experimental design, an equivalent numbers of cells are combined. The proteins are then extracted from this single sample and digested into tryptic peptides. The peptides are normally fractionated using SCX chromatography and these fractions are analyzed using LC–MS^2. The peptides are identified and their relative abundance determined from these data.

An excellent example of the use of SILAC in cancer is a study to identify biomarkers of esophageal squamous cell carcinoma (ESCC) (42). In this study, a SILAC-based quantitative proteomic approach was used to compare the secretome of ESCC cells with that of nonneoplastic esophageal squamous epithelial cells. Once the proteomes of each cell type are combined, there are a number of fractionation options available prior to MS analysis. In this study, the proteins were resolved by SDS-PAGE, and individual bands cut from the gel were in-gel digested using trypsin. The peptide fractions were identified and quantitated using LC–MS^2 on a high accuracy quadrupole time-of-flight mass spectrometer. In total, 441 proteins were quantified, including 120 proteins with at least twofold upregulation in the ESCC secretome compared to the nonneoplastic esophageal squamous epithelial cells. Several previously identified potential ESCC protein biomarkers including matrix metalloproteinase 1, transferrin receptor, and transforming-growth-factor-beta-induced 68 kDa protein were observed in this study. Several novel proteins not previously associated with ESCC were also identified. Among these, protein disulfide isomerase family a member 3 (PDIA3), GDP dissociation inhibitor 2 (GDI2), and lectin galactoside binding soluble 3 binding protein (LGALS3BP) were further validated on 137 ESC cases using immunoblot analysis and IHC labeling using tissue microarrays. This analysis showed overexpression of PDIA3, GDI2, and LGALS3BP in 93%, 93%, and 87% of 137 ESCC cases, respectively.

With commercial reagents becoming broadly available, SILAC is now a very attractive method for conducting quantitative proteomic studies. It definitely requires much less effort than methods such as ICAT or iTRAQ and the cells can be easily prepared in basic research laboratories that are accustomed to culturing cells. One major historical criticism of SILAC for quantitative proteomics is that it was only useful for cells that could be cultured in media. Over the years, this criticism has been pushed aside by the introduction of SILAC-labeled *Drosophila* (43) and mice (44). It also bears mentioning at this point that prior to the SILAC mouse, John Yates's 3rd group developed a ^{15}N-labeled rat (45). While the ^{15}N-labeled rat did not end up being as popular as the isotope-labeled mouse, the concept that we could isotopically label higher mammalian species for quantitative proteomics originated with the creative thinking of Dr. John Yates 3rd. Even more convenient, however, is the fact that Cambridge Isotopes Limited has made specific tissues from a SILAC-labeled mouse available for purchasing; precluding the need for individual laboratories to maintain their own mouse colonies.

While it is likely that a SILAC-labeled human will never be produced, Matthias Mann's group recently developed the concept of super-SILAC for studying human samples (46). In this strategy, Mann and colleagues mixed labeled protein lysates from several previously established cancer-derived cell lines that are cultured in SILAC media to provide an internal standard that can be added to unlabeled human tissue samples. The cell lines are mixed together in equal amounts and are selected to represent the full complexity of a specific tissue proteome rather than a single cell line. An equal amount of the super-SILAC-labeled proteome is added to each human sample of interest, essentially allowing an unlimited number of samples to be compared.

While providing an analytical reagent to conduct SILAC studies on human tissue, super-SILAC also demonstrates greater quantitative power over regular SILAC that utilizes a single cell line. In a comparative study, Mathias Mann's group labeled the human breast cancer cell line HCC1599 with SILAC and mixed the lysate with that of mammary carcinoma tissue from an individual with grade II lobular carcinoma (47). The combined proteome mixture was analyzed via fractionation and MS resulting in the quantitation of just over 4000 proteins. Oddly, however, the quantitative ratio distribution was bimodal and showed 755 proteins with a greater than fourfold abundance level in the tumor tissue compared to the cell line. This experiment suggested that a single labeled cell line does not provide a useful standard to quantitate proteins in tumor tissue.

They then prepared a super-SILAC mixture comprising isotope-labeled proteomes extracted from four breast cancer cell lines differing in origin, stage, and estrogen receptor and ErbB2 expression, as well as a normal mammary epithelial cell type. Comparing the proteomes of the tissue and this super-SILAC mixture resulted in the quantitation of a similar number of proteins as observed in the previous experiment; however, 90% of the quantified proteins were within a fourfold abundance level of the tissue. Almost 80% of the quantified proteins were within a twofold ratio between the super-SILAC mixture and carcinoma tissue. The distribution of quantitation ratios was also unimodal. This experiment demonstrated the superiority in using a super-SILAC mixture to compare protein abundances in human tissue.

3.6 NONISOTOPIC SOLUTION-BASED QUANTITATION

While stable isotope labeling methods represent a tremendous advance in quantitative proteomics, these were limited in the number of samples that could be directly compared. While stable isotope labeling methods continue to become easier to use and cover a broader type of samples, science always migrates toward more convenient methods that work. Nonisotopic-solution-based methods represent an efficient method for quantitating protein abundances across an unlimited number of proteomic samples. The basic sample preparation and MS analysis methods required for these nonisotopic-solution-based methods are no different than direct identification of complex proteomic samples.

3.6.1 Subtractive Proteomics—Peptide Counting

One of the recent developments in nonisotopic-solution-based methods to compare protein abundances termed subtractive proteomics (48) does not require any type of isotope labeling and has become a major tool in finding quantitative differences across proteomes. It is based on a very simple hypothesis: the number of peptides identified for a specific protein is a measure of its relative abundance. A very simple example is the comparison of wild-type serum and serum that has been treated to remove albumin (either via immunodepletion, filtration, dye binding, etc.). If both samples are analyzed by LC–MS^2 after being digested with trypsin, the number of albumin peptides in the wild-type serum will be at least an order of magnitude greater than the sample in which albumin was depleted. Probably, the earliest major report showing the use of the subtractive quantitative approach involved Dr. John Yates 3rd's laboratory in the search for novel nuclear envelope (NE) proteins (48). In this study, mouse liver NEs and microsomal membranes (MMs) were isolated via centrifugation. The isolated NEs are contaminated with membranes from the peripheral endoplasmic reticulum and mitochondria. The MMs, on the other hand, can be purified devoid of NE because the intact nuclei sediment readily via centrifugation. After isolating these two fractions, the proteins were extracted and analyzed using MudPIT. The proteins identified in the MM fraction were subtracted from the list of proteins identified in the NE (and associated contaminants) fraction to produce a dataset of proteins that were found exclusively in the NE. This final list contained 67 previously unknown putative integral NE proteins as well as the 13 proteins that were known at the time of publication to be part of the NE. Of eight proteins validated using immunofluorescence, all were found to localize to the NE. This result suggested that the remaining 57 were also NE proteins. Many of these proteins were associated with various dystrophies.

This study utilized the most extreme form of subtractive proteomics: to be considered a NE protein, they had to be identified uniquely in these fractions. The authors even admit that using such a stringent requirement may have resulted in some of the proteins they identified not being recognized as being part of the NE. It is readily apparent, however, that the same strategy could be applied to infer abundance differences between proteins that are identified in different samples. The major advantages of this method over other quantitative approaches are its simplicity, its universal applicability, and the number of samples that can be compared. As far as simplicity goes, subtractive proteomics requires no isotope labeling step and the data collection using MudPIT has been done routinely in proteomic laboratories for well over a decade. Quantitative comparison of the data can be done with Microsoft® Excel or Access. Since it requires no special labeling, subtractive proteomics can be applied to any type of sample in which proteins can be extracted. While isotope labeling methods can conceivably be used to compare an unlimited number of samples (e.g., use an isotope labeled reference proteome), none of them do it as neatly and easily as subtractive proteomics.

Owing to its simplicity and ability to compare unlimited numbers of samples, subtractive proteomics has been used widely in biomarker discovery. A study illustrating the broad use of subtractive proteomics is an attempt to find biomarkers of head and neck squamous cell carcinoma (HNSCC) by comparing the proteomes extracted from formalin-fixed paraffin-embedded normal epithelial tissues along with poorly (PD), moderately (MD), and well differentiated (WD) HNSCC tumors (49). About 20,000 cells were collected from each of these different samples using laser capture microdissection (LCM). The proteomes were separately extracted, digested with trypsin, and analyzed using LC–MS^2. There was not enough protein available to conduct SCX fractionation prior to LC–MS^2 analysis. Using the subtractive approach, the number of peptides identified per individual protein was used to quantitate differences in the relative abundances of proteins in the four cell types. Cytokeratin 4 was found to be more abundant in normal epithelial tissue compared with tumor cells, whereas cytokeratin 16, vimentin, and desmoplakin were more abundant in the tumor cells compared with normal epithelial tissue. Desmoplakin was found to be particularly more abundant in WD tumor cells. Unfortunately, approximately 120 additional proteins were found to be unique to the PD, MD, and WD HNSCC tumors. These additional proteins, which included proteins involved in DNA synthesis, metabolism, and signaling, were identified by at least four total peptides and about 33% were identified by at least 10.

With verification and validation being absolutely required in biomarker discovery, the above-mentioned study illustrates a hurdle that is consistently seen in global quantitative studies: which of the observed differences is worth the time and expense required for verification and validation studies? Regardless of the quantitative approach taken (i.e., SILAC, ICAT, iTRAQ, subtractive, 2D-PAGE, etc.), investigators typically find that 15–30% of the quantitated proteins show a difference in their relative abundance compared to their control cohort (50). With the proteome coverage afforded with modern day mass spectrometers, this number can easily represent hundreds of proteins.

Obviously, it is going to be critical to develop methods to recognize proteins that may survive verification and validation studies to graduate to useful clinical biomarkers. Accomplishing this goal is going to require higher confidence in the quantitative results

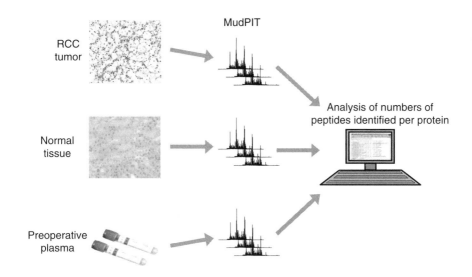

Figure 3.9. Quantitative comparison of healthy and tumor tissue obtained from a single patient suffering from renal cell carcinoma (RCC). Plasma obtained from this patient prior to surgery was also analyzed using liquid chromatography–tandem mass spectrometry (LC–MS²). The results obtained from the three samples were compared using a subtractive proteomics approach that compared the number of peptides identified for each specific protein.

afforded using MS as well as novel methods of analyzing the data. My laboratory conducted a simple subtractive proteomics study to determine whether a combined tissue–blood analysis could narrow the list of potential plasma-based tumor markers (51). The basic methodology behind the study is presented in Figure 3.9. Tumor and adjacent healthy tissue, along with preoperational plasma, was obtained from a single patient diagnosed with nonmetastatic renal cell carcinoma (RCC). As discussed in detail in Chapter 8, we compared the healthy and tumor tissue as we hypothesized that this comparison would provide the greatest chance of finding a protein that was unique to the tumor. We also examined the plasma sample, as circulating biomarkers are far more useful for diagnosing cancer than are tissue-based biomarkers. The proteomes extracted from each of these three samples were analyzed using shotgun proteomics.

To develop a list of potentially useful biomarkers using a subtractive quantitative approach, the data were filtered using the following rules:

1. Protein had to be identified uniquely in tumor and not in healthy tissue.
2. Protein identified uniquely in tumor had to be identified in the plasma sample.
3. Number of peptides identified for protein had to be higher in tumor than plasma.

Over 1000 unique proteins were identified in the combined analysis; however, only eight proteins passed these data-filtering criteria (Figure 3.10a). Eight of these proteins have been associated with some form of cancers, while three (cadherin 11, vascular cell adhesion molecule-1, and pyruvate kinase) have been specifically connected with

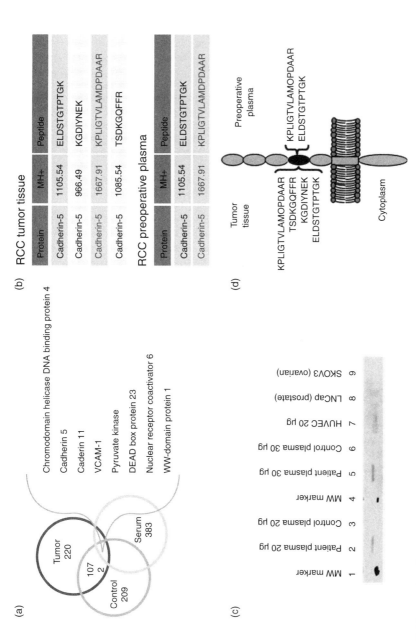

Figure 3.10. Results of comparative analysis of healthy tissue, tumor, and preoperation plasma from single renal cell carcinoma (RCC) patient. (a) After filtering the liquid chromatography–tandem mass spectrometry (LC–MS²) data to locate peptides that were unique to the tumor and plasma, and of higher relative abundance in the tumor, only eight proteins remained. (b) Cadherin was identified by four and two unique peptides in the tumor and plasma, respectively. (c) Western blotting showed the presence of cadherin in plasma of the RCC patient and HUVEC cells, but not in plasma from a healthy individual or LnCAP and SKOV3 cells. (d) All of the peptides identified for cadherin originate from the extracellular domain of the protein.

RCC (52–54) and are located within the membrane, suggesting they could be shed into circulation. Another encouraging result is what was not on this final list of proteins. Those proteins not on this final list included well known circulatory proteins (i.e., complement factors, acute-phase response proteins, etc.) that are commonly found to be differentially abundant in biomarker studies comparing blood samples taken from healthy and disease-affected individuals.

The identification of cadherin 5 as a potential biomarker for RCC serves as a useful illustration of how the subtractive method works. The MS data identified four and two unique peptides for cadherin 5 in the tumor and plasma samples, respectively (Figure 3.10b). While not strictly a part of a subtractive proteomics approach, the extracted ion chromatogram exhibited a greater intensity for the cadherin 5 peptide, KPLIGTVLAMDPDAAR, found in the tumor compared to plasma. Western blot analysis confirmed the presence of cadherin 5 in the preoperative plasma sample and in HUVEC cell lysates (positive control). Cadherin was not found in a healthy individual's sample or in ovarian (SKOV-3) and prostate cancer (LNCaP) cell lysates (Figure 3.10c). The cadherin 5 peptides identified all reside within the extracellular domain of this transmembrane protein, suggesting they could be readily shed into circulation enhancing their detection via a blood-based assay (Figure 3.10d). While these very preliminary findings do not prove that cadherin is a biomarker for RCC, these do demonstrate the utility of a combined blood–tissue analysis strategy that permits the detection of tumor proteins in blood.

For the results of this subtractive proteomics study to be enhanced, a greater number of subjects need to be studied to confirm their value. This study also used a stringent data-filtering strategy such as that used to identify NE proteins. This stringent filtering potentially eliminates proteins identified within the healthy tissue but were detected in the tumor tissue by a much greater number of peptides. Since none of the data are eliminated, it is easy to change the filtering criteria to search for additional putative biomarkers. For example, setting the filtering criteria to reveal proteins that were detected in both tissue types, but the number of peptides detected were n-fold greater (where may be set to >3) in the tumor may be more suitable.

3.6.2 Subtractive Proteomics—Peak Intensity

While using the number of peptides identified for specific proteins has proven useful, the field of proteomics always seems to have alternative strategies for every procedure. While a MS spectrum contains peaks of various intensities, these peaks are not considered as useful measures of a molecules quantity. If the peaks accurately reflected a molecule's abundance, the peaks in a spectrum of a tryptically digested protein would all be of equal intensity. Anyone who has generated a peptide map will tell you that they are not. There are a number of factors that affect the intensity of a MS signal produced by a protein or peptide. The two most commonly cited are ionization efficiency and matrix effects. Ionization efficiency refers to the ratio of the number of molecular ions formed compared to the number of electrons or photons used in the ion source. The number of ions formed is highly dependent on the chemistry of the analyte being ionized. This number is also affected by the environment (or matrix) of the molecule, as it may have to compete with other analytes for electrons or photons. In a complex proteome sample of the type analyzed in a quantitative study, the environment of a specific peptide can change between samples. Nevertheless, recent studies have developed methods that compare the intensity of MS signals for specific peptides found within comparative samples. In 2004, Nathan Yates published differential mass spectrometry (dMS), a fully automated method for detecting differences between LC–MS profiles (55). The dMS method directly compares peak intensities at each retention time and m/z ratio, returning a ranked list of differences between samples in two conditions. A key to the accuracy of dMS is that it analyzes multiple LC–MS runs to take into account the variability of intensity measurements. By incorporating multiple LC–MS runs, small but statistically significant differences in low abundance peptides are captured while large, but statistically insignificant, differences in peptides present at much greater concentrations are eliminated.

In a study illustrating the use of dMS in not only quantitating native peptides but also posttranslationally modified peptides, an experiment was performed to identify biomarkers of histone deacetylase (HDAC) inhibitors (56). Compounds that inhibit class 1 and class 2 HDAC enzymes have shown antitumor activity and there is continuing interest in developing specific HDAC inhibitors that retain anticancer activity but have fewer and less severe side effects. Recognizing HDAC1 inhibitors that are class-specific requires identifying pharmacodynamic markers that correlate closely with HDAC1-inhibition *in vitro* and *in vivo*. Unfortunately, existing histone markers of HDAC enzyme activity utilize pan-HDAC inhibitors and do not always delineate isoform-specific inhibitors. To identify biomarkers for class-specific HDAC enzymes, histones were isolated from HCT116 human colon cancer cell lines that had been treated with multiple classes of compounds that are specific for HDAC1 (MRLB-38489); HDAC1 and 3 (MRLB-32353); and HDAC1, 3, and 6 (vorinostat) enzymes. The isolated histones were digested using trypsin and analyzed using LC–MS^2 on a Fourier transform mass spectrometer. Data-dependent spectra were analyzed using SEQUEST against a human histone database from NCBI. Most of the peptides were identified as histones and showed significant differences in their acetylation status. The dMS analysis combined with peptide identification using Sequest revealed that histone H3 is acetylated at lysine residues 18 and 23 when treated with vorinostat compared to control. Following dMS analysis, HDAC1-selective features were found that were common to MRLB-38489, MRLB-32353, and vorinostat, while HDAC3-specific features that were shared by vorinostat and MRLB-32353 (but not MLRB-38489) were also recognized. The net result was 135 putative HDAC1-selective features and 29 HDAC3-selective features.

The number of compounds tested as HDAC enzyme inhibitors was increased to gain confidence in the peptide features found in the first stage of this study. The second dMS experiment was conducted using cell lines that were treated with two and three compounds, in addition to vorinostat, that were shown to be HDAC1- and HDAC 1/3-specific, respectively. The experiment included compounds with similar structures, but possessing different amounts of HDAC selectivity. For example, MRLB-07266 (HDAC1-selective) and MRLB-63397 (HDAC1/3-selective) both possess a benzothiophene core; however, these compounds are selective for HDAC1 or HDAC1/3, respectively. The dMS data were analyzed to reveal HDAC1-selective features as those features that are shared by all compounds, and HDAC3-selective features as those that are shared by vorinostat and MS-275, MRLB-63397, and HC-Toxin, but not by MRLB-88183 or MRLB-07266. In this comparative analysis, 31 HDAC1- and 14 HDAC3-selective features were identified. Of the 31 HDAC1-specific features, 24 were repeatedly identified in replicate experiments. Sequencing seven of these features revealed acetylated peptides originating from histones H4/O, H2B.B, H2B.N, H2B.E, and H3 as markers of HDAC1 enzyme inhibition. Unfortunately, owing to the data-dependent mode of operation only seven of the 24 features were successfully sequenced by the mass spectrometer. This study provides an excellent illustration of how dMS can be used in a targeted, quantitative mode to identify markers of the effect of a specific drug and may allow physicians to follow the pharmacological effect of patient treatment.

3.7 CONCLUSIONS

Basic research and clinical diagnostics both heavily rely on experiments that are capable of measuring the quantity of specific molecules. In basic research, detecting an increase in the abundance of a protein can indicate an increase in DNA transcription or mRNA translation resulting from some perturbation of the cell. In clinical diagnostics, the increase in a protein's abundance beyond a normal range can indicate the presence of a disease. Proteomics has taken the concept of quantitation beyond a specific molecule and focused it on quantifying as many proteins as possible in a single study. This leap from hypothesis- to discovery-driven studies is necessary if the dream of predictive systems biology and translation medicine is ever to become a reality.

Admittedly this chapter does not cover all of the different methods used for quantitative proteomics. The most obvious omission is protein arrays; however, this technology is covered within a dedicated chapter later in this book. Also missing is the large number of isotope labeling approaches that were developed and applied over the past several years. Probably, the most commonly used method that was not discussed in this chapter is trypsin-mediated $^{16}O/^{18}O$ labeling. In this approach, one proteome sample is digested using trypsin in normal isotopic abundance water. The comparative proteome sample is tryptically digested in water in which the ^{16}O atom has been replaced with an ^{18}O isotope (57, 58). The action of trypsin results in the exchange of the two O atoms at the C-terminus of the resulting peptide with two ^{18}O atoms from the water solvent. Other isotope labeling techniques using acrylamide (59), sulfur-36 (60), isothiocyanate (61), and dimethyl groups (62) have also been developed and utilized to a lesser extent than other methods highlighted in this chapter.

It is interesting to see how quantitative proteomics has evolved over the past decade. While mass spectrometers' forte has been the identification of proteins via peptide surrogates, the peptide signals were never considered to be quantitative due to unpredictable ionization efficiencies and matrix effects. Therefore, starting with ICAT, isotope labeling methods became all the rage. The belief (and correctly so) was that isotope labeling methods would provide an internal standard for every peptide in the proteome of interest while eliminating differences in ionization efficiency, minimizing matrix effects, and variation in chromatographic retention times. Owing to the limitations of ICAT, alternative methods such as enzyme-mediated $^{16}O/^{18}O$ labeling and iTRAQ were invented.

While both of these methods proved effective, metabolic labeling methods became increasingly popular as the belief was that they provide greater accuracy and ease of use. Metabolic labeling methods took on many different forms culminating with the introduction of a fully isotopically labeled mouse.

Then, a strange thing happened. Investigators started to examine whether the information provided by a data-dependent LC–MS^2 experiment could provide quantitative information. This method did not require any type of chemical or metabolic isotope labeling and was simply MudPIT analysis, which had become commonplace in proteomic laboratories. This strategy seems to have brought the field full circle to where it was prior to the invention of ICAT. Fortunately, study after study has shown that comparing signal intensities of specific peptides or counting the number of peptides identified for specific proteins between samples can provide reasonable quantitative results. It may be that the information for conducting global quantitative studies was in front of us all along.

There is still room for improvement in quantitative proteomics. The most pressing need is to develop better accuracy and precision in the measurements. Too much variability exists in the data acquired using MS, particularly in the field of biomarker discovery wherein validation is often impossible due to the unreasonable number of "potential" biomarkers that are reported in almost every study. Decreasing variability will provide fewer and stronger leads as to which observed differences between samples are physiological and which are random. To reach the necessary levels of accuracy and precision is going to require standardized sample acquisition and preparation as well as data acquisition and analysis. Continuing to increase the throughput of sample analysis will also permit greater numbers of samples to be compared helping to provide reliable statistical analysis to the data.

The next breakthrough in quantitative proteomics will be the ability to measure the absolute amounts of proteins. While it is currently possible to measure the absolute amount of specific proteins within complex mixtures, making this measurement for every protein in a proteome is still not routinely possible. Recent studies, however, have shown that this step may be possible in the next few years. Two recent studies by Matthias Selbach's (63) and Ruedi Aebersold's (64) laboratories proposed methods to measure the absolute abundance of proteins across complex proteome samples and compared these levels to their corresponding mRNAs. Dr. Selbach's laboratory used stable isotope incorporation and MS to quantitate pulse-labeled proteins and 4-thiouridine incorporation to quantitate newly synthesized RNA in mouse cells (63). These results showed that genes, such as transcription factors and kinases, that respond quickly to stimuli have short protein and mRNA half lives. Proteins and mRNA involved in constitutive processes such as translation and metabolism were more stable and longer lived. Dr. Aebersold's group proposed a method for measuring the absolute quantity of approximately 7000 proteins in a human cell line during log and M-phase growth (64). This study also found that proteins related to translation were high in abundance, and those involved in transcription and signaling were present in very low abundance. It was quite interesting that both studies suggested that the correlation between mRNA and protein levels is greater than that previously reported (65, 66). If quantitative proteomics continues to evolve as quickly as it has in the past decade, absolute quantitation of proteins across the entire proteome may be possible in the following decade.

REFERENCES

1. O'Farrell PH. High resolution two-dimensional electrophoresis of proteins. J. Biol. Chem. 1975;250:4007–4021.

2. Geisow MJ. Proteomics: One small step for a digital computer, one giant leap for humankind. Nat. Biotechnol. 1998;16:206.

3. Jagannadham MV, Abou-Eladab EF, Kulkarni HM.Identification of outer membrane proteins from an Antarctic bacterium *Pseudomonas syringae* Lz4W. Mol. Cell. Proteomics 2011;10:M110.004549.

4. Devraj K, Geguchadze R, Klinger ME, Freeman WM, Mokashi A, Hawkins RA, Simpson IA.Improved membrane protein solubilization and clean-up for optimum two-dimensional electrophoresis utilizing GLUT-1 as a classic integral membrane protein. J. Neurosci. Methods 2009;184:119–123.

5. Lee K, Pi K.Proteomic profiling combining solution-phase isoelectric fractionation with two-dimensional gel electrophoresis using narrow-pH-range immobilized pH gradient gels with slightly overlapping pH ranges. Anal. Bioanal. Chem. 2010;396:535–539.

6. Lee K, Pi K.Effect of separation dimensions on resolution and throughput using very narrow-range IEF for 2-DE after solution phase isoelectric fractionation of a complex proteome. J. Sep. Sci. 2009;32:1237–1242.

7. Semaan SM, Sang QX.Prefractionation enhances loading capacity and identification of basic proteins from human breast cancer tissues. Anal. Biochem. 2011;411:80–87.

8. Fitzgerald A, Walsh BJ.New method for prefractionation of plasma for proteomic analysis. Electrophoresis 2010;31:3580–3585.

9. Unlü M, Morgan ME, Minden JS. Difference gel electrophoresis: A single gel method for detecting changes in protein extracts. Electrophoresis 1997;18:2071–2077.

10. Ihling C, Sinz A.Proteome analysis of *Escherichia coli* using high-performance liquid chromatography and Fourier transform ion cyclotron resonance mass spectrometry. Proteomics 2005;5:2029–2042.

11. Dowsey AW, English JA, Lisacek F, Morris JS, Yang GZ, Dunn MJ.Image analysis tools and emerging algorithms for expression proteomics. Proteomics 2010;10:4226–4257.

12. Silva E, O'Gorman M, Becker S, Auer G, Eklund A, Grunewald J, Wheelock AM. In the eye of the beholder: Does the master see the SameSpots as the novice?. J. Proteome Res. 2010;9:1522–1532.

13. Patton WF. Detection technologies in proteome analysis. J. Chromatogr. B Analyt. Technol. Biomed. Life Sci. 2002;771:3–31.

14. O'Dwyer D, Ralton LD, O'Shea A, Murray GI. The proteomics of colorectal cancer: Identification of a protein signature associated with prognosis. PLoS One 2011;6:e27718.

15. Peng XC, Gong FM, Zhao YW, Zhou LX, Xie YW, Liao HL, Lin HJ, Li ZY, Tang MH, Tong AP. Comparative proteomic approach identifies PKM2 and cofilin-1 as potential diagnostic, prognostic and therapeutic targets for pulmonary adenocarcinoma. PLoS One 2011;6: e27309.

16. Unlü M, Morgan ME, Minden JS. Difference gel electrophoresis: A single method for detecting changes in protein extracts. Electrophoresis 1997;18:2071–2077.

17. Arruda SC, Barbosa Hde S, Azevedo RA, Arruda MA. Two-dimensional difference gel electrophoresis applied for analytical proteomics: Fundamentals and applications to the study of plant proteomics. Analyst 2011;136:4119–4126.

18. Minden JS, Dowd SR, Meyer HE, Stühler K. Difference gel electrophoresis. Electrophoresis 2009;30 (Suppl 1):S156–S161.

19. Coombs KM. Quantitative proteomics of complex mixtures. Expert Rev. Proteomics 2011;8:659–677.

20. Xiao H, Zhang L, Zhou H, Lee JM, Garon EB, Wong DT. Proteomic analysis of human saliva from lung cancer patients using two-dimensional difference gel electrophoresis and mass spectrometry. Mol. Cell. Proteomics 2012;11:M111.012112.

21. Washburn MP, Wolters D, Yates JR 3rd. Large-scale analysis of the yeast proteome by multidimensional protein identification technology. Nat. Biotechnol. 2001;19:242–247.

22. Xie F, Liu T, Qian WJ, Petyuk VA, Smith RD. Liquid chromatography–mass spectrometry-based quantitative proteomics. J. Biol. Chem. 2011;286:25443–25449.

23. Gygi SP, Rist B, Gerber SA, Turecek F, Gelb MH, Aebersold R. Quantitative analysis of complex protein mixtures using isotope-coded affinity tags. Nat. Biotechnol. 1999;17:994–999.

24. Hansen KC, Schmitt-Ulms G, Chalkley RJ, Hirsch J, Baldwin M A, Burlingame AL. Mass spectrometric analysis of protein mixtures at low levels using cleavable ^{13}C-isotope-coded affinity tag and multidimensional chromatography. Mol. Cell. Proteomics 2003;2:299–314.

25. Yu LR, Conrads TP, Uo T, Issaq HJ, Morrison RS, Veenstra TD. Evaluation of the acid-cleavable isotope-coded affinity tag reagents: Application to camptothecin-treated cortical neurons. J. Proteome Res. 2004;3:469–477.

26. Chen QR, Yu LR, Tsang P, Wei JS, Song YK, Cheuk A, Chung JY, Hewitt SM, Veenstra TD, Khan J. Systematic proteome analysis identifies transcription factor YY1 as a direct target of miR-34a. J. Proteome Res. 2011;10:479–487.

27. Peurala H, Greco D, Heikkinen T, Kaur S, Bartkova J, Jamshidi M, Aittomäki K, Heikkilä P, Bartek J, Blomqvist C, Bützow R, Nevanlinna H. miR-34a expression has an effect for lower risk of metastasis and associates with expression patterns predicting clinical outcome in breast cancer. PLoS One 2011;6:e26122.

28. Faca V, Coram M, Phanstiel D, Glukhova V, Zhang Q, Fitzgibbon M, McIntosh M, Hanash S. Quantitative analysis of acrylamide labeled serum proteins by LC–MS/MS. J. Proteome Res. 2006;5:2009–2018.

29. Wang H, Wong CH, Chin A, Kennedy J, Zhang Q, Hanash S. Quantitative serum proteomics using dual stable isotope coding and nano LC–MS/MSMS. J. Proteome Res. 2009;8:5412–5422.

30. Shadforth IP, Dunkley TP, Lilley KS, Bessant C. i-Tracker: For quantitative proteomics using iTRAQ. BMC Genomics 2005;6:145.

31. DeSouza L, Diehl G, Rodrigues MJ, Guo J, Romaschin AD, Colgan TJ, Siu KW. Search for cancer markers from endometrial tissues using differentially labeled tags iTRAQ and cICAT with multidimensional liquid chromatography and tandem mass spectrometry. J. Proteome Res 2005;4:377–386.

32. Unwin RD. Quantification of proteins by iTRAQ. Methods Mol. Biol. 2010;658:205–215.

33. Phanstiel D, Unwin R, McAlister GC, Coon JJ. Peptide quantification using 8-plex isobaric tags and electron transfer dissociation tandem mass spectrometry. Anal. Chem. 2009;81:1693–1698.

34. Ong SE, Blagoev B, Kratchmarova I, Kristensen DB, Steen H, Pandey A, Mann M. Stable isotope labeling by amino acids in cell culture, SILAC, as a simple and accurate approach to expression proteomics. Mol. Cell. Proteomics 2002;1:376–386.

35. Veenstra TD, Martinović S, Anderson GA, Pasa-Tolić L, Smith RD. Proteome analysis using selective incorporation of isotopically labeled amino acids. J. Am. Soc. Mass Spectrom. 2000;11:78–82.

36. Martinović S, Veenstra TD, Anderson GA, Pasa-Tolić L, Smith RD. Selective incorporation of isotopically labeled amino acids for identification of intact proteins on a proteome-wide level. J. Mass Spectrom. 2002;37:99–107.

37. Xia B, Pikus JD, Xia W, McClay K, Steffan RJ, Chae YK, Westler WM, Markley JL, Fox BG. Detection and classification of hyperfine-shifted ^{1}H, ^{2}H, and ^{15}N resonances of the Rieske ferredoxin component of toluene 4-monooxygenase. Biochemistry 1999;38:727–739.

38. Shindo K, Masuda K, Takahashi H, Arata Y, Shimada I. Backbone ^{1}H, ^{13}C, and ^{15}N resonance assignments of the anti-dansyl antibody Fv fragment. J. Biomol. NMR 2000;17:357–358.

39. Conrads TP, Alving K, Veenstra TD, Belov ME, Anderson GA, Anderson DJ, Lipton MS, Pasa-Tolić L, Udseth HR, Chrisler WB, Thrall BD. Smith RD. Quantitative analysis of bacterial and mammalian proteomes using a combination of cysteine affinity tags and ^{15}N-metabolic labeling. Anal. Chem. 2001;73:2132–2139.

40. Ong SE, Mann M. Stable isotope labeling by amino acids in cell culture for quantitative proteomics. Methods Mol. Biol. 2007;359:37–52.

41. Basu SS, Mesaros C, Gelhaus SL, Blair IA. Stable isotope labeling by essential nutrients in cell culture for preparation of labeled coenzyme A and its thioesters. Anal. Chem. 2011;83:1363–9136.

42. Kashyap MK, Harsha HC, Renuse S, Pawar H, Sahasrabuddhe NA, Kim MS, Marimuthu A, Keerthikumar S, Muthusamy B, Kandasamy K, Subbannayya Y, Prasad TS, Mahmood R, Chaerkady R, Meltzer SJ, Kumar RV, Rustgi AK, Pandey A. SILAC-based quantitative proteomic approach to identify potential biomarkers from the esophageal squamous cell carcinoma secretome. Cancer Biol. Ther. 2010;10:796–810.

43. Sury MD, Chen JX, Selbach M. The SILAC fly allows for accurate protein quantification in vivo. Mol. Cell. Proteomics 2010;9:2173–2183.

44. Zanivan S, Krueger M, Mann M. In vivo quantitative proteomics: The SILAC mouse. Methods Mol. Biol. 2012;757:435–450.

45. Wu CC, MacCoss MJ, Howell KE, Matthews DE, Yates JR 3rd. Metabolic labeling of mammalian organisms with stable isotopes for quantitative proteomic analysis. Anal. Chem. 2004;76:4951–4959.

46. Geiger T, Wisniewski JR, Cox J, Zanivan S, Kruger M, Ishihama Y, Mann M. Use of stable isotope labeling by amino acids in cell culture as a spike-in standard in quantitative proteomics. Nat. Protoc. 2011;6:147–157.

47. Geiger T, Cox J, Ostasiewicz P, Wisniewski JR, Mann M. Super-SILAC mix for quantitative proteomics of human tumor tissue. Nat. Methods 2010;7:383–385.

48. Schirmer EC, Florens L, Guan T, Yates JR 3rd, Gerace L. Nuclear membrane proteins with potential disease links found by subtractive proteomics. Science 2003;301:1380–1382.

49. Patel V, Hood BL, Molinolo AA, Lee NH, Conrads TP, Braisted JC, Krizman DB, Veenstra TD, Gutkind JS. Proteomic analysis of laser-captured paraffin-embedded tissues: A molecular portrait of head and neck cancer progression. Clin. Cancer Res. 2008;14:1002–1014.

50. Blonder J, Issaq HJ, Veenstra TD. Proteomic biomarker discovery: It's more than just mass spectrometry. Electrophoresis 2011;32:1541–1548.

51. Johann DJ Jr, Wei BR, Prieto DA, Chan KC, Ye X, Valera VA, Simpson RM, Rudnick PA, Xiao Z, Issaq HJ, Linehan WM, Stein SE, Veenstra TD, Blonder J. Combined blood/tissue

analysis for cancer biomarker discovery: Application to renal cell carcinoma. Anal. Chem. 2010;82:1584–1588.

52. Shimazui T, Giroldi LA, Bringuier PP, Oosterwijk E, Schalken JA. Complex cadherin expression in renal cell carcinoma. Cancer Res. 1996;6:3234–3237.

53. Hegele A, Varga Z, Kosche B, Stief T, Heidenreich A, Hofmann R. Pyruvate kinase type tumor M2 in urological malignancies. Urol. Int. 2003;70:55–58.

54. Vasselli JR, Shih JH, Iyengar SR, Maranchie J, Riss J, Worrell R, Torres-Cabala C, Tabios R, Mariotti A, Stearman R, Merino, M, Walther MM, Simon R, Klausner RD, Linehan WM. Predicting survival in patients with metastatic kidney cancer by gene-expression profiling in the primary tumor. Proc. Natl. Acad. Sci. U.S.A. 2003;100:6958–6963.

55. Wiener MC, Sachs JR, Deyanova EG, Yates NA. Differential mass spectrometry: A label-free LC–MS method for finding significant differences in complex peptide and protein mixtures. Anal. Chem. 2004;76:6085–6096.

56. Lee AY, Paweletz CP, Pollock RM, Settlage RE, Cruz JC, Secrist JP, Miller TA, Stanton MG, Kral AM, Ozerova ND, Meng F, Yates NA, Richon V, Hendrickson RC. Quantitative analysis of histone deacetylase-1 selective histone modifications by differential mass spectrometry. J. Proteome Res. 2008;7:5177–5186.

57. Yao X, Freas A, Ramirez J, Demirev PA, Fenselau C. Proteolytic ^{18}O labeling for comparative proteomics: Model studies with two serotypes of adenovirus. Anal. Chem. 2001;73:2836–2842.

58. Ye X, Luke B, Andresson T, Blonder J. ^{18}O stable isotope labeling in MS-based proteomics. Brief. Funct. Genomic Proteomic 2009;8:136–144.

59. Sechi SA. Method to identify and simultaneously determine the relative quantities of proteins isolated by gel electrophoresis. Rapid Commun. Mass Spectrom. 2002;16:1416–1424.

60. Jehmlich N, Kopinke FD, Lenhard S, Vogt C, Herbst FA, Seifert J, Lissner U, Völker U, Schmidt F, von Bergen M. Sulfur-36S stable isotope labeling of amino acids for quantification (SULAQ). Proteomics 2012;12:37–42.

61. Leng J, Wang H, Zhang L, Zhang J, Wang H, Cai T, Yao J, Guo Y. Integration of high accuracy N-terminus identification in peptide sequencing and comparative protein analysis via isothiocyanate-based isotope labeling reagent with ESI ion-trap TOF MS. J. Am. Soc. Mass Spectrom. 2011;22:1204–1213.

62. Song C, Wang F, Ye M, Cheng K, Chen R, Zhu J, Tan Y, Wang H, Figeys D, Zou H. Improvement of the quantification accuracy and throughput for phosphoproteome analysis by a pseudo triplex stable isotope dimethyl labeling approach. Anal. Chem. 2011;83:7755–7762.

63. Schwanhäusser B, Busse D, Li N, Dittmar G, Schuchhardt J, Wolf J, Chen W, Selbach M. Global quantification of mammalian gene expression control. Nature 2011;473:337–342.

64. Beck M, Schmidt A, Malmstroem J, Claassen M, Ori A, Szymborska A, Herzog F, Rinner O, Ellenberg J, Aebersold R. The quantitative proteome of a human cell. Mol. Syst. Biol. 2011;7:549–556.

65. Maier T, Guell M, Serrano L. Correlation of mRNA and protein in complex biological samples. FEBS Lett. 2009;583:3966–3973.

66. de Sousa Abreu R, Penalva LO, Marcotte EM, Vogel C. Global signatures of protein and mRNA expression levels. Mol. Biosyst. 2009;5:1512–1526.

4

PROTEOMIC ANALYSIS OF POSTTRANSLATIONAL MODIFICATIONS

4.1 INTRODUCTION

The simplest definition of a protein is a string of amino acids covalently bound together that fold in a defined three-dimensional structure to form a functional biomolecule. As protein chemists know; proteins are definitely much more than the sum of the amino acid parts. Although the function of a protein is highly dependent on its primary structure, there are additional, much more subtle factors that have a profound effect on their function. These factors can involve molecules as small as four atoms or as large as hundreds of atoms. These factors, commonly known as posttranslational modifications (PTMs), are instrumental in determining the activity of a protein to the point of changing its entire function within the cell. Most eucaryotic proteins are modified at some point during their lifetime and PTMs are believed to regulate the activity of most of these proteins (1). Therefore, to fully understand a protein it is essential to characterize its PTMs.

While there are estimated to be over 200 different kinds of PTMs (Table 4.1) (2), the most important that occur within eucaryotic cells are presently thought to be phosphorylation and glycosylation (3–5). The importance of phosphorylation and glycosylation of proteins to the overall function of the cell cannot be underestimated. A keyword search of PubMed using "phosphorylation" and "glycosylation" results in

Proteomic Applications in Cancer Detection and Discovery, First Edition. Timothy D. Veenstra.
© 2013 John Wiley & Sons, Inc. Published 2013 by John Wiley & Sons, Inc.

TABLE 4.1. Comparison of Properties of Genes and Proteins

	Genome	Proteome
Subunits	Four nucleotide bases	At least 20 amino acid residues
Modifications	Few	~300
Biophysical properties	Homogeneous	Heterogeneous
Form	Static	Dynamic
Function	One main function	Unknown number of functions

approximately 36,000 and 5500 papers, respectively, published during the year 2010. Between 30% and 50% and 50% and 70% of human proteins are believed to be phosphorylation and glycosylated, respectively (6–8). Phosphorylation, which occurs primarily on seryl (Ser), threonyl (Thr), and tyrosyl (Tyr) residues (Figure 4.1), is a key signaling event that historically has been conceptualized as an on/off switch that regulates a protein's activity. As evidence has accumulated, phosphorylation is now recognized as

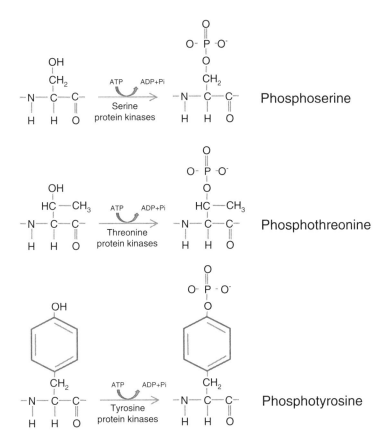

Figure 4.1. Examples of phosphorylation sites on threonine, serine, and tyrosine residues.

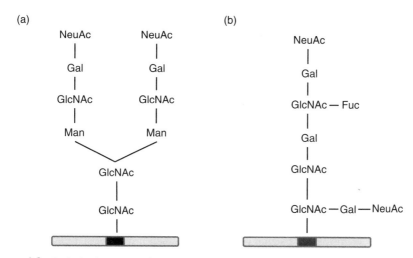

Figure 4.2. Carbohydrates on glycoproteins occur as either (a) N-linked or (b) O-linked to specific amino acid residues. N-linked glycosylation occurs on asparagines residues, whereas O-linked occurs on serine or threonine residues. The examples shown in the figure are typically found in mammalian cells and serve to illustrate the complexity of glycosylation modifications.

a dynamic and precise tuner of the potential function of a protein, rather than being a static on/off switch (9). For example, many transcription factors posses multiple phosphorylation sites that enable several different effects to occur within a single factor, finely tuning the transcriptional activity of the protein (10–12). While a least one-third of all of the proteins in a eucaryotic cell may be phosphorylated at any given time, it is hypothesized that not every phosphorylation site is central to a protein's function (7). Glycosylation, which occurs as both N-linked (to arginine residues) and O-linked (to serine or threonine residues; Figure 4.2) alters the physiochemical properties of a protein to change such parameters as its cell location, solubility, stability, overall structure, and immunogenicity (13). Glycoproteins are known to function in cell signaling, provide structural integrity to the cell, and act as adhesion molecules coupling neighboring cells together (14). Membrane proteins are commonly glycosylated and this PTM can act as recognition sites on different receptors (15). One of the major differences between glycosylation and phosphorylation is that glycosylation at a particular site is heterogeneous, making its absolute characterization very challenging using existing technologies. This heterogeneity can be either *macroheterogeneity*, in which a glycan is present or absent at a particular site, or *microheterogeneity*, in which the glycan at a particular site varies in composition. While a major part of this chapter will focus on phosphorylation and glycosylation, other important modifications such as acetylation, methylation, and ubiquitination will also be discussed.

Many methods have been developed prior to the "omics-era" to determine the PTM state of a protein; however, these techniques have been geared toward analyzing specific sites with targeted proteins. Affinity reagents such as antiphosphoamino acid-specific monoclonal antibodies (mAb) have been a popular choice for determining the phosphorylation status of specific proteins (16, 17). For glycoproteins, antibodies and lectins (i.e., sugar binding proteins) have been widely used to identify the glycosylation status of proteins (18, 19). Unfortunately, the antibodies and lectins are often insufficient in accurately describing the modified site. In a majority of cases, these affinity techniques will determine the presence of a PTM but not the specific site unless an antibody with specificity to a specific site has been manufactured. To have such an antibody, however, requires a hypothesis-driven approach wherein the site is already suspected to be modified. For a purely discovery-driven approach (which has propelled the development of omics technologies), affinity-based methods do not provide enough detailed information.

Presently, the best available technology (and one that has tremendous untapped potential) to characterize both known and unknown PTMs within a proteome is mass spectrometry (MS). The highly accurate masses and tandem MS capabilities of modern mass spectrometers provide a high degree of accuracy in the characterization of site-specific PTMs. While MS is excellent at identifying their presence on a specific residue, novel methods are enabling proteome-wide quantitation of specific PTMs across hundreds of proteins within complex mixtures (20, 21). In this chapter, I will summarize the techniques used to determine the presence of PTMs and describe examples on how these methods are being used to increase our knowledge of cancers.

4.2 PHOSPHORYLATION

Arguably, the most important and best-understood modification used to modulate protein activity and propagate signals for cell homeostasis is phosphorylation (7, 22, 23). Cell-cycle progression, differentiation, development, peptide hormone response, and adaptation are just a few of the processes that are regulated by protein phosphorylation (24, 25). This critical importance has translated into a lot of effort being put forth by different proteomic labs around the world. This level of effort is best illustrated by the number of publicly available databases that provide lists of identified phosphoproteins and phosphopeptides as well as software tools for their analysis (Table 4.2). No other modification has this number of available database resources.

Historically, the main method used to study protein phosphorylation required radiolabeling with ^{32}P inorganic phosphate (^{32}P$_i$) *in vivo* or [γ^{32}P]ATP *in vitro* (26). Radioactivity provided the signal that enabled the tracing of any biomolecule that incorporated the phosphate group. After labeling, proteins are separated using techniques such as two-dimensional polyacrylamide gel electrophoresis (2D-PAGE) or high-performance liquid chromatography (HPLC). The presence of radioactive-labeled proteins is measured either by autoradiography or scintillation counting. The radiolabeled protein is then treated in one of two ways depending on the research question being asked. If the purpose is to determine the types of amino acids that are modified, the phosphoprotein is completely hydrolyzed and the phosphoamino acid content determined. If the specific site(s) of phosphorylation is needed, the phosphoprotein is digested into peptides, which are separated (e.g., by HPLC). The radiolabeled peptides are collected and sequenced using Edman degradation to determine the phosphorylated site(s). The challenges posed

TABLE 4.2. Partial List of Publicly Available Phosphoproteomic Tools and Databases

Name	URL	Description
Gene ontology	http://www.geneontology.org/GO.tools.shtml	Gene ontology tools
Global proteome machine database	http://gpmdb.thegpm.org/	Public repository of proteomics data
SH2 domain sida	http://sh2.uchicago.edu/p://www.phosida.de/	Mouse and human SH2-domain proteins Sphoprotein data
PhosphoElm	http://phospho.elm.eu.org/	Database of S,T,Y Sphorylation sites
Phosphomouse	https://gygi.med.harvard.edu/phosphomouse/index.php	Mouse protein and phosphorylation data
PhosphoSite	http://phosphosite.org	Phosphorylation site database
STRING database	ttp://string-db.org/	Protein association networks database
Tranche	https://trancheproject.org/	Peptide data

by both methods are obvious; they are laborious, require large amounts of protein, and possess all of the difficulties associated with working with radioactive compounds. A few years ago, our group developed an *in vitro* labeling method that utilized stable isotope labeled adenosine triphosphate (ATP) and MS to identify phosphorylation residues within proteins (27). The commercially available ATP molecule is synthesized with four ^{18}O atoms on the terminal phosphate group. When this molecule is combined in an enzymatic reaction with a kinase and its target protein, the phosphate group transferred to the amino acid(s) retains three of the ^{18}O atoms (Figure 4.3a). To determine phosphorylated residues, the reaction is performed in duplicate using wild-type (i.e., ^{16}O-labeled ATP) and stable isotope-labeled ATP. The samples are combined, digested with trypsin and analyzed using MS. Peptides containing phosphorylated residues will be manifested in the mass spectrum by peaks separated by 6.02 Da (Figure 4.3b). Unlike radioactivity, which provides a single signal for peptides containing multiple sites of phosphorylation, the stable isotope method provides a direct signal for each phosphorylated version of a peptide allowing multiple sites to be identified in a single peptide (Figure 4.3c).

Quantitative differences in phosphorylation status of specific proteins have also been conducted by fractionating ^{32}P-labeling proteomes using 2D-PAGE and comparing the relative spot intensities observed using autoradiography (28). Unfortunately, the inconvenience of radioactivity has prevented ^{32}P-labeling from being an effective tool for measuring proteome-wide changes in phosphorylation. While dedicating a set of tools such as pippettemen to experiments involving radioactivity is appropriate, setting aside an entire mass spectrometer is impractical and unnecessary.

A more popular and universally applicable approach to studying phosphoproteins that is highly sensitive is immunostaining. Since phosphoproteins are typically of low

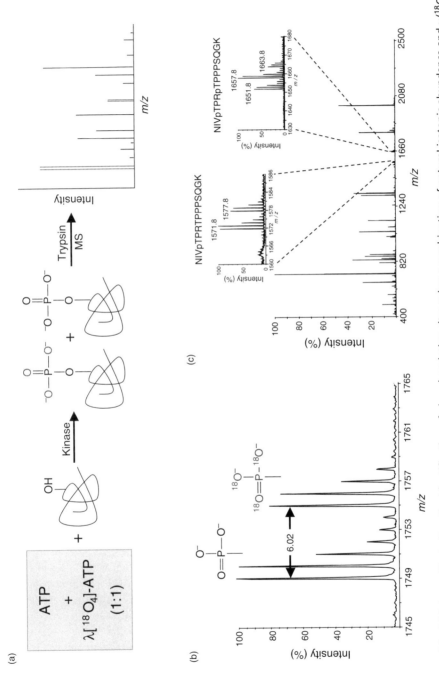

Figure 4.3. (a) Scheme illustrating the identification of phosphorylation sites using a combination of natural isotopic abundance and $\gamma(^{18}O_4)$-ATP. (b) Mass spectrum of peptide labeled with both natural isotopic abundance and $\gamma(^{18}O_4)$-ATP. (c) Peptide mapping of *in vitro* kinase reaction products of MAP kinase and MBP in the presence of a 1:1 mixture of ATP and $\gamma(^{18}O_4)$-ATP showing the identification of a mono- and diphosphorylated version of a peptide.

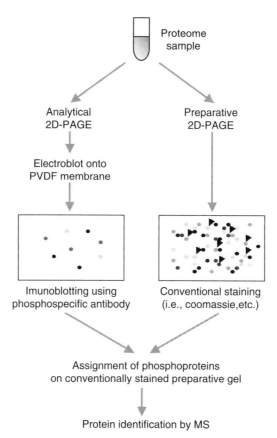

Figure 4.4. The use of 2D-PAGE and immunoblotting to identify phosphoproteins. Due to the low abundance of most phosphoproteins, the sample a preparative and analytical gel of the proteome sample can be run. Obviously, a greater amount of sample is loaded onto the preparative gel. After separation, the analytical gel is immunoblotted using a phospho-specific antibody to reveal the location of phosphoproteins within the gel. Spots visualized in this manner are aligned with their spot on the preparative gel that has been stained using a conventional staining method. The phosphoprotein can then identified using standard MS-based proteome methods.

abundance within a proteome, finding and identifying them using immunostaining requires a two-pronged approach as illustrated in Figure 4.4 (29). The proteome of interest is split into two fractions and both are separated on a 2D-PAGE gel. One gel is considered analytical and the other preparative. A larger amount of the proteome is separated on the preparative gel to facilitate downstream protein identification. The proteins within the analytical gel are transferred to a polyvinylidene fluoride (PVDF) membrane, which is then probed using an antibody that detects phosphorylated resides within proteins. The preparative gel is simply stained with a colorimetric dye such as

Coomassie to reveal the location of proteins. The analytical and preparative gels are aligned to correlate any protein spots that were recognized by the antibody with those stained on the preparative gel. Spots that match between the analytical and preparative gel are cored from the preparative gel, in-gel digested, and the protein identified using MS.

This affinity-based detection method, however, only determines if a protein is phosphorylated and does not directly identify the modified residue, unless an antibody that targets a specific phosphorylated residue is available. Knowing the exact site that is modified is important since identical modifications at different sites within the same protein can have widely different effects on protein activity. In addition, a single protein can be phosphorylated by different kinases resulting in disparate functions. For example, the multifunctional transcription factor TFII-I has broad biological roles in transcription and signal transduction. Depending on the cell type, the protein has been shown to contain Src-dependent tyrosine phosphorylation sites (30). The protein, however, interacts both physically and functionally with Bruton's tyrosine kinase (31, 32) and inducible tyrosine kinase (Itk) (33). In addition, the α-amino-3-hydroxy-5-methyl-4-isoxazolepropionate (AMPA) receptor is phosphorylated by cAMP-dependent protein kinase II and protein kinases A and C. Each phosphorylation event results in a different AMPA receptor function (34, 35). While site-specific mAbs can be used to map specific sites of phosphorylation, preparing these antibodies is very laborious and expensive. It is unlikely that enough antibodies with the required specificity will be manufactured in the near future to cover the tens of thousands of predicted phosphorylated sites within the human proteome.

While there are a number of reasons that MS is quickly supplanting all other techniques as the method of choice to characterize phosphoproteins, the main one is specificity. Unlike an antibody, the mass spectrometer can produce a highly confident signal that directly identifies a phosphorylated site within the protein. While MS can find phosphorylated proteins by comparing the calculated mass of a protein to its accurate mass measurement before and after phosphatase treatment (36), this type of experiment is rarely done. There are two fundamental reasons for this: (1) obtaining the mass of an intact protein requires the protein to be highly purified prior to MS measurement and (2) there are a number of PTMs that occur to a protein that cannot be factored into the calculated difference between the experimental and calculated masses. The bigger reason, however, is that knowing a protein *is* phosphorylated is not as important as knowing *where it is* phosphorylated.

4.2.1   Identification of Phosphorylated Proteins

The most common use of MS in characterizing phosphorylation sites is to enzymatically digest the protein into peptides and to analyze the fragments using MS^2 or MS^3 (37–39). If a peptide is known to contain only a single phosphorylated residue, MS is sufficient; however, MS^2 is always recommended as it provides more definitive peptide identification. If the peptide contains multiple phosphorylation sites, or multiple sites that can potentially be phosphorylated (i.e., many Ser, Thr, or Tyr residues) MS^2 is necessary to establish the specific phosphorylated site(s).

4.2.2 Phosphopeptide Mapping

The most basic method for identifying a phosphorylation site within a protein is phosphopeptide mapping. In phosphopeptide mapping, a purified (or at minimum, highly enriched) phosphoprotein is digested into peptides and their mass-to-charge (m/z) ratios are recorded (40,41). If the protein being studied is known, phosphorylated peptides will have an experimental mass 79.96 Da higher than the calculated mass of the unmodified peptide. If the protein is unknown, the experimental masses of unmodified peptides can be used to identify the protein and peaks with m/z ratios 79.96 units (or integral units of) higher than the masses of predicted unmodified peptides can be assumed to be phosphorylated. Obviously using a higher resolution mass spectrometer, such as a time-of-flight (TOF) or Fourier transform ion cyclotron resonance (FTICR), aids in the confidence of the identification. Additionally, the protein can be analyzed prior to and after treatment with a phosphatase enzyme (42). Peaks that show a shift of an integral value of 79.96 Da can be assumed to be phosphorylated.

4.2.3 Collision-Induced Dissociation

As with protein identification, collision-induced dissociation (CID) (or MS^2) is still the most common method of identifying phosphorylation sites within proteins (43,44). The main advantage of MS^2 over peptide mapping is that it is capable of providing the exact site of phosphorylation within a peptide. Phosphopeptide mapping only provides the exact site of modification if there is a single residue within the peptide that can be phosphorylated. Many different MS^2 strategies exist; however, they all rely on the lability of the phosphoester bonds of pTyr, pThr, and pSer residues. The phosphoester bond is easily fragmented within a collision cell, electrospray ionization ion source, or during postsource decay when using a matrix-assisted laser desorption ionization (MALDI)-based mass spectrometer. The loss of the phosphate group as HPO_3 or H_3PO_4 is a favored fragmentation event, and generally dominates over amide backbone cleavages, especially when the molecular ion is singly charged. It has been observed that phosphate tends to be lost from pSer more readily than pThr and from pThr more readily than from pTyr in MS^2 experiments (45). Phosphate groups in smaller peptides are also generally more labile than those in larger peptides, since approximately the same collisional energy is spread throughout a smaller number of covalent bonds.

The lability of the phosphodiester bond can make identifying phophosphorylated peptides challenging. Ideally, a significant percentage of the phosphopeptides retain the phosphate group on the modified amino acid and the presence of this modified residue is identified by the increase in the apparent MW at a particular point in the sequence. This situation is illustrated in Figure 4.5 for a phosphorylated and nonphosphorylated version of a peptide. The MS^2 spectrum of the phosphorylated version of the peptide

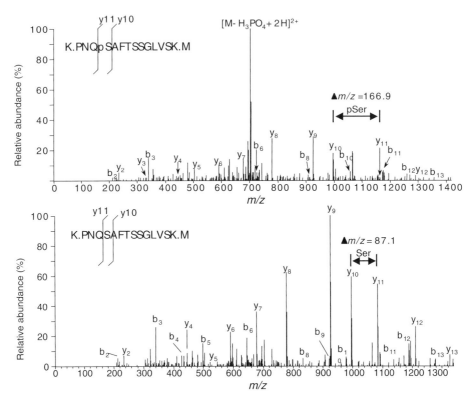

Figure 4.5. A comparison of tandem MS^2 spectra of phosphorylated (top) and unmodified (bottom) form of the same peptide.

shows a large peak representing the loss of H_3PO_4; however, the signals from the other fragments are strong enough to show the difference between the y^{10} and y^{11} ions as 166.9 Da, representing a phosphorylated Ser residue. The difference in the y^{10} and y^{11} fragment ions in the nonphosphorylated version of the peptide are separated by only 87.1 Da; the mass of an unmodified Ser residue. Unfortunately, what happens all too frequently is CID energy localizes to the phosphodiester bond resulting in a MS^2 spectrum showing an intense single peak correlating to the loss of H_3PO_4. While this MS^2 spectrum indicates the presence of a phosphorylated peptide, there is insufficient signal from the other fragments to determine the sequence of the peptide and hence its identity. For example, Figure 4.6 shows the MS spectrum of a proteome mixture in which the peak at m/z 923.8 was selected for MS^2. The MS^2 spectrum shows a single large peak (m/z 874.8) that is characteristic of a phosphorylated peptide in which the phosphate group has been lost. Unfortunately, not enough fragment ions provide signals that can identify the peptide. The key is to use the ion-trap's capability of isolating fragment ions and perform another round of CID (i.e., MS^3) (46). All of the sequence information to identify the peptide is present within the peak at m/z 874.8,

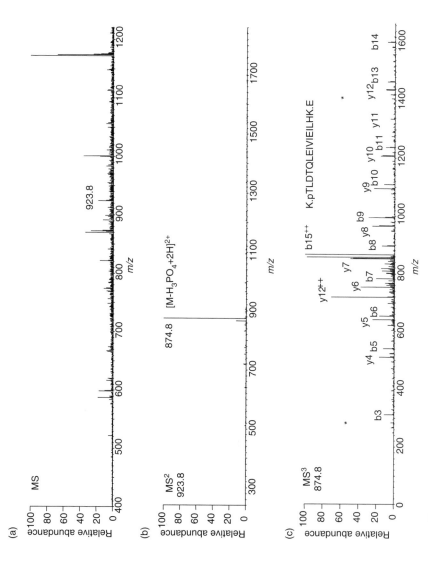

Figure 4.6. Use of MS^3 for identification of phosphorylated peptides. In the top panel, the peak at m/z 923.8 was selected for CID. The MS^2 spectrum is dominated by a single intense peak indicative of the loss of the phosphate group; however, insufficient information is available within the spectrum to identify the peptide's sequence. Isolation of the fragment ion at m/z 874.8 and subjecting this ion to another round of CID (i.e., MS^3) provides sufficient information to identify the peptide.

97

which is the doubly charged version of the peak at 923.8 minus H_3PO_4. When m/z 874.8 is fragmented the resulting MS^3 spectrum shows signals originating from the now unmodified peptide. These signals are now prominent enough to conclusively identify the peptide.

In a study comparing the utility of MS^2 and MS^3 for identifying phosphorylation sites within a proteome, Yu et al. analyzed the phosphoproteome of HeLa cells that was extracted using titanium dioxide (TiO_2) chromatography (47). Over 850 phospho-peptides corresponding to 1034 distinct phosphorylated residues were identified using data-dependent neutral loss nano-RPLC–MS^2–MS^3analysis. Of these, 41% and 35% of the phosphopeptides were identified only by MS^2 and MS^3, respectively, while 24% were identified by both MS^2 and MS^3. Cross-validation by MS^2 and MS^3 scans pro-vided a confidence level of 99.5% for the identification of phosphopeptides. This study illustrated the value of conducting both rounds of CID in producing high quality datasets of phosphopeptides.

4.2.4 Electron Capture and Electron Transfer Dissociation

As mentioned the lability of the phosphodiester linkage makes identifying phospho-peptides challenging. The ion fragmentation technique, electron capture dissociation (ECD), appears to be a promising method for identifying phosphopeptides as well as sequencing large protein fragments (48, 49). In ECD, multiply protonated peptide (or protein) ions are trapped and energetic electrons are introduced. The energy released from the electrons causes fragmentation almost exclusively along the backbone of the peptide to occur, resulting in c and z ions (50). Labile modifications such as phospho-rylation (and O-glycosylation), however, are retained. In comparison to CID spectra of peptide ions, ECD spectra show considerably less side-chain losses of PTMs, and much higher numbers of backbone cleavages preserving both sequence and PTM information.

4.2.5 Electron Transfer Dissociation

Though ECD has shown great promise in its utility for characterizing PTMs, due to the physical constraints of the technique it is restricted solely to FTICR instruments, which require a great deal of technical expertise to operate and are the most expensive MS instruments. It remains to be seen, therefore, how broadly utilized this technique will be in proteomics. Very recently, Hunt and coworkers have developed an alternative to ECD where fragmentation is accomplished through a gas phase reaction after coincidentally trapping peptide/protein molecular ions with anions formed by a chemical ionization source (51). The entire process can be accomplished on millisecond timescales and utilizes a commercially available and affordable QLT.

In the initial demonstration of the utility of ETD, a synthetic doubly phosphory-lated peptide (LPISASHpSpSKTR) was subjected to fragmentation by reaction with anthracene anions in the QLT. After 50 ms of reaction time, all predicted c- and z-type fragment ions were observed, including both of the intact phosphoseryl residues. An experiment coupling ETD with nanoflow LC separation showed that ten synthetic pep-tides present at a concentration in the low femtomolar range, including a phosphopeptide, could be readily sequenced from the observed fragment ions. When the same mixture was analyzed using CID, the phosphopeptide fragment ion spectrum was dominated by ions corresponding to the loss of phosphoric acid with insufficient information to identify the peptide sequence.

4.2.6 Enrichment of Phosphopeptides

Divide and conquer is a basic tenet of proteomic analysis. Whether the aim is to identify membrane proteins, a specific protein complex, or modified proteins, the ability to prepare a sample that is enriched with the target proteins is critical in obtaining useful MS data. While techniques such as ECD and ETD continue to mature, most phosphorylated residues will still be identified through MS^2 and MS^3 CID analysis of phosphopeptides produced through a digest with trypsin or some other proteolytic enzyme. What this digestion does is produces an enormous pool of peptides of which only a small percentage are phosphorylated. Subjecting the entire pool for data-dependent LC–MS^2 (or MS^3) analysis will result in most of the identifications being from unmodified peptides owing to the relative high abundance compared to phosphopeptides. Detecting and characterizing phosphopeptides can be enhanced by increasing the percentage of peptides within a mixture that are phosphorylated or eliminating the background of unmodified peptides (two sides of the same coin).

4.2.7 Immunoaffinity Chromatography

One of the best methods for extracting phosphorylated proteins from a sample is immunoaffinity (52–54). Phosphorylated protein(s) that are immunoprecipitated from a complex mixture are then digested and analyzed using the MS strategies described above. The selection of the type of antibody to use is dependent on the goal of the study. If a specific protein is of interest, an antibody directed toward any epitope within that protein may be used and the residue(s) that is phosphorylated is determined using MS. If the study is a global phosphoproteomic study, there are a number of pan-phosphoseryl (pSer), -phosphothreonyl (pThr), and -phosphotyrosyl (pTyr) antibodies available. While the phosphoamino acid is part of the recognition epitope, it is not the sole recognition factor between the antibody and phosphoprotein as neighboring residues (i.e., the consensus sequence) surrounding the phosphorylated residue contributes to the specificity between the antibody and specific phosphoproteins.

In most global phosphoproteomic studies, an antibody directed toward a specific type of phosphorylated residue is utilized. Such a strategy was used to identify differences in the phosphorylation state of proteins within resting platelets and those that had been stimulated with thrombin (55). The platelet proteome was separated using 2D and 1D-PAGE and phosphotyrosine-containing proteins were identified using an antiphosphotyrosine antibody. Proteins that reacted positively with the antibody were analyzed by MALDI-TOF and LC–MS^2. A study to measure changes in the phosphorylation states of proteins extracted from murine fibroblasts exposed to tumor necrosis factor-α (TNF-α) utilized a similar strategy (56). Two separate (but identical) 2D-PAGE resolved fibroblast proteomes were either electroblotted onto a PVDF membrane and immunoblotted with a phosphotyrosine-specific antibody or visualized within the gel using silver stain. Proteins that showed a positive interaction with the antibody were aligned with their corresponding spots within the silver stained gel and selected for MS analysis. Twenty-one different phosphoproteins were identified within the fibroblasts, including eight that showed a time-dependent change in their phosphorylation state after treatment with TNF-α.

As MS methods have evolved so has the evolution of using antibodies to decipher the phosphoproteome. To provide the most definite result that a phosphorylation event has been identified in a proteomic study it is important to observe the phosphorylated residue in the MS data. Many studies (including those described above) often use an antibody to enrich for phosphoproteins but the exact phosphorylated residue is not observed in the MS analysis. The fact that the observed protein is phosphorylated is simply presumed. To increase the confidence in these types of analysis, Rush et al. immunoprecipitated phosphotyrosine peptides from a complex proteome by digesting the proteome prior to the immunoprecipitation step (Figure 4.7) (57,58). In this strategy, a tryptically digested proteome samples was passed over a protein G agarose column to which a phosphotyrosine monoclonal antibody had been coupled. The captured phosphopeptides were analyzed using LC–MS^2. By immunoprecipitating the pTyr peptides, this method ensures that the mixture analyzed by MS is highly enriched with peptides containing phosphotyrosine residues and much of the background of unmodified peptides is eliminated during sample preparation. In a proof-of-performance study, 194 and 185 phosphotyrosine sites were identified in pervanadate-treated Jurkat cells and NIH-3T3 cells constitutively expressing active c-Src. About 70% of the identified sites had not previously been shown to be phosphorylated.

4.2.8 Immobilized Metal Affinity Chromatography

One of the most popular affinity-based methods to extract phosphoproteins or phosphopeptides is immobilized metal affinity chromatography (IMAC) (59–61). In IMAC, trivalent cations such as Fe^{3+} or Ga^{3+} are coupled to a chromatographic solid support through chelation to iminodiacetic or nitrilotriacetic. Phosphopeptides are extracted from a proteome sample owing to the greater affinity of phosphate groups than unmodified residues. As with other methods, nonspecifically bound peptides are washed from the column and the remaining bound material is analyzed using MS. IMAC can be done both at the protein and peptide level, however, it is more effective with downstream analysis to conduct it at the peptide level since much of the background of unmodified peptides is removed. Unfortunately nonspecific binding of unmodified peptides is a major problem with IMAC. In particular, acidic peptides containing carboxylate groups (e.g., Asp and Glu) will also bind to IMAC columns (62). To increase the selectivity of IMAC, one group added methanolic-HCl to a tryptically digested proteome to convert all of the carboxylate groups to methyl esters, while leaving the phosphate groups intact (63). When this strategy was applied to a whole yeast lysate more than 1000 phosphopeptides were identified. While this does not sound like a large number compared to today's

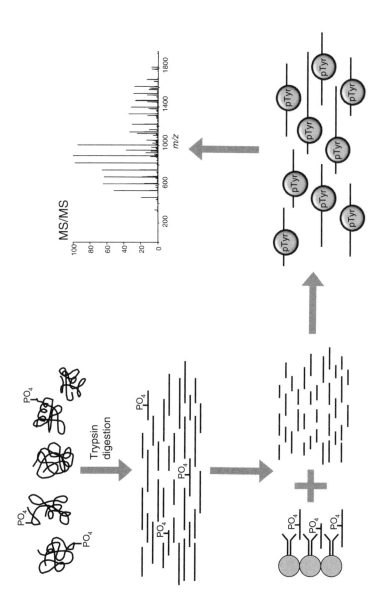

Figure 4.7. After digesting the proteome into peptides, it is passed across an antiphosphotyrosine monoclonal antibody column to extract pTyr-containing peptides. This enriched mixture is then analyzed by liquid chromatography coupled on-line with tandem MS[2] to identify the exact sequence of the pTyr peptides.

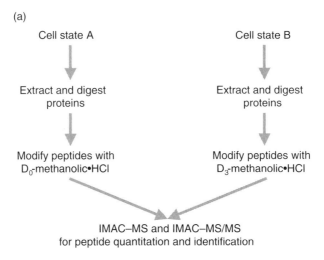

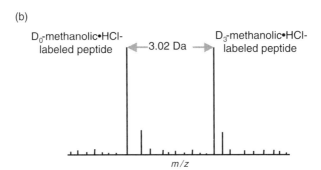

Figure 4.8. (a) Schematic representation of phosphopeptide quantitation using D_0- and D_3-methanolic·HCl derivatization of peptides followed by IMAC enrichment of phosphopeptides and quantitation and identification by MS and MS/MS. (b) Representative MS spectrum of a pair of phosphopeptides derivatized with isotopically different versions of methanolic·HCL.

standards, this number was achieved almost a decade ago with mass spectrometers far less powerful than currently available.

Since the above-mentioned method utilized a chemical labeling approach, it was straightforward to extend it to a quantitative phosphorylation project by utilizing stable isotope labeled reagents (Figure 4.8). Two different cell systems were compared by esterifying one with D_0-methanolic·HCl and the other with D_3-methanolic·HCl (63). The phosphorylation state of proteins within mammalian sperm as they underwent capacitation was studied. Phosphopeptides isolated from sperm at some $t = 0$ were esterified using D_0-methanolic·HCl and enriched using IMAC. Phosphopeptides within sperm that were isolated during capacitation was then labeled with D_3-methanolic·HCl and also enriched with IMAC. The two differentially labeled samples were mixed and analyzed using LC–MS^2. Several changes in the phosphorylation sites of proteins were identified including that of valosin-containing protein, a key protein involved in sperm capacitation.

4.2.9   Metal Oxide Affinity Chromatography

Metal oxide affinity chromatography (MOAC) is presently one of the most often utilized strategies for enriching for phosphopeptides (38, 64, 65). TiO$_2$ is especially popular, however, other compounds such as zirconium oxide (ZrO$_2$), have been utilized. Both of these materials have better selectivity towards phosphopeptides than IMAC and also have higher capacity. To improve the binding selectivity of phosphopeptides on TiO$_2$, substances such as 2,5-dihydrobenzoic acid, glutamic acid, and phthalic acid have been added to the loading buffer to minimize the binding of unmodified peptides (66–69). While IMAC and MOAC are generally used singularly, there is nothing preventing them from being used in tandem. Studies have shown that TiO$_2$ is able to separated monophosphorylated peptides more efficiently than multiphosphorylated peptides (70). Conversely, IMAC is most effective for enriching for multiphosphorylated peptides. Applying IMAC and MOAC in tandem enables non-, mono-, and multiphosphorylated peptides to be resolved into distinct aliquots.

4.3   GLYCOSYLATION

Along with phosphorylation, glycosylation is one of the most common modifications that occur within the cell (71–73), but unfortunately determining the exact structure of this modification at any specific residue is immensely more challenging. A significant percentage of proteins, especially those that are secreted or associated with the cell membrane, have been shown to be glycosylated. Among its many perceived functions, glycosylation is believed to increase a protein's solubility and permit specific interactions with other molecules (74). Carbohydrate groups are commonly found on classes of proteins such as immunoglobulins, proteases, cytokines, and cell surface receptors, which function primarily through interactions with other proteins.

The two main types of glycosylation are N-linked (covalently attached to the side chain nitrogen atom of an Asn residue) and O-linked (attached to the side chain oxygen atoms of primarily Ser and Thr residues) (75, 76). Consensus sequences can be used to predict whether a specific residue has the potential to be N-glycosylated as this modification typically occurs within the sequence -Asn–Xaa–Ser-, -Asn–Xaa–Thr-, or -Asn–Xaa–Cys- (where Xaa is any residue except Pro). There are no readily apparent consensus sequences to predict sites of O-linked glycosylation although algorithms are continuing to be developed that may assist in the prediction of Ser and Thr residues that are glycosylated (77).

The reason that characterizing glycosylation sites within proteins and peptides is important is simply because like phosphorylation, the modification plays a major role in the function of the protein. This importance is adequately illustrated by a study conducted in our laboratory in which a biomarker for interstitial cystitis was discovered

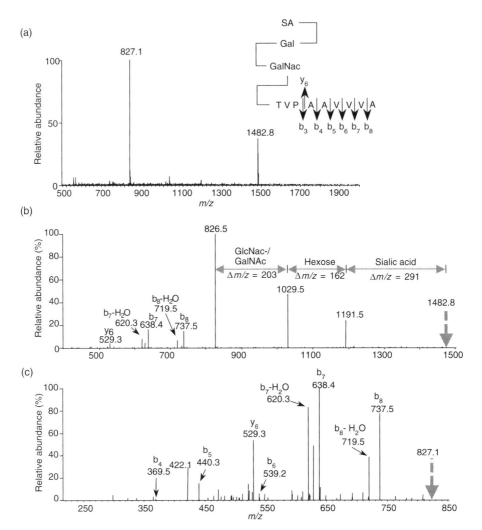

Figure 4.9. (a) Tandem MS analysis of APF showing the identification of the glycosylation group (upper trace) and peptide sequence (lower trace). (b) Inhibition of bladder epithelial T24 cells. Native APF isolated from urine of patients with interstitial cystitis and the synthetic version had a significant negative effect on cell proliferation, while the synthetic peptide alone (i.e., minus the glycosylation modification) had little, if any, effect.

(78). The presence of an small, bioactive peptide (termed antiproliferative factor or APF) in the urine of IC patients was first postulated over a decade ago (79). Using a combination of chromatographic methods and cell-based assays, APF was subsequently identified as a nine-residue glycosylated peptide, as shown in Figure 4.9 (78). The peptide sequence was determined to be TVPAAVVVA while the glycosylation group was composed of sialic acid (SA), galactose (Gal), and N-acetylgalactosamine (GalNac) bound to the N-terminal Thr residue. A synthetic version of this sialoglycopeptide had the same physiological affect on bladder epithelial cells as the wild-type peptide; however, removal of the glycosylation group inhibited the peptide's activity.

The challenges in determining the structures of glycans are due to their multi-branched organization, the variety of different carbohydrates that can be present, the isobaric nature of many of these moieties, and the different linkages that couple the groups together. Techniques such as X-ray crystallography and nuclear magnetic resonance spectroscopy may not provide structural information about the carbohydrate constituency due to their nonrigid structure. Determining the structure of N-linked eucaryotic glycosylation groups is somewhat simplified since these groups contain a pentasaccharide core made up of three mannose (Man) and two N-acetylglucosamine (GlcNac) units (80, 81). Additional carbohydrate residues are attached to this element via two outer mannose residues. High-mannose type modifications contain additional mannose residues that are linked to the pentasaccharide core (82). In complex glycosylation, GlcNac, Gal, SA, and L-fucose residues may be present with the glycan chains typically terminating with SA. There are also hybrid-type glycans that possess the features of high Man and complex glycans. Unfortunately, N-linked glycans represent the simpler of the two major types as O-linked glycans possess a wider variety of core structures whose makeup can range from monosaccharide modifications to large sulphated polysaccharides.

To characterize glycosylated proteins various pieces of information are required such as identification of the site(s) of glycosylation, structural characterization of the glycolytic side chain, quantitation of the extent of glycosylation at each site, and identification of the number of different glycoforms of each protein. MS contributes significantly to the identification of the sites of glycosylation and structural characterization of the glycolytic side chain using both MS and MS^2 measurements. The complete characterization of the different glycoforms is highly dependent on the ability to fractionate the various glycoforms and identify the glycolytic composition of each. To completely characterize a glycosylated protein can require a three-pronged approach as shown in Figure 4.10. The protein sample is often split into two separate aliquots, with one being enzymatically digested (trypsin being most commonly used) and the other being deglycosylated. The digested fraction is then analyzed directly using MS^2 to identify the glycosylated peptides. Unfortunately the MS^2 data may not be sufficient to accurately describe the glycan structure or may result in spectra too complicated to accurately identify the protein backbone. To provide additional data, the deglycosylated intact protein is also digested into peptides, which are identified to provide the backbone structure of the protein. The liberated glycans are also analyzed using MS^2 to identify their structure. In all three cases, the data are analyzed to not only identify the peptides and glycans but also correlate which peptides were modified by which glycans.

4.3.1   Mass Spectrometry Characterization

A subtractive approach can be used to locate an N- or O-linked glycosylated residue within a protein. In this strategy, a peptide map of the protein is obtained prior to and after

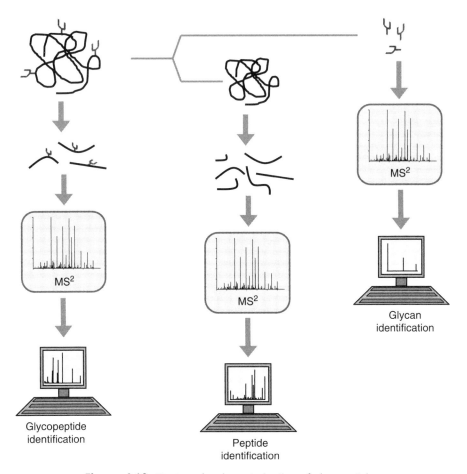

Figure 4.10. Strategy for characterization of glycoproteins.

treatment using a deglycosylation agent (83, 84). O-linked and N-linked carbohydrates can be removed using base-catalyzed β-elimination and N-glycanases, respectively (85). The peptide maps of the glycoprotein before and after deglycosylation are acquired and the appearance (or disappearance) of new molecular ions enables the identification of those peptides that were originally glycosylated. N-linked carbohydrates can be digested with endoglycosidase H resulting in a glycopeptides containing a GlcNAc attached to the residue (86, 87). A peptide signal 203 Da higher than its deglycosylated counterpart is observed by MS. Another useful marker is that fact that digestion of N-linked glycopeptides with N-glycanase converts the Asn residue to an aspartyl (Asp) residue resulting in a 1 Da increase in the mass of the deglycosylated peptide compared to its calculated unmodified mass (88).

As mentioned above, the complex branching present in many oligosaccharides, makes structural elucidation of the carbohydrate group using MS extremely difficult.

Accurately determining the carbohydrate group's structure requires identifying fragment ions indicative of the intersaccharide linkages and cross-ring products; however, these ions are notoriously of low intensity in MS^2 spectra (89, 90). These molecular ions may be entirely absent from the CID spectrum of complex glycans, resulting in the need for additional derivatization methods or other analytical techniques to be employed, such as methylation analysis, to obtain data to determine the carbohydrate structure.

4.3.2 Electron Capture and Electron Transfer Dissociation

As with phosphoprotein analysis described above, ECD and ETD hold tremendous promise in the characterization of glycoproteins (91). The selective fragmentation along the peptide backbone, allows ECD to determine the site of glycosylation since the labile modification is retained on the peptide backbone. In general, the fragmentation characteristics of ETD are similar to ECD. ETD provides an almost complete c- and z-type ion series without any identifiable loss of the glycosylation group from N- and O-linked glycopeptides. Therefore, ECD and ETD allow information about the peptide sequence, glycan attachment sites, and glycan mass to be retained. Typical neutral losses that occur upon ETD that are indicative of N-linked glycosylation are low molecular weight losses like 43 Da, which originate from the radical neutral loss *N*-acetyl moieties of the HexNAc units (92). Importantly for the characterization of the specific site of modification, ETD spectra of glycopeptides contain ions resulting from glycosidic bond cleavages while retaining those related to the intact peptide.

4.3.3 Targeted Identification of Glycoproteins

An excellent example of the use of ECD to characterize a glycoprotein is in the analysis of a lectin from the coral tree, *Erythrina corallodendron* (93). The FTICR mass spectrum of a tryptic digest of this protein revealed two major peaks that corresponded to the masses of known glycopeptide structures. The ECD fragment spectrum obtained from the N-glycosylated peptide of *m/z* 1005.5 (residues 100–116) was dominated by N-terminal c-type ions that provided enough information to allow the amino acid sequence and glycan structure of this peptide to be identified. Fragment ions corresponding to cleavages at 10 of 15 peptide backbone amide bonds were observed. Interestingly, the amino acid residues that did not fragment during ECD were located close to the glycosylated Asn residue. This result suggests that the glycan sterically hinders access to the proximal backbone carbonyl oxygens. Three glycosylated fragment ions were observed in the ECD spectrum of this glycopeptide. All three contained the entire glycan structure with no fragmentation of glycosidic bonds occurring.

However, the structure of this glycan modification was still incomplete. While ECD effectively located the modification, the lack of fragmentation of the carbohydrate group left its structure unknown. To characterize the modification further, infrared multiphoton dissociation (IRMPD) (94) was employed (93). As expected from a complex modification, a complicated fragmentation pattern was observed. Fortunately, however, the fragmentation pattern showed several ions originating from the parent glycopeptide with the loss of one or more sugars. IRMPD resulted in fragmentation at each glycosidic bond, as well as the loss of the entire glycan. The observed monosaccharide losses revealed the presence of multiple branch points in the structure of the glycan. A doubly protonated fragment at m/z 1097.0 ($[M - Man_3XylGlcNAc + 2H]^{2+}$) indicated the position of fucosylation within the inner GlcNAc residue. The IRMPD spectrum did not provide sufficient information to deduce the peptide sequence, but when used in combination with ECD (or ETD) the entire glycopeptide can be extensively characterized.

4.3.4 Proteome-Wide Identification of Glycoproteins

As described in previous chapters, much of proteomics is focused on characterizing as many proteins within a proteome as possible (i.e., global proteomics). As described in this chapter, there are a number of techniques available for doing global analysis of phosphoproteins. Unfortunately, conducting global analysis of glycoproteins is not nearly as mature. As mentioned before, the complexity and heterogeneity of glycosylation makes characterizing these modifications very challenging, yet their importance in cell physiology remains. Like any complex mixture, fractionation is crucial for analyzing the glycome. One of the best methods utilizes 2D-PAGE fractionation of the proteome following by detection of glycosylated proteins using fluorescent dyes that are specific for glycosylation groups (95, 96). With the movement away from 2D-PAGE gels, lectin affinity chromatography makes an effective means to enrich mixtures for glycoproteins that will eventually be analyzed using solution-based MS techniques (97–99). While lectin chromatography can enrich glycomic samples, they have a variety of different affinities; therefore, there is no one single lectin that will extract all glycoproteins from a proteome. Hydrazide chemistry has also been utilized to enrich glycoproteins and glycopeptides. In this method, a covalent hydrazone bond is formed between the glycans and the hydrazide groups that are coupled to a solid support (100). This method has been applied mostly for N-linked glycoprotein analysis since enzymes such as PNGase F can effectively remove the covalently attached glycoprotein from the support (101).

Quantitative glycoproteomics has become an important area in the attempt to find circulating biomarkers that can diagnose disease conditions such as tumorigenesis and inflammation. The results of several large scale glycoproteomic studies of human plasma and serum have contributed to the development of blood glycoproteomic databases containing hundreds of known glycoproteins (102, 103). These studies have used many of the aforementioned techniques including hydrazide chemistry and lectin affinity chromatography. Although not as comprehensively characterized as serum or plasma, the glycoproteome has also been analyzed for biofluids such as cerebrospinal fluid, tears, urine, and pancreatic juice illustrating the universal importance of glycosylation throughout the human body (104).

Most of the isotopic labeling techniques utilized for general proteomics can be utilized for quantitative glycoproteomics to compare diseased samples to controls, for example. Probably, the only one that would limit the coverage of glycoproteins would be ICAT since it specifically labels cysteinyl residues. One recent study used iTRAQ labeling combined with different glycoprotein enrichment approaches to study hepatocellular carcinoma. In this study, N-linked glycoproteins were enriched from hepatocellular carcinoma patients and controls using multilectin chromatography followed by labeling with the iTRAQ reagents and LC–MS analysis. In addition to iTRAQ, ^{18}O-labeling can be readily introduced into N-glycopeptides through trypsin- or PNGase F-mediated reactions depending on the desired site of labeling. The SILAC method is also effective for quantitative glycoproteomics since the technique incorporates isotope labels along the peptide backbone. Currently, the most popular method of quantitative glycoproteomics is the label-free approach owing to its simplicity and low cost compared to other methods.

4.4   OTHER POSTTRANSLATIONAL MODIFICATIONS

While this chapter has focused on the characterization of two major types of PTMs, over 200 different types are known. Many of these are rare, yet their presence is still critical to the protein's function. While the characterization of phosphorylation sites dominate the field of proteomics and identifying sites of glycosylation continues to be thought of as a critical area, other modifications have also been detected by MS. Detection of ubiquitinated sites using MS is becoming more common as tryptically digesting a protein leaves a residual glycyl-glycyl dipeptide covalently attached to the modified lysyl residue of the target protein. This signature modification with a mass of 114.1 Da is easily detected owing to its consistency. Early application of this rule to whole yeast cell lysates resulted in the identification of 110 ubiquitination sites from 72 different proteins (105). The ability to detect ubiquitination sites globally have progressed to the point where over 750 unique lysine ubiquitination sites on 471 proteins have been reported in eucaryotic cells (106). A very recent study has claimed to identify 11,054 endogenous putative ubiquitination sites on 4273 human proteins; equivalent to about two-thirds of the putative ubiquitination sites within the human proteome (107). The specificity afforded with MS continues to enhance its reputation as the most confident method for identifying ubiquitinated residues within single proteins as well.

Other specific modifications that are gaining in importance and are increasingly being characterized using MS are acetylation and methylation. The identification of acetylation sites has traditionally followed a straightforward approach in which a modification of 42 Da on lysyl or N-terminal residues while a modification of 14 Da on lysyl or arginyl residues is used to indicate a methylated residue. In addition, multiples of 14 Da can indicate multiple methylations on the same residue. One of the earliest examples of a study that characterized both types of modifications was conducted on proteins from human lens tissues (108). Lens tissue proteins were digested using three different enzymes to increase coverage and identify overlapping peptide sequences. The three peptide mixtures were characterized using MudPIT and the resulting data was analyzed for modifications including acetylation and methylation. Multiple proteins containing either single or multiple forms of these modifications were identified, at a higher percentage than most scientists probably expected. The ability to map acetylation sties has progressed to a point where the quantitative change in 3600 lysine acetylation sites on 1750 proteins in response to the deacetylase inhibitors has been reported (109).

4.5 CONCLUSIONS

One of the major goals of proteomics is to characterize the protein complement of cells enabling the interactions between constituents of various pathways and networks to be deciphered. Identifying the proteins involved is critical; however, this information does not provide enough information to achieve the ultimate goal. Without knowing the signals that contribute to protein interactions, the dynamics within the model cannot be ascertained. The advances made in identifying proteins in complex biological mixtures have also benefited the characterization of PTMs. Obviously identifying certain PTMs are easier than others, yet one thing is common to methods for characterizing all of them; incorporating a sample preparation method that enriches the mixture for the specific type of modification being studied results in higher confidence MS results. Proteomics has only begun to scratch the surface based on the number of known, and possibly yet undiscovered, PTMs. Fortunately, ways to analyze these modifications are always improving beginning with sample preparation through to bioinformatic analysis of the data.

REFERENCES

1. Mann M, Jensen ON. Proteomic analysis of post-translational modifications. Nature 2003;21:255–261.
2. Krishna RG, Wold F. Post-translational modifications of proteins. Adv. Enzymol. Relat. Areas Mol. Biol. 1993;67:265–298.
3. Mok J, Zhu X, Snyder M Dissecting phosphorylation networks: Lessons learned from yeast. Expert Rev. Proteomics 2011;8:775–86.
4. Taylor AD, Hancock WS, Hincapie M, Taniguchi N, Hanash SM. Towards an integrated proteomic and glycomic approach to finding cancer biomarkers. Genome Med 2009;1:57.
5. Griffin TJ, Goodlett DR, Aebersold R. Advances in proteome analysis by mass spectrometry. Curr. Opin. Biotechnol 2001;12:607–612.
6. Pinna LA, Ruzzene M. How do protein kinases recognize their substrates?. Biochim. Biophys. Acta 1996;1314:191–225.
7. Cohen P. The origins of protein phosphorylation. Nat. Cell Biol 2002;4:E127–130.
8. An HJ, Froehlich JW, Lebrilla CB. Determination of glycosylation sites and site-specific heterogeneity in glycoproteins. Curr. Opin. Chem. Biol 2009;13:421–46.
9. Holmberg CI, Tran SE, Eriksson JE, Sistonen L. Multisite phosphorylation provides sophisticated regulation of transcription factors. Trends Biochem. Sci. 2002;27:619–627.
10. Frank CL, Ge X, Xie Z, Zhou Y, Tsai LH Control of Activating Transcription Factor 4 (ATF4) persistence by multisite phosphorylation impacts cell cycle progression and neurogenesis. J. Biol. Chem. 2010;285:33324–33337.
11. Gardner KH, Montminy M. Can you hear me now? regulating transcriptional activators by phosphorylation. Sci. STKE 2005;301:pe44.
12. Foster KG, Acosta-Jaquez HA, Romeo Y, Ekim B, Soliman GA, Carriere A, Roux PP, Ballif BA, Fingar DC. Regulation of mTOR Complex 1 (mTORC1) by Raptor Ser863 and multisite phosphorylation. J. Biol. Chem. 2010;285:80–94.

13. Haltiwanger RS, Lowe JB. Role of glycosylation in development. Annu. Rev. Biochem. 2004;73:491–537.

14. Puri A, Neelamegham S. Understanding glycomechanics using mathematical modeling: A review of current approaches to simulate cellular glycosylation reaction networks. Ann. Biomed. Eng. 2012;40:816–827.

15. Bertozzi CR, Kiessling LL. Chemical glycobiology. Science 2001;291:2357–2364.

16. Archuleta AJ, Stutzke CA, Nixon KM, Browning MD. Optimized protocol to make phospho-specific antibodies that work. Methods Mol. Biol. 2011;717:69–88.

17. Brumbaugh K, Johnson W, Liao WC, Lin MS, Houchins JP, Cooper J, Stoesz S, Campos-Gonzalez R. Overview of the generation, validation, and application of phosphosite-specific antibodies. Methods Mol. Biol. 2011;717:3–43.

18. Rawling J, Melero JA. The use of monoclonal antibodies and lectins to identify changes in viral glycoproteins that are influenced by glycosylation: The case of human respiratory syncytial virus attachment (G) glycoprotein. Methods Mol. Biol. 2007;379:109–125.

19. Nagai R, Horiuchi S, Unno Y. Application of monoclonal antibody libraries for the measurement of glycation adducts. Biochem. Soc. Trans. 2003;31:1438–1440.

20. Wu J, Warren P, Shakey Q, Sousa E, Hill A, Ryan TE, He T. Integrating titania enrichment, iTRAQ labeling, and orbitrap CID-HCD for global identification and quantitative analysis of phosphopeptides. Proteomics 2010;10:2224–2234.

21. Ge F, Xiao CL, Bi LJ, Tao SC, Xiong S, Yin XF, Li LP, Lu CH, Jia HT, He QY. Quantitative phosphoproteomics of proteasome inhibition in multiple myeloma cells. PLoS One 2010;5:e13095.

22. Cohen P. Targeting protein kinases for the development of anti-inflammatory drugs. Curr. Opin. Cell. Biol. 2009;21:317–324.

23. Cohen, P. Protein kinases—the major drug targets of the twenty-first century? Nat. Rev. Drug Discov. 2002;1:309–315.

24. Pawson T, Kofler M. Kinome signaling through regulated protein-protein interactions in normal and cancer cells. Curr. Opin. Cell. Biol. 2009;21:147–153.

25. Pawson T, Linding R. Network medicine. FEBS Lett. 2008;582:1266–1270.

26. Manning DR, DiSalvo J, Stull JT. Protein phosphorylation: Quantitative analysis in vivo and in intact cell systems. Mol. Cell. Endocrinol. 1980;19:1–19.

27. Zhou M, Meng Z, Jobson AG, Pommier Y, Veenstra TD. Detection of in vitro kinase generated protein phosphorylation sites using gamma[18O4]-ATP and mass spectrometry. Anal. Chem. 2007;79:7603–7610.

28. Chu G, Egnaczyk GF, Zhao W, Jo SH, Fan GC, Maggio JE, Xiao RP, Kranias EG. Phospho-proteome analysis of cardiomyocytes subjected to beta-adrenergic stimulation: identification and characterization of a cardiac heat shock protein p20. Circ. Res. 2004;94:184–193.

29. Ducret A, Desponts C, Desmarais S, Gresser MJ, Ramachandran CA. General method for the rapid characterization of tyrosine-phosphorylated proteins by mini two-dimensional gel electrophoresis. Electrophoresis 2000;21:2196–2208.

30. Bu Y, Gao L, Gelman IH. Role for transcription factor TFII-I in the suppression of SSeCKS/Gravin/Akap12 transcription by Src. Int. J. Cancer 2011;128:1836–1842.

31. Caraveo G, van Rossum DB, Patterson RL, Snyder SH, Desiderio S. Action of TFII-I outside the nucleus as an inhibitor of agonist-induced calcium entry. Science 2006;314:122–125.

32. Sacristan C, Tussie-Luna MI, Logan SM, Roy AL. Mechanism of Bruton's tyrosine kinase-mediated recruitment and regulation of TFII-I. J. Biol. Chem. 2004;279:7147–7158.

33. Sacristán C, Schattgen SA, Berg LJ, Bunnell SC, Roy AL, Rosenstein Y. Characterization of a novel interaction between transcription factor TFII-I and the inducible tyrosine kinase in T cells. Eur. J. Immunol. 2009;39:2584–2595.

34. McDonald BJ, Chung HJ, Huganir RL. Identification of protein kinase C phosphorylation sites within the AMPA receptor GluR2 subunit. Neuropharmacology 2001;41:672–679.

35. Zheng Z, Keifer J. PKA has a critical role in synaptic delivery of GluR1- and GluR4-containing AMPARs during initial stages of acquisition of in vitro classical conditioning. J. Neurophysiol 2009;101:2539–2549.

36. Han JM, Kim JH, Lee BD, Lee SD, Kim Y, Jung, YW, Lee S, Cho W, Ohba M, Kuroki T, Suh PG, Ryu SH. Phosphorylation-dependent regulation of phospholipase D2 by protein kinase Cdelta in rat pheochromocytoma PC12 Cells. J. Biol. Chem. 2002;277:8290–8297.

37. Reinders J, Sickmann A. State-of-the-art in phosphoproteomics. Proteomics 2005;5:4052–4061.

38. Eyrich B, Sickmann A, Zahedi RP Catch me if you can: Mass spectrometry-based phospho-proteomics and quantification strategies. Proteomics 2011;11:554–570.

39. Sarg B, Chwatal S, Talasz H, Lindner HH. Testis-specific linker histone H1t is multiply phosphorylated during spermatogenesis. identification of phosphorylation sites. J. Biol. Chem. 2009;284:3610–3618.

40. Hasan N, Wu HF. Highly selective and sensitive enrichment of phosphopeptides via NiO nanoparticles using a microwave-assisted centrifugation on-particle ionization/enrichment approach in MALDI-MS. Anal. Bioanal. Chem. 2011;400:3451–3462.

41. Schilling B, Gafni J, Torcassi C, Cong X, Row RH, LaFevre-Bernt MA, Cusack MP, Ratovitski T, Hirschhorn R, Ross CA, Gibson BW, Ellerby LM. Huntingtin phosphorylation sites mapped by mass spectrometry. modulation of cleavage and toxicity. J. Biol. Chem. 2006;281:23686–23697.

42. Wu HY, Tseng VS, Chen LC, Chang YC, Ping P, Liao CC, Tsay YG, Yu JS, Liao PC. Combining alkaline phosphatase treatment and hybrid linear ion trap/orbitrap high mass accuracy liquid chromatography-mass spectrometry data for the efficient and confident identification of protein phosphorylation. Anal. Chem. 2009;81:7778–7787.

43. Kim MS, Zhong J, Kandasamy K, Delanghe B, Pandey A. Systematic evaluation of alternating CID and ETD fragmentation for phosphorylated peptides. Proteomics 2011;11:2568–2572.

44. Palumbo AM, Smith SA, Kalcic CL, Dantus M, Stemmer PM, Reid GE. Tandem mass spectrometry strategies for phosphoproteome analysis. Mass Spectrom. Rev. 2011;30:600–625.

45. Sickmann A, Mreyen M, Meyer HE. Identification of modified proteins by mass spectrometry. IUBMB 2002;54:51–57.

46. Macek B, Waanders LF, Olsen JV, Mann M. Top-down protein sequencing and MS^3 on a hybrid linear quadrupole ion trap-orbitrap mass spectrometer. Mol. Cell. Proteomics 2006;5:949–958.

47. Yu LR, Zhu Z, Chan KC, Issaq HJ, Dimitrov DS, Veenstra TD. Improved titanium dioxide enrichment of phosphopeptides from hela cells and high confident phosphopeptide identification by cross-validation of MS/MS and MS/MS/MS spectra. J. Proteome Res. 2007;6:4150–4162.

48. Stensballe A, Jensen ON, Olsen JV, Haselmann KF, Zubarev RA. Electron capture disso-ciation of singly and multiply phosphorylated peptides. Rapid Commun. Mass Spectrom. 2000;14:1793–1800.

49. Palumbo AM, Smith SA, Kalcic CL, Dantus M, Stemmer PM, Reid GE. Tandem mass spectrometry strategies for phosphoproteome analysis. Mass Spectrom. Rev. 2011;30:600–625.

50. Chalmers MJ, Hakansson K, Johnson R, Smith R, Shen J, Emmett MR, Marshall AG. Protein kinase a phosphorylation characterized by tandem fourier transform ion cyclotron resonance mass spectrometry. Proteomics 2004;4:970–981.

51. Syka JEP, Coon JJ, Schroeder MJ, Shabanowitz J, Hunt DF. Peptide and protein sequence analysis by electron transfer dissociation. Proc. Natl. Acad. Sci. U.S.A. 2004;101:9528–9533.

52. Pelech S. Tracking cell signaling protein expression and phosphorylation by innovative proteomic solutions. Curr. Pharm. Biotechnol. 2004;5:69–77.

53. Carrascal M, Gay M, Ovelleiro D, Casas V, Gelpí E, Abian J. Characterization of the human plasma phosphoproteome using linear ion trap mass spectrometry and multiple search engines. J. Proteome Res. 2010;9:876–884.

54. Boersema PJ, Foong LY, Ding VM, Lemeer S, van Breukelen B, Philp R, Boekhorst J, Snel B, den Hertog J, Choo AB, Heck AJ. In-depth qualitative and quantitative profiling of tyrosine phosphorylation using a combination of phosphopeptide immunoaffinity purification and stable isotope dimethyl labeling. Mol. Cell. Proteomics 2010;9:84–99.

55. Marcus K, Moebius J, Meyer HE. Differential analysis of phosphorylated proteins in resting and thrombin-stimulated human platelets. Anal. Bioanal. Chem. 2003;376:973–993.

56. Yanagida M, Miura Y, Yagasaki K, Taoka M, Isobe T, Takahashi N. Matrix assisted laser desorption/ionization-time of flight-mass spectrometry analysis of proteins detected by anti-phosphotyrosine antibody on two-dimensional-gels of fibrolast cell lysates after tumor necro-sis factor-alpha stimulation. Electrophoresis 2000;21:1890–1898.

57. Rush J, Moritz A, Lee KA, Guo A, Goss VL, Spek EJ, Zhang H, Zha XM, Polakiewicz RD, Comb MJ. Immunoaffinity profiling of tyrosine phosphorylation in cancer cells. Nat. Biotechnol. 2005;23:94–101.

58. Conrads TP, Veenstra TD. An enriched look at tyrosine phosphorylation. Nat. Biotechnol. 2005;23:36–37.

59. Ye J, Zhang X, Young C, Zhao X, Hao Q, Cheng L, Jensen ON. Optimized IMAC-IMAC protocol for phosphopeptide recovery from complex biological samples. J. Proteome Res. 2010;9:3561–3573.

60. Eyrich B, Sickmann A, Zahedi RP. Catch me if you can: Mass spectrometry-based phos-phoproteomics and quantification strategies. Proteomics 2011;11:554–570.

61. Thingholm TE, Jensen ON, Larsen MR. Enrichment and separation of mono- and multiply phosphorylated peptides using sequential elution from IMAC prior to mass spectrometric analysis. Methods Mol. Biol. 2009;527:67–78.

62. Seeley EH, Riggs LD, Regnier FE. Reduction of non-specific binding in Ga(III) immobilized metal affinity chromatography for phosphopeptides by using endoproteinase Glu-C as the digestive enzyme. J. Chromatogr. B Analyt. Technol. Biomed. Life Sci. 2005;817:81–88.

63. Ficarro SB, McCleland ML, Stukenberg PT, Burke DJ, Ross MM, Shabanowitz J, Hunt DF, White FM. Phosphoproteome analysis by mass spectrometry and its application to *Saccharomyces cerevisiae*. Nat. Biotechnol. 2002;20:301–305.

64. Colby T, Röhrig H, Harzen A, Schmidt J. modified metal-oxide affinity enrichment combined with 2D-PAGE and analysis of phosphoproteomes. Methods Mol. Biol. 2011;779:273–286.

65. Mazanek M, Roitinger E, Hudecz O, Hutchins JR, Hegemann B, Mitulović G, Taus T, Stingl C, Peters JM, Mechtler K. A new acid mix enhances phosphopeptide enrichment on titanium- and zirconium-dioxide for mapping of phosphorylation sites on protein complexes. J. Chromatogr. B Analyt. Technol. Biomed. Life Sci. 2010;878:515–524.

66. Larsen MR, Thingholm TE, Jensen ON, Roepstorff P, Jorgensen TJ. Highly selective enrichment of phosphorylated peptides from peptide mixtures using titanium dioxide microcolumns. Mol. Cell. Proteomics 2005;4;873–886.

67. Wu J, Shakey Q, Liu W, Schuller A, Follettie MT. Global profiling of phosphopeptides by titania affinity enrichment. J Proteome Res. 2007;6:4684–4689.

68. Mazanek M, Mituloviae G, Herzog F, Stingl, C, Hutchins JR, Peters JM, Mechtler K. Titanium dioxide as a chemo-affinity solid phase in offline phosphopeptide chromatography prior to HPLC–MS/MS analysis. Nat. Protoc. 2007;2:1059–1069.

69. Bodenmiller B, Mueller LN, Mueller M, Domon B, Aebersold R. Reproducible isolation of distinct, overlapping segments of the phosphoproteome. Nat. Methods 2007;4:231–237.

70. Thingholm TE, Jensen ON, Robinson PJ, Larsen MR. SIMAC (Sequential Elution from IMAC), a phosphoproteomics strategy for the rapid separation of monophosphorylated from multiply phosphorylated peptides. Mol. Cell. Proteomics 2008;7:661–671.

71. Schwarz F, Aebi M. Mechanisms and principles of N-Linked protein glycosylation. Curr. Opin. Struct. Biol. 2011;21:576–582.

72. Stalnaker SH, Stuart R, Wells L. Mammalian O-Mannosylation: unsolved questions of structure/function. Curr. Opin. Struct. Biol. 2011;21:603–609.

73. Nishimura S. Toward automated glycan analysis. Adv. Carbohydr. Chem. Biochem. 2011;65:219–271.

74. Solá RJ, Griebenow K. Effects of glycosylation on the stability of protein pharmaceuticals. J. Pharm. Sci. 2009;98:1223–1245.

75. Nothaft H, Szymanski CM. Protein glycosylation in bacteria: Sweeter than ever. Nat. Rev. Microbiol. 2010;8:765–778.

76. Zaia J. Mass spectrometry and glycomics. OMICS 2010;14:401–418.

77. Chen YZ, Tang YR, Sheng ZY, Zhang Z. Prediction of mucin-type O-glycosylation sites in mammalian proteins using the composition of k-spaced amino acid pairs. BMC Bioinformatics 2008;9:101.

78. Keay SK, Szekely Z, Conrads TP, Veenstra TD, Barchi JJ Jr, Zhang CO, Koch KR, Michejda CJ. An antiproliferative factor from interstitial cystitis patients is a frizzled 8 protein-related sialoglycopeptide. Proc. Natl. Acad. Sci. U.S.A. 2004;101:11803–11808.

79. Keay SK, Zhang CO, Shoenfelt J, Erickson DR, Whitmore K, Warren JW, Marvel R, Chai T. Sensitivity and specificity of antiproliferative factor, heparin-binding epidermal growth factor-like growth factor, and epidermal growth factor as urine markers for interstitial cystitis. Urology 2001;57:9–14.

80. Tsai HY, Boonyapranai K, Sriyam S, Yu CJ, Wu SW, Khoo KH, Phutrakul S, Chen ST. Glycoproteomics analysis to identify a glycoform on haptoglobin associated with lung cancer. Proteomics 2011;11:2162–2170.

81. Williams DC Jr, Lee JY, Cai M, Bewley CA, Clore GM. Crystal structures of the HIV-1 inhibitory cyanobacterial protein MVL free and bound to Man3GlcNAc2: Structural

basis for specificity and high-affinity binding to the core pentasaccharide from N-Linked oligomannoside. J. Biol. Chem. 2005;280:29269–29276.

82. Lederkremer GZ. Glycoprotein folding, quality control and ER-associated degradation. Curr. Opin. Struct. Biol. 2009;19:515–523.

83. Wilson N, Simpson R, Cooper-Liddell C. Introductory glycosylation analysis using SDS-PAGE and peptide mass fingerprinting. Methods Mol. Biol. 2009;534:205–212.

84. Hagglund P, Bunkenborg J, Elortza F, Jensen ON, Roepstorff P. A new strategy for identification of N-Glycosylated proteins and unambiguous assignment of their glycosylation sites using HILIC enrichment and partial deglycosylation. J. Proteome Res. 2004;3: 556–566.

85. Fryksdale BG, Jedrzejewski PT, Wong DL, Gaertner AL, Miller BS. Impact of deglycosylation methods on two-dimensional gel electrophoresis and matrix assisted laser desorption/ionization-time of flight-mass spectrometry for proteomic analysis. Electrophoresis 2002;23:2184–2193.

86. Henriksson H, Denman SE, Campuzano ID, Ademark P, Master ER, Teeri TT, Brumer H III. N-Linked glycosylation of native and recombinant cauliflower xyloglucan endotransglycosylase 16A. Biochem. J. 2004;375:61–73.

87. Segu ZM, Hussein A, Novotny MV, Mechref Y. Assigning N-Glycosylation sites of glycoproteins using LC/MSMS in conjunction with endo-m/exoglycosidase mixture. J. Proteome Res. 2010;9:3598–3607.

88. Henning S, Peter-Katalinić J, PohlentzG. Structure analysis of N-glycoproteins. Methods Mol Biol. 2009;492:181–200.

89. Sheeley DM, Reinhold VN. Structural characterization of carbohydrate sequence, linkage, and branching in a quadrupole ion trap mass spectrometer: Neutral oligosaccharides and N-Linked glycans. Anal. Chem. 1998;70:3053–3059.

90. Weiskopf AS, Vouros P, Harvey DJ. Characterization of oligosaccharide composition and structure by quadrupole ion trap mass spectrometry. Rapid Commun. Mass Spectrom. 1997;11:1493–1504.

91. Bakhtiar R, Guan Z. Electron capture dissociation mass spectrometry in characterization of peptides and proteins. Biotechnol. Lett. 2006;28:1047–1059.

92. Wiesner J, Premsler T, Sickmann A. Application of Electron Transfer Dissociation (ETD) for the analysis of posttranslational modifications. Proteomics 2008;8:4466–4483.

93. Hakansson K, Cooper HJ, Emmett MR, Costello CE, Marshall AG, Nilsson CL. Electron capture dissociation and infrared multiphoton dissociation MS/MS of an N-Glycosylated tryptic peptic to yield complementary sequence information. Anal. Chem. 2001;73:4530–4536.

94. Little DP, Speir JP, Senko MW, O'Connor PB, McLafferty FW. Infrared multiphoton dissociation of large multiply charged ions for biomolecule sequencing. Anal. Chem. 1994;66:2809–2815.

95. Kanninen K, Goldsteins G, Auriola S, Alafuzoff I, Koistinaho J. Glycosylation changes in Alzheimer's disease as revealed by a proteomic approach. Neurosci. Lett. 2004;367:235–240.

96. Chiang YH, Wu YJ, Lu YT, Chen KH, Lin TC, Chen YK, Li DT, Shi FK, Chen CC, Hsu JL. Simple and specific dual-wavelength excitable dye staining for glycoprotein detection in polyacrylamide gels and its application in glycoproteomics. J. Biomed. Biotechnol. 2011;2011:780108.

97. Choi E, Loo D, Dennis JW, O'Leary CA, Hill MM. High-throughput lectin magnetic bead array-coupled tandem mass spectrometry for glycoprotein biomarker discovery. Electrophoresis 2011;32:3564–3575.

98. Wang SH, Wu SW, Khoo KH. MS-based glycomic strategies for probing the structural details of polylactosaminoglycan chain on N-glycans and glycoproteomic identification of its protein carriers. Proteomics 2011;11:2812–2829.

99. Abbott KL, Pierce JM. Lectin-based glycoproteomic techniques for the enrichment and identification of potential biomarkers. Methods Enzymol. 2010;480:461–76.

100. Zhang H, Li XJ, Martin DB, Aebersold R. Identification and quantification of N-Linked glycoproteins using hydrazide chemistry, stable isotope labeling and mass spectrometry. Nat. Biotechnol. 2003;21:660–666.

101. Liu T, Qian WJ, Gritsenko MA, Camp DG II, Monroe ME, Moore RJ, Smith RD. Human plasma N-Glycoproteome analysis by immunoaffinity subtraction, hydrazide chemistry, and mass spectrometry. J. Proteome Res. 2005;4:2070–2080.

102. Yang Z, Hancock WS, Chew TR, Bonilla L. A study of glycoproteins in human serum and plasma reference standards (HUPO) using multilectin affinity chromatography coupled with RPLC–MS/MS. Proteomics 2005;5:3353–3366.

103. Jung K, Cho W, Regnier FE. Glycoproteomics of plasma based on narrow selectivity lectin affinity chromatography. J. Proteome Res. 2009;8:643–650.

104. Jimenez CR, Piersma S, Pham TV. High-throughput and targeted in-depth mass spectrometry-based approaches for biofluid profiling and biomarker discovery. Biomark. Med. 2007;1:541–565.

105. Peng J, Schwartz D, Elias JE, Thoreen CC, Cheng D, Marsischky G, Roelofs J, Finley D, Gygi SP. A proteomics approach to understanding protein ubiquitination. Nat. Biotechnol. 2003;21:921–926.

106. Danielsen JM, Sylvestersen KB, Bekker-Jensen S, Szklarczyk D, Poulsen JW, Horn H, Jensen LJ, Mailand N, Nielsen ML. Mass spectrometric analysis of lysine ubiquitylation reveals promiscuity at site level. Mol. Cell. Proteomics 2011;10:M110.003590.

107. Wagner SA, Beli P, Weinert BT, Nielsen ML, Cox J, Mann M, Choudhary C. A proteome-wide, quantitative survey of in vivo ubiquitylation sites reveals widespread regulatory roles. Mol. Cell. Proteomics 2011;10:M111.013284.

108. MacCoss MJ, McDonald WH, Saraf A, Sadygov R, Clark JM, Tasto JJ, Gould KL, Wolters D, Washburn M, Weiss A, Clark JI, Yates JR 3rd. Shotgun identification of protein modifications from protein complexes and lens tissue. Proc. Natl. Acad. Sci. U.S.A. 2002;99:7900–7905.

109. Choudhary C, Kumar C, Gnad F, Nielsen ML, Rehman M, Walther TC, Olsen JV, Mann M. Lysine acetylation targets protein complexes and co-regulates major cellular functions. Science 2009;325:834–40.

5

CHARACTERIZATION OF PROTEIN COMPLEXES

5.1 INTRODUCTION

Frederick Douglass, one of the most prominent leaders in the American abolitionist movement of the mid-1800s, would have easily grasped cell biology. Mr. Douglass remarked that "Properly speaking, there are in the world no such men as self-made men" (1). Mr. Douglass's views of people would have enabled him to readily understand proteins. No protein is able to make itself and neither is it able to function without the aid of other proteins or biomolecules. Even proteases may require the presence of metal ions in addition to its substrate to carry out its function. Biomolecular interactions drive every process within the cell including structural integrity (e.g., actin filaments), molecular transport (e.g., hemoglobin), signal propagation (e.g., kinases and phosphatases), DNA replication, protein translation, energy production, and cell division. This chapter will focus primarily on protein–protein interactions; however, proteins interact with a diverse range of biomolecules including metals (e.g., calcium transport), DNA (e.g., transcription factors), RNA (e.g., small nuclear ribonucleoproteins), lipids (e.g., apolipoproteins), ATP (e.g., kinases), etc.

For cells to function properly, protein interactions must occur with exact spatial and temporal accuracy. For example, signal transduction is initiated by the binding of a biomolecule to a cell surface receptor, which activates an internal protein pathway.

Proteomic Applications in Cancer Detection and Discovery, First Edition. Timothy D. Veenstra.
© 2013 John Wiley & Sons, Inc. Published 2013 by John Wiley & Sons, Inc.

The signal is propagated along the pathway through the simple addition or removal of a phosphate group via a kinase or phosphatase, respectively. The involved proteins must be proximal to each other and interact in a precise order for the signal to reach the nucleus. As far as timing, proteins interactions can occur over a long (e.g., actin/myosin filaments for maintaining the cell's structure) or a short period of time (e.g., the addition of a phosphate group by a kinase). Another obvious example is cell division. In controlled cell division, prophase, prometaphase, metaphase, anaphase, and telophase all occur sequentially. Below this layer, however, there are a number of coordinated microinteractions that drive each phase. The complete repertoire of protein interactions that must occur for cell division are not yet completely understood; however, it is known that imperfection in the timing and order result in uncontrolled cell growth and contribute to conditions such as cancer.

In diseases such as cancer, the spatial and temporal regulation of protein interactions goes awry. For example, interactions within the mitogen-activated protein (MAPK) and extracellular-signal-regulated (ERK) kinases regulate almost all processes initiated by the binding of an extracellular ligand. The most prominent of the MAPK pathways involved in cancers is the Ras-Raf-MEK1/MEK2-ERK1/ERK2 cascade (2,3). This pathway regulates cell proliferation, differentiation and survival, cell adhesion and motility, and embryonal development (4). Deregulation of the ERK signaling cascade contributes to cancer initiation and progression of a majority of human carcinomas (5–7). Disrupting ERK signaling using kinase inhibitors (e.g., gefitinib and erlotinib) or monoclonal antibodies (e.g., herceptin) represents one of the main treatment regiments for non–small cell lung cancer (8).

The importance of investigating protein–protein interactions is underscored by reading high impact journals such as *Science*, *Nature*, and *Cell*. It is difficult to find an edition of these journals that does not contain a research article describing a novel protein interaction. Basic research has long recognized the importance of identifying protein interactions; however, modern proteomic methods have brought fundamental changes into how these discoveries are made. Historically, hypothesis-driven methods have been used to identify protein interactions. Hypothesis-driven methods utilized an affinity device (such as an antibody or aptamer) to bind a targeted protein under nondenaturing conditions that preserve protein–protein interactions within the sample (Figure 5.1). After conducting a series of washing steps to try to eliminate nonspecifically bound proteins while retaining as many specifically bound proteins as possible, the isolated complex is separated using polyacrylamide gel electrophoresis (PAGE) and transferred to a polyvinylidene fluoride (PVDF) membrane. The PVDF membrane is probed with an antibody directed toward a protein that is hypothesized to bind to the target protein. If a band is observed near the anticipated molecular weight of the hypothesized protein, it is concluded that this protein interacts with the target protein.

There are several deficiencies when using this hypothesis-driven approach. It is costly, in both time and money, to prove a hypothesis incorrect. This approach only identifies a single interacting protein per experiment unless the blot is probed with multiple antibodies. Probably most disconcerting is that this strategy is extremely prone to false negative and false positive results. For example, identifying a true interaction depends on how well the protein complex was isolated. If the washing steps performed

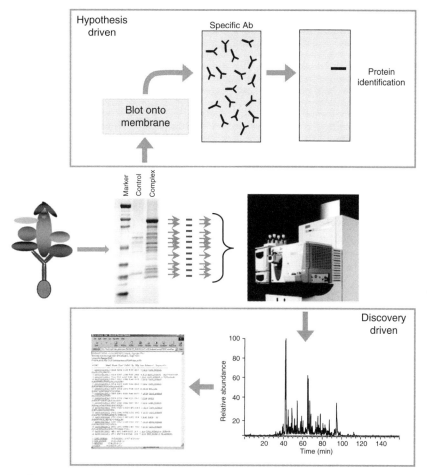

Figure 5.1. Discovery and hypothesis-driven methods of identifying protein–protein interactions. In a discovery-driven approach (bottom panel), a target protein and its binding partners are extracted from cells using techniques such as IP. The extracted proteins are separated by gel electrophoresis and stained bands are excised and processed for analysis by peptide mapping or liquid chromatography coupled on line with tandem MS (LC–MS^2). In a hypothesis-driven approach (upper panel), the proteins from the gel are blotted onto a membrane, which is interrogated with an antibody specific for a protein that is hypothesized to interact with the target protein.

during the complex isolation are too stringent, a protein that is actually part of the complex can be dissociated prior to its isolation. If the washing steps are too relaxed, a protein may be identified as part of the complex, when in reality it is not. Over the years, our laboratory has seen a large number of supposed isolated protein complexes that resemble cell lysates when separated on a PAGE gel. Regardless of these deficiencies, however, hypothesis-driven approaches continue to play a major role in basic research (9).

The development of proteomic technologies, specifically mass spectrometry (MS), has made a tremendous impact on how protein interactions are identified. Although biomarker discovery and systems biology gets a lot of the headlines, it is arguable that MS's biggest impact has been in the characterization of protein complexes. MS has shifted protein–protein interaction studies from hypothesis to discovery driven. The sample preparation steps for a discovery-driven approach are identical to those used in a hypothesis-driven method (Figure 5.1). Where they diverge is after separation of the complex using PAGE. In a discovery-driven approach proteins are not transferred to a PVDF membrane. The protein bands that have been visualized using coomassie or silver stain are cut from the gel and processed for MS analysis. The major advantage of this method is that it requires no hypothesis regarding interacting proteins. Obviously, identifying proteins that have been previously shown to interact with the protein of interest can increase the confidence in novel findings. The rate of discovery using this approach is dramatically increased since the entire complement of proteins within the isolated complex can be analyzed. The discovery rate of potential interacting proteins makes validation of the functional importance of these interactions the primary bottleneck in these studies.

5.2 METHODS FOR ISOLATING PROTEIN COMPLEXES

The most critical step in identifying protein interactions is the isolation of the complex. Garbage-in, garbage-out is a fitting adage for this type of experiment. Current proteomic technologies, particularly MS, are proficient at protein identification. This proficiency can be both a benefit and detriment to the analysis of protein complexes. Operating a mass spectrometer in a data-dependent mode results in the selection of peaks based solely on their signal intensity. The instrument cannot determine *a priori* the peaks that originate from specifically and nonspecifically bound proteins. If the protein complex extraction is not optimized to minimize abundance of nonspecifically bound proteins, these will be identified at the expense of those that are specifically bound to the protein of interest. Conversely, if the complex is isolated in a manner that removes specifically bound proteins, those that functionally interact with the target protein will not be available for identification.

5.2.1 Optimizing Protein Complex Isolation

Selecting the conditions to isolate a protein complex depends on a number of issues. At the forefront is the ability to isolate sufficient amounts of material for downstream analysis. Consequently, most complex isolations target an overexpressed protein that is transfected into a cell line. This strategy increases the amount of protein per cell and allows the cells used in the study to be amplified to a number necessary for isolating enough material. Endogenous proteins can be targeted; however, a much greater amount of starting material (cell number, total cell lysates, etc.) is typically required depending on its abundance in the cell. Depending on the protein expression efficiency in an overexpressed mammalian cell system, approximately 5–10 million cells are usually required for LC–MS^2 analysis; however, some studies can require 10 times this amount. Targeting an endogenous protein may require over 100 million cells.

The next consideration is how to isolate the protein complex. The most popular method in use today is immunoprecipitation (IP). IP uses an antibody directed against the target protein. With any technology that uses antibodies, the specificity of its interaction with the target protein should be as high as possible, and the off-rate should be slower than the on-rate. Specificity is important so that the antibody does not cross-react with other proteins, decreasing the efficiency of binding to the intended target. It needs to be remembered that an antibody that is useful for doing Westerns or immunohistochemistry may not be useful for performing IPs. It is important to test the antibody's capability for performing IPs or obtain the necessary data from the commercial vendor.

To obviate the need for antibodies directed against specific proteins, a number of IPs are performed targeting overexpressed chimeric proteins (Figure 5.2). These chimeras contain a "tag" that serves as an epitope for isolating the target protein and its associated partners (10). The tags commonly used today can range from a short peptide to an entire protein. Some of the most commonly used peptide epitopes are Flag (DYKDDDDK), c-Myc (EQKLISEEDL), hemagglutinin (HA; YPYDVPDYA), and hexa-histidine

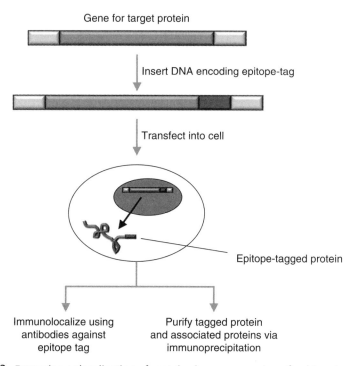

Figure 5.2. Extraction or localization of protein via overexpression of a chimeric protein. The sequence of the protein of interest is cloned into a vector that expresses the protein with a tag attached to it. The vector is transfected into a cell where the tagged-protein is expressed and isolated or localized using an antibody directed to the tag.

(His$_6$; HHHHHH). These tags can be utilized in solo or in combination. For example, many studies in our laboratory have utilized three FLAG tags in sequence at the N- or C-terminus of the protein of interest. In tandem affinity purification (TAP) tagging, two different tags are expressed as part of the target protein (11). Different combinations of tags have been used including FLAG, His$_6$, calmodulin-binding peptide (CBP), protein A, maltose-binding protein, etc. (12). Using a tagged protein eliminates the need for an antibody specific to the protein of interest. Instead, well-characterized, commercially available antibodies that are already coupled to resins can be used. Covalent coupling of the antibody to the column support prevents it from being eluted with the complex and interfering with the downstream fractionation and MS analysis. For other tags, customized resins can be used to bind the tagged protein. For example, calmodulin columns are used for CBP tagged proteins, immobilized metal affinity chromatography for His$_6$ tags, etc. A schematic of isolation of a protein complex through TAP-tagging utilizing Protein A and CBP is shown and described in Figure 5.3.

5.2.2 Importance of Optimizing Isolation Conditions

Taking the time to optimize the conditions for isolating a protein complex is definitely time well spent. The aim is to reduce the number of nonspecifically bound proteins without removing specifically bound ones. This balance is difficult to achieve and requires cell lysis, protein binding, and column washing conditions that preserve the protein complex as much as possible. Many studies are doomed at the cell lysis step as the conditions used are too harsh and break apart any noncovalent complexes that were within the cell. For example, sonication, high levels of detergent, and high salt concentrations should never be used to lyse cells when the objective is to isolate a protein complex.

It has been suggested that IP methods such as TAP tagging allows for a single purification method, enabling parallel characterization of multiple complexes (13, 14). Essentially, the belief was that once the method was optimized for a single protein, it could be used for all other proteins. Unfortunately, empirical evidence has not supported this assumption. While epitope tags have simplified the IP of protein complexes, they still do not allow for a single set of conditions to be used to isolate every protein complex. Proteins are like children; they may seem the same, yet each individual child has different needs. A large number of global interactome studies have utilized a single set of isolation conditions for every protein (15, 16). Unfortunately, as discussed later, this strategy has resulted in difficult to reproduce data and proposed interactions that are a challenge to functionally verify. It needs to be remembered that each protein within the cell is unique with its own characteristic interactions. These interaction parameters not only include the proteins it interacts with but also the affinities and kinetics of these proteins. While it requires an investment of time, optimizing the isolation conditions is well worth it as it makes it much easier to recognize specifically and nonspecifically bound proteins based on the MS data.

Isolating a protein complex via IP requires mild cell lysis conditions so that the integrity of the protein complex is preserved (17). Proper lysis conditions include, up to 0.5% Triton X-100, PBS or Tris buffer at neutral pH, physiological ionic strength

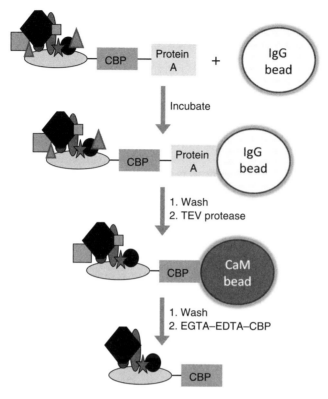

Purified complex

Figure 5.3. Extraction of a protein complex using TAP-tagging utilizing CBP and Protein A. In this example, the protein of interest is expressed as a chimera containing sequences corresponding to CBP and protein A. In the first extraction round, IgG is used to bind to protein A and the bound complex is washed to remove nonspecifically bound proteins. The complex is cleaved from the IgG support using tobacco etch virus protease that cuts at a specific TEV-protease site that is engineered between the two affinity tags. The complex is then bound to a calmodulin column through interaction with CBP. After another washing step to remove more nonspecifically bound proteins, the final complex is removed from the complex by introducing ethylenediaminetetraacetic acid, or ethylene glycol tetraacetic acid, or an excess of CBP (which competes for calmodulin binding sites with the tagged protein of interest). The purified complex can then be analyzed using MS, etc.

$(I = 150 \text{ mM})$, 5–10% glycerol, etc. Protease inhibitors should always be added during cell lysis to avoid enzymatic protein degradation. Phosphatase inhibitors should also be added to prevent dephosphorylation since the binding of a protein can be directly dependent on its phosphorylation state. Another necessary optimization step is covalently conjugating the antibody to Protein G Sepharose beads (18). Covalently coupling the antibody to the beads increases the amount of protein complex that can be isolated. Importantly for downstream MS analysis, covalently coupling the antibody to the beads prevents it from being eluted when the complex is being washed off the column. The antibody heavy and light chains are observed as two intense bands on the sodium dodecyl sulphate (SDS)-PAGE gel at molecular weights 50,000 and 25,000 Da, respectively. These proteins will mask a large part of the gel and smear throughout the entire lane. Even if a nongel separation is used, peptides from the antibody will be in greater abundance than any other protein in the sample. Theses peptides will be selected for identification by the mass spectrometer at the expense of proteins within the complex.

The column-bound protein complex needs to be washed several times using optimized salt and detergent concentrations. As always, the goal is to maximize the number of specifically bound proteins while minimizing background species. This requirement begs the question: "If we do not know what the complex is, how can I tell when the washing conditions are optimized?" It is impossible to knowingly get to the point at which all nonspecific proteins are removed and only specifically bound proteins are retained. Having a proper negative control is critical. A good negative control is invaluable for identifying a large percentage of nonspecifically binding proteins. A useful negative control for an IP includes cells that are not transfected or are transfected with the empty expression vector. For endogenously expressed proteins, using cells that do not express the target protein can serve as a negative control. The negative control needs to be treated identically to the experimental IP with respect to cell number, lysis, binding, washing, and elution conditions.

The negative control can be used at two different steps in the analysis. If the protein complex is analyzed by first separating it on a SDS-PAGE gel, the negative control should be separated on the gel as well. Once visualized, only protein bands that are observed in the experimental lane can be cut out for MS analysis. If the protein complex is analyzed directly using LC–MS^2, proteins identified in both the experimental and negative control should be considered as nonspecifically bound.

The importance of using a negative control cannot be underestimated. In my experience in the laboratory, I have witnessed IP samples that looked exactly like the corresponding cell lysate. If this IP was analyzed using Western blotting, every hypothesis would give a false positive result. It is likely that a lot of novel protein interactions that have been published over the last couple of decades are false positives owing to the fact that the complex isolation conditions were not optimized or a negative control was not analyzed.

5.2.3 Oligoprecipitation

While protein complex isolation has been dominated by IP, this mode of extraction may not reflect the environment of the target protein within the cell. Proteins have specific functions, but are not always involved in that function and not continuously bound in a complex. For example, phosphorylation status can dictate a protein's function. The phosphorylated and nonphosphorylated forms of a protein generally bind to different sets of proteins. Using an IP to extract a complex bound to the phosphorylated form will generally also isolate the nonphosphorylated form of the protein and its binding partners unless an antibody that only targets the phosphorylated version of the protein is used.

The identified proteins will be a hybrid of proteins bound to both the phosphorylated and nonphosphorylated forms of the target protein.

Transcription factors also have both active and inactive forms within a cell. In most studies, investigators are interested in the complex that forms around a transcription factor when it is bound to DNA. Using an antibody-based IP will isolate both the transcriptionally active and inactive forms of the transcription factor. This method makes it difficult to determine which proteins that bind to the targeted transcription factor are part of the transcriptional complex and which are part of the transcriptionally inactive complex. Isolating an active transcriptional complex requires using a double-stranded (ds) DNA response element as the bait or affinity resin or at least making this part of the isolation mixture (19–21). For several decades now, identifying interactions between DNA and proteins such as transcription factors has been done using electrophoretic mobility shift assays (EMSA), as illustrated in Figure 5.4. EMSA have also been used to identify RNA–protein interactions (22). Conducting an EMSA requires a synthetically prepared dsDNA binding site, which is radioactively labeled (i.e., ^{32}P). This piece of dsDNA is incubated with a cell or nuclear lysate and the mixture is separated on

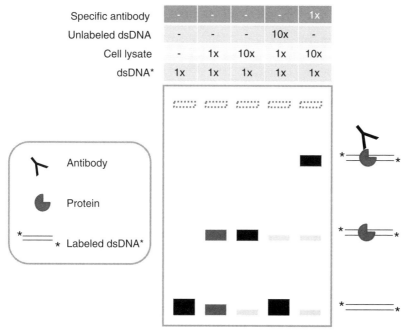

Figure 5.4. Principles of EMSA for measuring DNA–protein interactions. In an EMSA, various combinations of radioactively labeled ds DNA (dsDNA*), cell lysate, unlabeled dsDNA, and an antibody directed to a specific protein are mixed and run on a polyacrylamide gel. The binding of a protein to the dsDNA* is visualized through autoradiography. If the band is shifted to a higher molecular weight when the antibody is present, the conclusion is that the protein that the antibody binds to is binding to the dsDNA*.

a polyacrylamide gel. The location of the DNA–protein complex is visualized using autoradiography. To identify the transcription factor bound to the dsDNA strand, an antibody targeting a specific transcription factor is added to the mixture prior to its separation. If the position of the complex has shifted to a higher molecular weight owing to the binding of the antibody to the DNA–protein complex, then the identity of the protein within the complex is concluded based on the antibody selected. To confirm an interaction, the dsDNA:protein ratio may be changed or unlabeled (i.e., "cold") dsDNA may be added. If a binding event is occurring, decreasing the dsDNA–protein ratio will result in a more intense band. Adding "cold" dsDNA should result in a less intense band when measured using autoradiography.

Although EMSA have been successfully used for several decades now, they are severely limited in the information they provide. These types of studies are purely hypothesis driven. Both the transcription factor and DNA-binding site must be determined hypothetically, since the dsDNA strand must be designed based on the suspected interaction. The identity of the bound protein is inferred by its interaction with a chosen antibody; however, false positive results can result from cross-reactivity. The net result of an EMSA is a single protein binding to a DNA sequence: other proteins within the complex are not revealed. The radioactive labeling does not allow the complex to be cored from the gel and immediately analyzed using MS. Indeed, there are very few instances of where EMSA-derived complexes have been successfully characterized using MS.

Ideally, DNA–protein complexes could be discovered similar to what has been described above for protein–protein interactions. Several years ago, the concept of oligoprecipitation (OP) as a means of capturing the "active form" of a transcription factor when it is bound to its DNA response element was developed (20, 21). In this method, a synthetically prepared dsDNA containing a biotin group is coupled to streptavidin beads via a biotin group (Figure 5.5). A cell or nuclear extract is passed over the column and proteins with an affinity for the DNA are allowed to bind. While cell lysates can be used, it is better to use a nuclear extract since it will reduce the complexity of the mixture and have a higher chance of containing the protein complex in its transcriptionally active form. After washing the column to remove as many nonspecifically bound proteins as possible while preserving the transcriptional complex, the proteins are eluted from the column. For this type of experiment, the negative control can be either a cell lysate that does not express the protein of interest or dsDNA in which the sequence of the protein binding site is scrambled so that it no longer binds the transcriptional complex.

Investigators have utilized the OP-extraction approach to study transcriptional complexes that are activated in the presence of elevated inorganic phosphate (P_i) as found in differentiating osteoblasts (21). While the mechanism of osteoblast differentiation is unclear, the transcription factor, early growth response protein 1 (EGR1), is involved in the early response of osteoblasts to P_i. An OP experiment was conducted to determine proteins involved in the EGR1 transcriptional complex in response to elevated Pi levels. The oligonucleotide sequences 5′-biotin-GGA TCC A**GC GGG GGC G**AG CGG GGG CGA-3′ and 5′-TCG CCC CCG CT**C GCC CCC GC**T GGA TCC-3′ containing the EGR1 binding DNA sequence (bold) were heated to 95°C and cooled to ambient temperature to allow the two strands to anneal. The dsDNA was bound to a streptavidin column and a nuclear lysate was passed over the column. After a series of washing

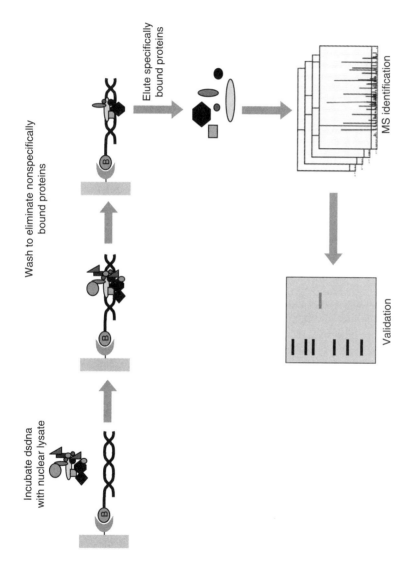

Figure 5.5. Principal of OP for extracting DNA-protein complexes. In contrast to IP, OP uses a biotin-labeled ds DNA (dsDNA) coupled to a solid avidin support. The dsDNA is incubated with a nuclear extract and after incubation, is washed to remove nonspecifically bound proteins. The protein complex is eluted and analyzed using tandem MS. As with any discovery-driven approach, orthogonal experiments to validate the results are critical.

Incubate dsdna
with nuclear lysate

Wash to eliminate nonspecifically
bound proteins

Elute specifically
bound proteins

MS identification

Validation

steps, the bound proteins were eluted, digested with trypsin, and analyzed using LC–MS^2. Forty-four proteins were identified as being associated with transcriptionally active EGR1. While the specificity of all of these proteins was not validated, several were and a majority of the others had previously been shown to be associated with nuclear function. The proteins identified in the EGR1 complex were from classes including other transcription factors, splicing factors, regulatory enzymes, and ribonuclear proteins. While EMSA was used to confirm the EGR1 interaction with it dsDNA consensus sequence, OP has the distinct advantage in being able to identify several proteins within a complex using a purely discovery-driven approach.

A previous study using OP added a stable isotope labeling step to assist in determining those proteins that specifically bound to the target dsDNA (Figure 5.6) (20). The study was interested in identifying the protein factor that bound to the transcriptional regulatory element X (Trex), which is a positive control site within the muscle creatine kinase (MCK) enhancer. The Trex site is important for *MCK* expression in skeletal and cardiac muscle. To isolate the Trex binding protein, wild-type and mutant dsDNAs were prepared and coupled to magnetic beads. Each oligoprobe was incubated with a HeLa cell nuclear extract and placed in a magnetic field to extract the particles containing the dsDNA/protein complexes. After washing each sample to remove nonspecifically bound proteins, the remaining proteins were eluted from their respective dsDNAs. Instead of taking each sample directly to MS analysis, the investigators labeled them with either the light or heavy version of the isotope-coded affinity tags (ICAT) reagents. While ICAT is discussed in the chapter on "Quantitative Proteomics," suffice it to say here that this method is used to measure the relative abundance of peptides in comparative samples. After extracting Trex-binding factor (TrexBF) using magnetic beads coupled to oligonucleotides containing either wild-type or mutant Trex sites to enrich the TrexBF, the two extracts were labeled with the different isotopic versions of the ICAT reagents. Finding a MS signal whose relative abundance was much greater in the extract using the wild-type oligonucleotide for Trex would lead to the identity of the TrexBF. This protein was identified as Six4, a homeodomain transcription factor of the Six/sine oculis family. The importance of using an isotope tag to distinguish the BF is exemplified by the fact that the extracts contained ∼900 copurifying background proteins.

5.3 PROTEOME SCREENING USING TANDEM AFFINITY PURIFICATION

With the discovery-driven mentality that modern proteomics brought to the discovery of protein complexes, it was inevitable that ambitious groups would attempt to use this technology to determine the protein interaction "circuitry" of the cell. Imagine being able to have a direct blueprint of the wiring within a cell. This knowledge would immediately increase the number of viable drug targets an order of magnitude and save billions of dollars in research funding.

Within the previous decade, two studies attempted such a feat using *Saccharomyces cerevisiae* (*S. cerevisiae*) (15, 16). *S. cerevisiae* was an excellent choice as a model organism since it is easy to culture and replicates quickly. Conducting the studies

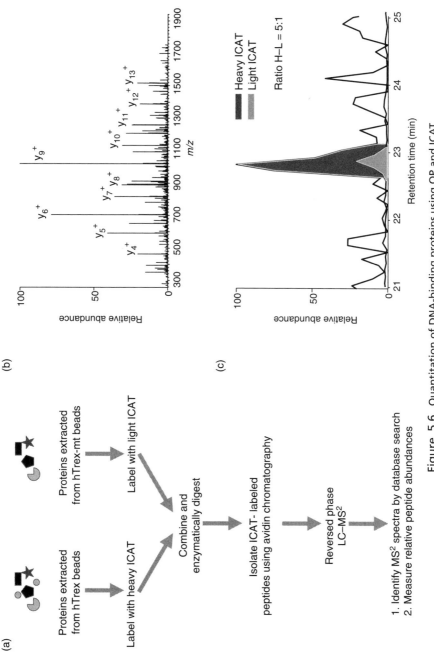

Figure 5.6. Quantitation of DNA-binding proteins using OP and ICAT.

129

reported by these two groups required a lot of material. As a eucaryote, *S. cerevisiae* has a similar internal cell structure to mammalian cells and indeed a number of human proteins were identified through the initial discovery of a homolog in yeast. In addition, *S. cerevisiae* can be easily genetically manipulated and transformed to express chimeric proteins necessary to identify hundreds of protein complexes.

While both groups used a tagged protein strategy, one utilized a FLAG-tag (16), while the other used a TAP method in which the tag is made up of Protein A and CBP (15). After isolation of the complexes, both studies used a similar strategy of one-dimensional-PAGE separation followed by band excision and MS identification. The study by Gavin et al. reported 491 complexes composed of 1483 proteins (15), while that by Krogan et al. reported 547 complexes containing 2702 proteins (16). Combining the two data sets and remove redundancy, results in the identification of 3033 proteins (i.e., almost 50% of the *S. cerevisiae* proteome) as part of the identified complexes (23). Only about 1/3 of the proteins (1152) were found to be common to both data sets. Only six of the complexes in the two data sets were identical. Almost 200 of the complexes reported by Krogan et al. do not share a single protein found in any complex reported by Gavin et al. Conversely, only 20 complexes reported by Gavin et al. do not contain any proteins within any of the complexes reported by Krogan et al. Approximately 80% of the complexes found in ether study have less than 50% overlap with the other. Over one-third of the studies reported by Krogan et al. had less than a 5% overlap with those reported by Gavin et al.

The overlap between these two studies gives one pause. From what is known about conducting protein complex studies, regardless of the strategy, this overlap should not be too surprising. It is prudent to first examine sources of potential differences within the two studies. The studies used their own specific set of algorithms to determine correctly identified proteins and determine probabilistic measurements of the proteins within the complexes. Beyond software, the most likely source of dissimilarity between these two studies is in the purification of the protein complexes themselves. While the advantages/disadvantages, preference/dislike of TAP and FLAG tags can be debated, the greatest source of error in the complexes is probably the number of nonspecifically bound proteins that were identified. In both studies, the investigators attempted to utilize a one-size fits all approach to isolating protein complexes. It has been my lab's (and countless others) experience that the conditions for each complex must be carefully optimized to maximize the number of specifically bound proteins while minimizing the number of nonspecifically bound ones.

I remember hearing Dr. Roger Brent give a seminar in 2000 at Pacific Northwest National Laboratories. He was describing his vision on identifying protein interactions on an "omics" scale, much like what the previous two studies were attempting. Probably, the statement I remember most about his presentation was something to the affect that "Maybe there is no shortcut to do this. It is just simply going to be a long and arduous task." I think Dr. Brent was exactly correct. Proteins are like children. Each child has their own personality and there is no single approach that works in getting them all to reach their maximum potential. Even if every complex is carefully isolated and identified, the data would only provide a baseline of the cell's circuitry as it must always be remembered that the proteome is dynamic and in constant flux.

5.4   YEAST TWO-HYBRID SCREENING

Beyond using MS, all of the methods described up to this point to characterize protein interactions have another commonality: the complex is removed from its cellular environment prior to identification. While the perfect *in vivo* interaction discovery study has yet to be designed, the yeast two-hybrid system provides a method to screen protein–protein interactions with living cells (Figure 5.7) (24,25). In yeast, two-hybrid screening

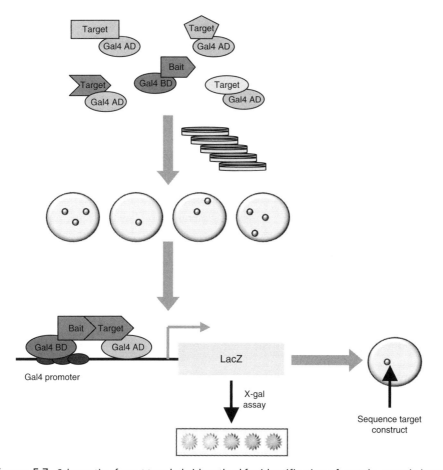

Figure 5.7. Schematic of yeast two-hybrid method for identification of protein–protein interactions. A plasmid containing a bait protein fused to the DNA-BD of a transcription factor (Gal4 in this example) is transformed into yeast cells that contain a library of plasmids expressing target proteins fused to the AD of the same transcription factor. The yeast cells are cultured on plates and interaction of the bait with a target protein brings the domains of the transcription factor together resulting in the expression of a LacZ reporter gene. Yeast colonies containing an activated reporter gene are selected and the plasmids sequence to identify the interacting target and bait proteins.

the "bait" protein (e.g., X) is expressed as a chimeric protein fused to the DNA-binding domain (DB) of a transcription factor (usually Gal4). The "target" proteins (i.e., Y) that the bait protein is screened against are expressed within the cell fused to the activation domain (AD) of the same transcription factor. If the bait and target proteins bind, the transcription factors BD and AD interact to form an active transcriptional complex. The transcriptional complex then activates the expression of a reporter gene permitting the yeast to proliferate on selected media. Without an active transcriptional complex, the yeast cells do not grow.

The biggest difference between identifying protein complexes using IP or OP and the yeast two-hybrid system is the extent of interactions that can be discovered. An IP or OP experiment is designed to extract all of the proteins present within a complex regardless of whether they interact with the target protein or not. The yeast two-hybrid system is a binary measurement. Only the protein that interacts directly with the bait protein will be identified.

Preparing a yeast-two hybrid screen requires transforming cultures of yeast cells with a clonal library expressing individual bait proteins. The culture suspension is then spread onto solid media that lacks the nutrient provided by the reporter gene that is expressed if the transcription factor is formed through a BD–AD interaction. Yeast colonies where the BD–AD does not interact will fail to grow. At least two selection criteria based on the activation of different reporter genes are used in two-hybrid screens. Colonies that grow (i.e., pass the first nutrient-based selection criteria) are tested in a colorimetric assay. The colorimetric assay is generally based on the expression of LacZ, an *Escherichia coli* gene that results in the yeast cells turning blue (26).

An advantage of the yeast two-hybrid system are that DNA, not proteins, is manipulated allowing the copious genomic resources that are available for a variety of different organisms to be utilized. Yeast two-hybrid screening is also high-throughput compared to IP. This throughput has allowed large-scale interaction studies for organisms such as *S. cerevisiae* (27, 28), *Helicobacter pylori* (29), *C. elegans* (30), *Drosophila melanogaster* (31), and *Homo sapiens* (32) to be completed. Ideally, a two-hybrid screen using all known proteins would identify all the binary interactions within known complexes, but this result rarely occurs. Owing to the high false negative rate of two-hybrid screening, only a small percentage of true protein interactions are identified (25). Unfortunately, the yeast two-hybrid system detects large numbers of interactions that have no validatable biological significance.

This high false discovery rate is illustrated in the comparison of data obtained by two independent groups that conducted two-hybrid screening using ~6000 yeast bait proteins (27, 28). There was an approximately 15% overlap in the interaction pairs that were detected in either study. While there have been many improvements minimizing the number of false positive identifications detected in yeast two-hybrid screening, the results still require independent validation before they can be considered as absolutely legitimate. As mentioned previously, two-hybrid screening produces binary interactions, whereas other methods such as IP identify multiple binding partners that are part of the complex with the targeted protein. It is not surprising that interactome data obtained in studies utilizing two-hybrid screening show very little overall agreement to affinity purification/MS-based studies. The obvious reason for this discrepancy is the differences in the two methods, but the fact that two-hybrid screening tends to detect transient interactions while protein complex purification methods identify stable interactions must also play a role.

Probably no investigator has utilized yeast two-hybrid screening to expand the knowledge of protein–protein interactions more than Dr. Marc Vidal. In a recent study, his group presented a yeast two-hybrid derived *Caenorhabditis (C.) elegans* interactome derived from testing a matrix of ~10,000 proteins (33). The authors claim that their data set (Worm Interactome 2007 or WI-2007) of protein interactions is similar in quality to low-throughput data (e.g., IP) already reported in peer-reviewed literature. They used previously filtered interactome data sets along with the data in WI-2007 to generate a consolidated map of *C. elegans* proteins interactions. Overall, the map estimates that the *C. elegans* interactome is composed of approximately 116,000 interactions. This result would suggest an average of between 11 and 12 interactions per protein based on the screening of a 10,000 × 10,000 matrix.

5.5   QUICK LC–MS METHOD TO IDENTIFY SPECIFICALLY BOUND PROTEINS

As mentioned previously, one of the biggest challenges in characterizing protein inter-actions is determining which of the identified proteins specifically interact with the target protein. Many different classes of proteins are isolated in the extraction procedure and it is virtually impossible to ensure the optimal conditions are being used at each step. Unfortunately, conventional MS identification does not provide any evidence of the functional significance of an identified protein: it only tells you that the protein is present. The only evidence that MS can provide is that a protein was identified more predominantly in the experimental sample but not in the negative control.

What does it mean that a protein is identified more predominantly in the experimental sample? Essentially it means that the abundance of the protein was greater in the experimental sample than the negative control. While MS signals are not themselves inherently quantitative (owing to factors such as differing ionization efficiencies among peptides), certain conclusions concerning the abundance of a particular protein can be made based on the recorded data. One of the most popular strategies is spectral counting (34). In spectral counting the number of peptides identified for a specific protein is compared between the experimental sample and negative control. The basic hypothesis is that the more abundant a protein, the greater chance that its peptides will be selected and identified using MS^2. This strategy works best for proteins that are present in the experimental sample and completely absent from the negative control. Unfortunately, if a protein (because of its size, sequence, or abundance) is identified by only one or two peptides in the experimental sample, it can be difficult to assess its specificity in the complex. In our laboratory, we prefer to see at least three peptides from a protein in the experimental sample versus zero in the negative control before considering this protein as a worthwhile candidate for subsequent functional validation. There are many instances where a protein will be identified in both the experimental sample and negative control. A good rule of thumb is that the ratio between the numbers of peptides identified

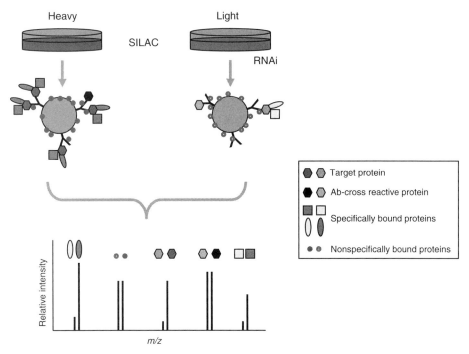

Figure 5.8. Principles of QUICK. In QUICK, proteins are differentially labeled using SILAC. The expression of the target protein is knocked down in one of the cultures using RNAi is used to knock down the expression of the target protein in the light isotope-labeled cells. IP is used to isolate the complexes and they are combined and characterized using liquid chromatography coupled directly on-line with tandem MS (LC–MS^2). Nonspecifically bound proteins will be represented in the mass spectrum by peaks of equal intensity. Proteins that interact specifically with the target protein will produce peaks of lesser relative intensity from the cells in which its expression was knocked down using RNAi.

in the experimental sample compared to the negative control must be at least 3 before considering the protein as a potential interactor.

While spectral counting has repeatedly proven to be a successful strategy, as mentioned it is not very effective for proteins identified by only one or two peptides. Fortunately, there are other strategies for comparing the abundances of proteins in complexes available. The laboratory of Matthias Mann developed a rapid, LC–MS^2 based, strategy that utilizes stable isotope labeling to discriminate specifically and nonspecifically bound members of an extracted protein complex (Figure 5.8). This strategy combines stable-isotope labeling with amino acids in cell culture (SILAC) (35) with IP and RNA interference (RNA$_i$) to knockdown protein expression (36). The strategy, termed QUICK (quantitative IP combined with knockdown), involves culturing cells in medium containing a heavy isotope-substituted amino acid (e.g., $^{13}C_6$ lysine), so that every protein contains a heavy version of that particular amino acid. A separate culture that is to be used in comparison is grown in normal isotopic abundance medium. One of the cultures is treated with an RNAi targeted to a specific protein to knockdown its expression. The targeted protein complex is then extracted from each culture separately using the exact same conditions. After combining the two IP samples, the mixture is proteolytically digested and analyzed using LC–MS^2. Proteins that interact nonspecifically with any of the components of the IP will ideally be represented in the mass spectra as a doublet of peaks with an area ratio of 1:1. Proteins that are specifically bound members of the complex will be represented by either a single peak from the non-RNAi-treated culture or a doublet in which the peptide peaks originating from the RNAi-treated culture are much lower in intensity.

To demonstrate the QUICK method, it was applied to identify specific binding partners of β-catenin and Cbl (36). T-cell-specific transcription factor 4, α-catenin, and β-catenin interacting protein all had significantly higher abundance ratios in the non-RNAi-treated sample in the study in which β-catenin was the target protein. In the Cbl study, nexin 18 along with three proteins known to interact with Cbl were identified. While it may not enhance the ability to identify more novel binding partners, the QUICK strategy is able to eliminate many nonspecifically binding proteins based on the quantitative MS data. In the case of β-catenin, well over 100 nonspecifically interacting proteins were recognizable based on the LC–MS^2 data. The QUICK strategy is also amenable to monitoring changes in the composition of protein complexes extracted using different methods and is not limited to RNAi-treated cells. As SILAC methods develop, it is now possible to use the basic QUICK strategy to analyze complexes extracted directly from organisms as complex as mouse.

5.6   PROTEIN ARRAYS

Protein expression and purification technologies have also seen enormous growth in the past decade. I remember as a post-doc the chore of expressing and purifying a single protein. Today large repositories of expression vectors containing open reading frames (ORFs) of almost every conceivable protein for most major organisms exist. These reagents have enabled massive screening of protein interactions using arrays or protein chips (Figure 5.9) (37,38). Protein chips that contain thousands of expressed proteins can be produced from having knowledge of the genomic sequence of an organism (39, 40). Each ORF identified in an organism's annotated genome is cloned in a suitable expression and purification vector. Individual flasks containing host cells are then transformed with the expression vectors and induced to express their specific protein. The expressed proteins are purified using automated high-throughput parallel methods. The isolated proteins are spotted onto a glass slide creating a huge matrix of spots each containing a specific protein. The entire protein chip is interrogated with a specific molecule to which a detectable label is attached. After a series of washing steps to remove labeled molecules that are bound nonspecifically or do not bind at all, the protein chip is dried and the labeled molecules detected photometrically. A list of proteins corresponding to the spots at which labeled molecules are found to bind is generated. The potential binding candidates are then validated using an orthogonal method to measure protein interactions.

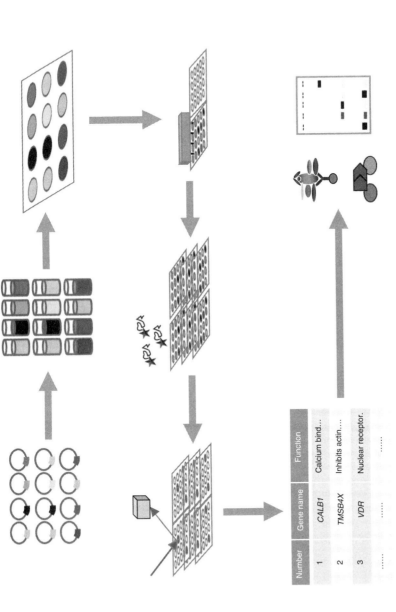

Number	Gene name	Function
1	CALB1	Calcium bind...
2	TMSB4X	Inhibits actin.....
3	VDR	Nuclear receptor.
......	......	......

Figure 5.9. Overview of identification of protein interactions using proteome microarrays. The array is constructed by cloning each ORF into a vector suitable for the expression and purification of the protein from the native organism. Cells are transformed with the addition of the plasmid(s). Cells transformed with the plasmids are grown and induced to overexpress the proteins, which are then purified using a single, high-throughput method. A piezoelectric spray arrayer or microcontact printer is used to transfer the purified proteins to slides. A labeled probe is introduced to the microarrays and after a certain incubation time, unbound proteins are washed away. After drying the microarrays, proteins that interact with the labeled probe are detected, typically using a fluorescence scanner. The fluorescent intensities from each spot are analyzed to generate a candidate interaction list, which are validated using an orthogonal technique (e.g., IP, gel-shift assay, and yeast two-hybrid screen).

Protein chips for conducting proteome screening of interactions have been produced for the approximately 6000 ORFs represented in the *S. cerevisiae* genome (41). The first complete *S. cerevisiae* collection included a plasmid library of ∼5800 individual ORFs that contained glutathione-*S*-transferase coding sequences fused to their 5' ends to assisted in their isolation (42). Obviously, printing protein chips of this nature requires careful optimization of many parameters to ensure that the proteins are of sufficient quantity and quality. Another critical parameter was found to be the surface chemistry of the slide that the proteins were printed on. The surface chemistry must be selected to optimize the binding affinity of the proteins that are printed onto the chip, and minimize nonspecific binding between the labeled probes and the slide surface. The surface of the slide must also not interfere with the imaging and detection of potential protein interactions. Purified proteins can be printed using either a microcontact printer or a piezoelectric spray arrayer (43). To circumvent having to purify and print proteins directly onto slides, alternative methods such as direct cell-free expression of proteins on chips have also been developed over the past few years (44, 45).

To identify protein interactions, the chips are incubated with a biomolecule of interest that is labeled with a probe that can be detected fluorometrically. After a specific incubation time, excess unbound and nonspecifically bound molecules are removed by washing the entire array. Excess buffer is removed by centrifuging the protein chip and the amount of probe bound to each spot is measured using a detection method such as a fluorescence slide scanner. Potential interactions are identified by locating protein spots that show higher binding with the labeled probe than most other spots. Analysis of the data in this manner provides a list of candidate interactions that require validation using an orthogonal assay such as an IP or EMSA.

5.7 FLUORESCENCE MICROSCOPY

Admittedly, most of the methods for characterizing protein interactions in this chapter have involved MS. It is important to keep in mind there are a number of other important methods for characterizing these interactions. These methods include Western blotting, EMSA, fluorescence spectroscopy, NMR spectroscopy, size-exclusion chromatography, etc. While it is impossible to cover them all, suffice it to say they can play an important role as either the primary method for identifying a protein interaction or as validation of a potential interaction found using a discovery driven approach. Over the past few years, there have been a number of improvements made in the fluorescent tagging of proteins and laser scanning confocal microscopy (LSCM) (46). Most LSCM methods use dual color images that are collected separately and superimposed to detect the presence of colocalizing signals. To localize a protein, the cell is irradiated with a laser corresponding to the specific absorption wavelength of the fluorophore of interest and the fluorophore's emission wavelength is detected. The process is repeated for a second compound that absorbs and emits light at a different wavelength. Combining the images is then performed to determine if the two signals overlap. For example, proteins are commonly tagged with green and red fluorophores. If the proteins colocalize, superimposing their images will result in orange pixels being observed. If orange pixels

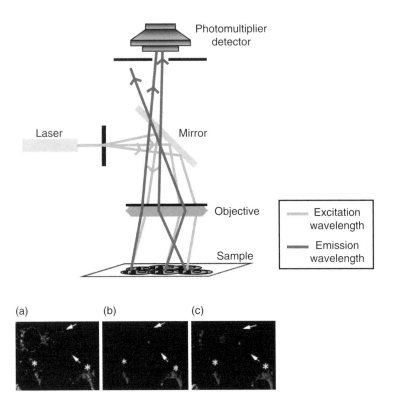

Figure 5.10. Principles of LSCM. (Top panel) The sample is irradiated via a laser with a wavelength of light that excites a fluorophore attached to the protein of interest. The emission wavelength of the excited fluorophore is recorded. (Bottom panel) Colocalization of proteins tagged with fluorophores having different excitation and emission wavelengths is determined by overlaying the two resultant images. In this example, the overlaying of the green and red images results in an orange color showing that the two proteins of interest occupy similar spaces in the cell and therefore, probably interact.

are observed it suggests that the proteins occupy the same space and are potentially interacting (Figure 5.10).

The field of protein colocalization has been tremendously advanced by the development of various fluorescent tags (47, 48). Fluorophores such as Cy2, Cy3, and Cy5 are used to covalently modify specific proteins or proteins can be expressed directly within transfected cells containing a fluorescent tag. Popular fluorescently labeled proteins include green, red, and yellow fluorescent protein. In the technique known as immunofluorescence, fluorescently tagged antibodies are also used to localize proteins within cells (47). The first step in any localization study is attaching the cells to a solid support, such as a microscope slide. Adherent cells can be grown on microscope slides while cells grown in suspension are generally centrifuged onto slides or bound to a solid support via covalent linkers, however, they can be analyzed while in suspension if neither of the previous methods is viable. Using immunofluorescence may necessitate fixing the cells either by using organic solvents such as alcohols and acetone or with paraformaldehyde. As in immunohistochemistry, paraformaldehyde cross-linking preserves the cellular architecture. Paraformaldehyde is better at maintaining cellular architecture than organic solvents but requires a permeabilization step that enables the antibody to enter the cell and access the antigen. Since the purpose of LSCM is to visualize a protein interaction in its native state, cells should be minimally perturbed prior to and during data acquisition.

Once the cells of interest have been prepared, the process of determining if two (or more) proteins colocalize is initiated. The fluorescent labels on the proteins are excited using lasers tuned to the respective tag's excitation wavelengths and the emission wavelengths are monitored (46, 49). If the images obtained from the different fluorophores overlap, it is assumed that the proteins interact by virtue of them occupying the same space. Since LSCM techniques require either labeling specific proteins or using antibodies, this method is hypothesis driven. One of its most valuable roles in basic research is validating biomolecular interactions that are found using discovery-driven methods.

5.8 MULTIEPITOPE LIGAND CARTOGRAPHY

As mentioned above, LSCM is not considered a discovery-driven method since specific proteins are targeted by incorporating fluorophores or using antibodies. In addition, LSCM has historically been limited to 2–3 proteins per study. A method termed multiepitope ligand cartography (MELC) has been developed that is capable of mapping the location of tens of proteins within the same cell or tissue sample (50–52). While it does not technically graduate LSCM to a discovery-driven method, MELC does expand the experimental space that can be interrogated in a single experiment. Similar to LSCM, this technology uses fluorescently labeled antibodies (or other affinity reagents) in a multiplexed mode to localize proteins. Unlike LSCM, however, MELC incorporates multiple rounds of introducing antibodies into the cell. After incorporating a single antibody into the cells and allowing it to bind to its intended target, the image is capture, the fluorescent signal is bleached, and the next fluorescently tagged antibody is introduced. MELC uses several iterations of this procedure to localize tens of different proteins within the same biological system.

While the technology behind the procedure appears to be straightforward, and can be likened to souped-up LSCM, interpretation and presentation of the data are anything but simple. In an early study illustrating MELC technology, fluorescently tagged antibodies were used to localize eighteen cell surface receptors on peripheral blood mononuclear cells (50). Fluorescently labeled antibodies were introduced to the cells in pairs in nine cycles composed of introduction of the antibodies to the cells, capturing of the individual fluorescent images, bleaching to remove any residual signal, and introducing the next pair of tagged antibodies. The proteins were localized by superimposing the intensities of the signals obtained for all 18 fluorescently tagged antibodies. Unfortunately, the images produced using this visualization method are very complex and it is challenging to provide an estimate of the protein complexes *in vivo*. The analysis is simplified by converting the data generated into a vector representing the intensity of the signal provided by each antibody at each pixel recorded in the image. These vectors are presented in a tabular view allowing numerical visualization of each antibody's fluorescence at individual locations within the cell.

The MELC method has been used to differentiate changes in protein complexes in skin biopsies obtained from patients with psoriasis, atopic dermatitis (a related but distinct condition) (53), and healthy controls. Forty-nine different proteins were localized in each of the clinical biopsies with the aim to identify proteins that have diagnostic value or could act as a therapeutic target for patients suffering from psoriasis. The data showed that the protein levels across the tissue section were generally similar in the psoriasis and atopic dermatitis samples. Protein complex motifs were identified that were unique to atopic dermatitis and could differentiate these samples from those obtained from patients with psoriasis. They could also distinguish both of these skin conditions from healthy controls. While this method is still not a purely discovery-driven approach, if the claim that hundreds of proteins can potentially be colocalized using MELC is realized, the range of hypotheses that can be tested is greatly increased.

5.9 CONCLUSIONS

The development of proteomics technology in the past decade has brought tremendous advances in how protein and biomolecular complexes are identified. While this chapter attempts to discuss many of the predominant proteomic methods for identifying protein interactions, even it is incomplete as novel developments for elucidating cellular interactomes are regularly being produced. While basic researchers used to be confined to confirming or refuting hypothetical interactions, tools that allow novel interactions to be discovered are now commonplace. The developments highlighted above span the entire range of complexity right from the characterization of proteins bound to a single protein target to entire proteome-wide scanning to measure entire interactomes. There are, however, many difficulties that state-of-the-art technology cannot yet overcome. For instance, MS is no different than any other "black-box" instrument and operates under the law of "garbage in, garbage out." The need for advanced sample preparation methods to limit the number of nonspecific entities identified within protein complexes will be crucial for obtaining interaction maps that truly reflect cell physiology. The lack of congruence amongst large-scale interactome studies suggests that each analytic method only provides part of the truth. Whether repeat analysis or continued comparison amongst large interactome data sets will provide an accurate view of the protein circuitry within the cell remains to be seen. At this time, many of the discovery-based methods for identifying protein interactions provide "possibilities" that require further validation before any certain biological function can be established for a novel protein–protein interaction.

REFERENCES

1. Douglass F. "Narrative of the Life of Frederick Douglass, an American Slave, Written by Himself". In: Baym N, editors. *The Norton Anthology of American Literature*. 6th ed. vol. B. London, New York: Norton; 2003. 2032–2097.

2. Seger R, Krebs EG. The MAPK signaling cascade. FASEB J. 1995;9:726–735.

3. Yoon S, Seger R. The Extracellular signal-regulated kinase: Multiple substrates regulate diverse cellular functions. Growth Factors 2006;24:21–44.

4. Burgermeister E, Seger R. PPARγ and MEK interactions in cancer. PPAR Res. 2008;2008:309–469.

5. Torii S, Yamamoto T, Tsuchiya Y, Nishida E. ERK MAP kinase in G_1 cell cycle progression and cancer. Cancer Sci. 2006;97:697–702.

6. Roberts PJ, Der CJ. Targeting the Raf-MEK-ERK mitogen-activated protein kinase cascade for the treatment of cancer. Oncogene 2007;26:3291–3310.

7. McCubrey JA, Steelman LS, Chappell WH, Abrams SL, Wong EW, Chang F, Lehmann B, Terrian DM., Milella M, Tafuri A, Stivala F, Libra M, Basecke J, Evangelisti C, Martelli AM, Franklin RA. Roles of the Raf/MEK/ERK pathway in cell growth, malignant transformation and drug resistance. Biochim. Biophys. Acta. 2007;1773:1263–1284.

8. Lurje G, Lenz HJ. EGFR signaling and drug discovery. Oncol. 2009;77:400–410.

9. Ramisetty SR, Washburn MP. Unraveling the dynamics of protein interactions with quantitative mass spectrometry. Crit. Rev. Biochem. Mol. Biol. 2001;46:216–228.

10. Shiio Y, Itoh M, Inoue J. Epitope tagging. Methods Enzymol. 1995;254:497–502.

11. Williamson MP, Sutcliffe MJ. Protein-protein interactions. Biochem. Soc. Trans. 2010;38: 875–878.

12. Collins MO, Choudhary JS. Mapping multiprotein complexes by affinity purification and mass spectrometry. Curr. Opin. Biotechnol. 2008;19:324–330.

13. Gingras AC, Aebersold, R, Raught B. Advances in protein complex analysis using mass spectrometry. J. Physiol. 2005;563:11–21.

14. Gingras AC, Gstaiger M, Aebersold R, Raught B. Advances of protein complex analysis using mass spectrometry. Nat. Rev. Mol. Cell Biol. 2005;8:645–654.

15. Gavin AC, Aloy P, Grandi P, Krause R, Boesche M, Marzioch M, Rau C, Jensen LJ., Bastuck S, Dümpelfeld B, Edelmann A, Heurtier MA, Hoffman V, Hoefert C, Klein K, Hudak M, Michon AM, Schelder M, Schirle M, Remor M, Rudi T, Hooper S, Bauer A, Bouwmeester T, Casari G, Drewes G, Neubauer G, Rick JM, Kuster B, Bork P, Russell RB, Superti-Furga G. Proteome survey reveals modularity of the yeast cell machinery. Nature 2006;440:631–636.

16. Krogan NJ, Cagney G, Yu H, Zhong G, Guo X, Ignatchenko A, Li J, Pu S, Datta N, Tikuisis AP, Punna T, Peregrín-Alvarez JM, Shales M, Zhang X, Davey M, Robinson MD, Paccanaro A, Bray JE, Sheung A, Beattie B, Richards DP, Canadien V, Lalev A, Mena F, Wong P, Starostine A, Canete MM, Vlasblom J, Wu S, Orsi C, Collins SR, Chandran S, Haw R, Rilstone JJ, Gandi K, Thompson NJ, Musso G, St Onge P, Ghanny S, Lam MH, Butland G, Altaf-Ul AM, Kanaya S, Shilatifard A, O'Shea E, Weissman JS, Ingles CJ, Hughes TR, Parkinson J, Gerstein M, Wodak SJ, Emili A, Greenblatt JF. Global landscape of protein complexes in the yeast saccharomyces cerevisiae. Nature 2006;440:637–643.

17. Dougherty MK, Ritt DA, Zhou M, Specht SI, Monson DM, Veenstra TD, Morrison DK. KSR2 is a calcineurin substrate that promotes ERK cascade activation in response to calcium signals. Mol. Cell 2009;34:652–662.

18. Ory S, Zhou M, Conrads TP, Veenstra TD, Morrison DK. Protein phosphatase 2A positively regulates ras signaling by dephosphorylating KSR1 and Raf-1 on critical 14-3-3 binding sites. Curr. Biol. 2003;13:1356–1364.

19. Jiang D, Jarrett HW, Haskins WE. Methods for proteomic analysis of transcription factors. J. Chromatogr. A 2009;1216:6881–6889.

20. Himeda CL, Ranish JA, Angello JC, Maire P, Aebersold R, Hauschka SD. Quantitative proteomic identification of six4 as the Trex-binding factor in the muscle creatine kinase enhancer. Mol. Cell. Biol. 2004;24:2132–2143.

21. Meng Z, Camalier CE., Lucas DA, Veenstra TD, Beck GR Jr, Conrads TP. Probing early growth response 1 interacting proteins at the active promoter in osteoblast cells using oligo-precipitation and mass spectrometry. J. Proteome Res. 2006;5:1931–1939.

22. Ryder SP, Recht MI, Williamson JR. Quantitative analysis of protein–RNA interactions by gel mobility shift. Methods Mol. Biol. 2008;488:99–115.

23. Goll J, Uetz P. The elusive yeast interactome. Genome Biol. 2006;7:223.

24. Gunsalus KC, Piano F. RNAi as a tool to study cell biology: Building the genome-phenome bridge. Curr. Opin. Cell Biol. 2005;17:3–8.

25. Walhout AJ, Boulton SJ, Vidal M. Yeast two-hybrid systems and protein interaction mapping projects for yeast and worm. Yeast 2000;17:88–94.

26. Sanchez-Ramos J, Song S, Dailey M, Cardozo-Pelaez F, Hazzi C, Stedeford T, Willing A, Freeman TB, Saporta S, Zigova T, Sanberg PR, Snyder EY. The X-gal caution in neural transplantation studies. Cell Transplant. 2000;9:657–667.

27. Uetz P, Giot L, Cagney G, Mansfield TA, Judson RS, Knight JR, Lockshon D, Narayan V, Srinivasan M, Pochart P, Qureshi-Emili A, Li Y, Godwin B, Conover D, Kalbfleisch T, Vijayadamodar G, Yang M, Johnston M, Fields S, Rothberg JM. A comprehensive analysis of protein–protein interactions in saccharomyces cerevisiae. Nature 2000;403:623–627.

28. Ito T, Chiba T, Ozawa R, Yoshida M, Hattori M, Sakaki Y. A comprehensive two-hybrid analysis to explore the yeast protein interactome. Proc. Natl. Acad. Sci. U.S.A. 2001;98:4569–4574.

29. Rain JC, Selig L, De Reuse H, Battaglia V, Reverdy C, Simon S, Lenzen G, Petel F, Wojcik J, Schachter V, Chemama Y, Labigne A, Legrain P. The protein–protein interaction map of *Helicobacter pylori*, Nature 2001;409:211–215.

30. Li S, Armstrong CM, Bertin N, Ge H, Milstein S, Boxem M, Vidalain PO, Han JD, Chesneau A, Hao T, Goldberg DS, Li N, Martinez M, Rual JF, Lamesch P, Xu L, Tewari M, Wong SL, Zhang LV, Berriz GF, Jacotot L, Vaglio P, Reboul J, Hirozane-Kishikawa T, Li Q, Gabel HW, Elewa A, Baumgartner B, Rose DJ, Yu H, Bosak S, Sequerra R, Fraser A, Mango SE, Saxton WM, Strome S, Van Den Heuvel S, Piano F, Vandenhaute J, Sardet C, Gerstein M, Doucette-Stamm L, Gunsalus KC, Harper JW, Cusick ME, Roth FP, Hill DE, Vidal M. A map of the interactome network of the Metazoan C. elegans. Science 2004;303:540–543.

31. Giot L, Bader JS, Brouwer C, Chaudhuri A, Kuang B, Li Y, Hao YL, Ooi CE, Godwin B, Vitols E, Vijayadamodar G, Pochart P, Machineni H, Welsh M, Kong Y, Zerhusen B, Malcolm R, Varrone Z, Collis A, Minto M, Burgess S, McDaniel L, Stimpson E, Spriggs F, Williams J, Neurath K, Ioime N, Agee M, Voss E, Furtak K, Renzulli R, Aanensen N, Carrolla S, Bickelhaupt E, Lazovatsky Y, DaSilva A, Zhong J, Stanyon CA, Finley RL Jr., White KP, Braverman M, Jarvie T, Gold S, Leach M, Knight J, Shimkets RA, McKenna MP, Chant J, Rothberg JM. A protein interaction map of drosophilia melanogaster. Science 2003;302:1727–1736.

32. Rual JF, Venkatesan K, Hao T, Hirozane-Kishikawa T, Dricot A, Li N, Berriz GF, Gibbons FD, Dreze M, Ayivi-Guedehoussou N, Klitgord N, Simon C, Boxem M, Milstein S, Rosenberg J, Goldberg DS, Zhang LV, Wong SL, Franklin G, Li S, Albala JS, Lim J, Fraughton C, Llamosas E, Cevik S, Bex C, Lamesch P, Sikorski RS, Vandenhaute J, Zoghbi HY, Smolyar A, Bosak S, Sequerra R, Doucette-Stamm L, Cusick ME, Hill DE, Roth FP, Vidal M. Towards a proteome-scale map of the human protein–protein interaction network. Nature 2005;437:1173–1178.

33. Simonis N, Rual JF, Carvunis AR, Tasan M, Lemmens I, Hirozane-Kishikawa T, Hao T, Sahalie JM, Venkatesan K, Gebreab F, Cevik S, Klitgord N, Fan C, Braun P, Li N, Ayivi-Guedehoussou N, Dann E, Bertin N, Szeto D, Dricot A, Yildirim MA, Lin C, de Smet AS, Kao HL, Simon C, Smolyar A, Ahn JS, Tewari M, Boxem M, Milstein S, Yu H, Dreze M, Vandenhaute J, Gunsalus KC, Cusick ME, Hill DE, Tavernier J, Roth FP, Vidal M. Empirically controlled mapping of the caenorhabditis elegans protein-protein interactome network. Nat. Methods 2009;6:47–54.

34. Carvalho PC, Hewel J, Barbosa VC, Yates JR 3rd. Identifying differences in protein expression levels by spectral counting and feature selection. Genet. Mol. Res. 2008;7:342–356.

35. Ong SE, Blagoev B, Kratchmarova I, Kristensen DB, Steen H, Pandey A, Mann M. Stable isotope labeling by amino acids in cell culture, SILAC, as a simple and accurate approach to expression proteomics. Mol. Cell. Proteomics 2002;1:376–386.

36. Selbach M, Mann M. Protein interaction screening by quantitative immunoprecipitation combined with knockdown (QUICK). Nat. Methods 2006;3:981–983.

37. Cormier CY, Mohr SE, Zuo D, Hu Y, Rolfs A, Kramer J, Taycher E, Kelley F, Fiacco M, Turnbull G, LaBaer J. Protein structure initiative material repository: An open shared public resource of structural genomic plasmids for the biological community. Nucleic Acids Res. 2010;38:D743–D749.

38. Cormier CY, Park JG, Fiacco M, Steel J, Hunter P, Kramer J, Singla R, LaBaer J. PSI:Biology-materials repository: A biologist's resource for protein expression plasmids. J. Struct. Funct. Genomics 2011;12:55–62.

39. Chen R, Snyder M. Yeast proteomics and protein microarrays. J. Proteomics 2010;73:2147–2157.

40. Kung LA, Snyder M. Proteome chips for whole-organism assays. Nat. Rev. Mol. Cell Biol. 2006;7:617–622.

41. Goffeau A, Barrell BG, Bussey H, Davis RW, Dujon B, Feldmann H, Galibert F, Hoheisel JD, Jacq C, Johnston M, Louis EJ, Mewes HW, Murakami Y, Philippsen P, Tettelin H, Oliver SG. Life with 6000 genes. Science 1996;274:563–547.

42. Zhu H, Bilgin M, Bangham R, Hall D, Casamayor A, Bertone P, Lan N, Jansen R, Bidlingmaier S, Houfek T, Mitchell T, Miller P, Dean RA, Gerstein M, Snyder M. Global analysis of protein activities using proteome chips. Science 2001;293:2101–2105.

43. Delehanty JB, Ligler FS. Method for printing functional protein microarrays. BioTechniques 2003;34:380–385.

44. Stoevesandt O, Taussig MJ, He M. Producing protein microarrays from DNA microarrays. Methods Mol. Biol. 2011;785:265–276.

45. Sitaraman K, Chatterjee DK. Protein–protein interactions: An application of Tus–Ter mediated protein microarray system. Methods Mol. Biol. 2011;723:185–200.

46. Hutter H. Fluorescent report methods. Methods Mol. Biol. 2006;351:155–173.

47. Giepmanns BN, Adams SR, Ellisman MH, Tsien RY. The fluorescent toolbox for assessing protein location and function. Science 2006;312:217–224.

48. Chan CP. Ingenious nanoprobes in bioassays. Bioanalysis 2009;1:115–133.

49. Miyashita T. Confocal microscopy for intracellular co-localization of proteins. Methods Mol. Biol. 2004;261:399–410.

50. Schubert W, Bonnekoh B, Pommer AJ, Philipsen L, Böckelmann R, Malykh Y, Gollnick H, Friedenberger M, Bode M, Dress AW. Analyzing proteome topology and function by automated multidimensional fluorescence microscopy. Nat. Biotechnol. 2006;24:1270–1278.

51. Berndt U, Philipsen L, Bartsch S, Hu Y, Röcken C, Bertram W, Hämmerle M, Rösch T, Sturm A. Comparative multi-epitope-ligand-cartography reveals essential immunological alterations in Barrett's metaplasia and esophageal adenocarcinoma. Mol. Cancer 2010;9:177

52. Pierre S, Scholich K. Toponomics: Studying protein–protein interactions and protein networks in intact tissue. Mol. Biosyst. 2010;6:641–647.

53. Boguniewicz M, Leung DY. Atopic dermatitis. J. Allergy Clin. Immunol. 2006;117:S475–S480.

6

GLOBAL PHOSPHORYLATION ANALYSIS

6.1 INTRODUCTION

In Chapter 4, over 2500 words were devoted to methods for measuring the phosphory-lation states of proteins. So the question may be why do we need another chapter on phosphorylation? The previous chapter provided a background on the importance of phosphorylation and methods, both at the sample preparation and mass spectrometry (MS) stage, for measuring these posttranslational modifications. What it did not provide, however, were examples of how measuring phosphoproteins can further our understand-ing of cancer. I am a big advocate that to fully understand something, it is important to see how it is done. Therefore, in this chapter, I am going to present examples on how global analysis of protein phosphorylation has been applied to issues in tumorigenesis, cancer progression, and treatment. This chapter will include studies in which the goal was to characterize a large number of phosphoproteins; however, I have chosen not to include studies that simply show the ability to identify phosphoproteins. Do not misunderstand me; developing and demonstrating the ability to identify thousands of phosphoryla-tion sites within complex proteomes was absolutely critical (1). However, I wanted to include studies in which the collected data were distilled to discover something impor-tant about the system being analyzed and could lead to logical hypotheses that could be readily tested.

Proteomic Applications in Cancer Detection and Discovery, First Edition. Timothy D. Veenstra.
© 2013 John Wiley & Sons, Inc. Published 2013 by John Wiley & Sons, Inc.

6.2   GLIOBLASTOMA MULTIFORME

The most aggressive form of adult human brain tumor is glioblastoma mutliforme (GBM), which has a median survival of <12 months (2–4). The low median survival is partially due to the lack of therapeutic treatments and reliable early detection biomarkers; however, epidermal growth factor receptor (EGFR)-targeted therapies are being pursued for GBMs that overexpress this protein (5, 6). About 50% of GBM tumors overexpress a mutant EGFR, termed EGFRvIII, which lacks exons 2–7 (7). While EGFRvIII lacks the extracellular domain and is incapable of binding EGFR ligands, it is constitutively tyrosine phosphorylated at a level roughly 10% of that of stimulated EGFR. EGFRvIII has been found exclusively in tumors and investigators have shown a correlation between its expression and poor prognosis for GBM patients (8). These last two observations suggest that EGFRvIII is critically important in GBM tumorigenesis and progression.

6.2.1   Effect of EGFRvIII Expression of Phosphorylation

The effect of EGFRvIII receptor expression level on phosphotyrosine-mediated cellular signaling networks was determined using many of the foundations of proteomics technology: sample preparation, stable isotope labeling, chromatographic fractionation, and MS. The samples analyzed were acquired from three different glioblastoma (UM87MG) cell lines that express various levels of EGFRvII;U87-M; 1.5 million copies EGFRvIII per cell: U87-H; 2 million copies EGFRvIII per cell: and U87-SH; 3 million copies EGFRvIII per cell (9). These different populations were acquired using fluorescence-activated cell sorting. A control population of U87MG cells (U87-DK for dead kinase) that express an inactive form of EGFRvIII, was used as a control. After the proteins were extracted from the cells, they were digested into peptides (Figure 6.1). Tyrosine-phosphorylated peptides were immunoprecipitated using a pan-specific antiphosphoty-rosine antibody, followed by immobilized metal affinity chromatography (IMAC). The peptides were finally labeled with the four-plex iTRAQ reagents just prior to liquid chromatography-tandem mass spectrometry (LC-MS^2) analysis. This method resulted in the quantitation of 99 phosphorylation sites within 69 proteins amongst the four cell lines. While not an overwhelming number, ultimately the data provided incredibly valuable information on the role of specific phosphorylation events with GBM tumors.

Eight phosphorylation sites were quantitated on EGFRvIII itself, with the degree of phosphorylation at each site being determined within each cell type (i.e., U87-M, U87-H, U87-SH, and U87-DK). While the increase in phosphorylation at Y845 and Y1086 increased proportionally with EGFRvIII expression level, phosphorylation of Y974, Y1068, Y1114, Y1148, and Y1173 increased more than twofold when U87-M (1.5 million copies EGFRvIII per cell) and U87-H (two million copies EGFRvIII per cell) cells were compared. If they had followed the trend of Y845 and Y1086, the increase in phosphorylation at these five residues would only have been expected to be about 1.3-fold. This disproportionate increase suggested that once EGFRvIII expression exceeds a certain threshold expression level, the receptor becomes activated and autophosphorylation increases significantly. The amount of phosphorylation at these

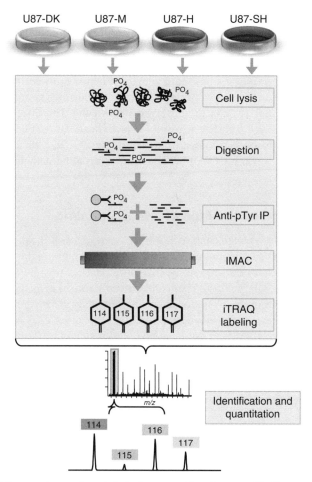

U87-DK U87-M U87-H U87-SH

Cell lysis

Digestion

Anti-pTyr IP

IMAC

iTRAQ labeling

Identification and quantitation

Figure 6.1. Cell lines and experimental strategy used for the quantitative analysis of phosphotyrosine peptides and proteins for the discovery of the mechanism of EGFRvIII in glioblastoma cells (U87MG). Four cell lines expressing different levels of active EGFRvIII (DK, dead kinase; M, medium; H, high; SH, super high) were generated. Phosphotyrosine peptides were extracted from these cell lines using a combination of immunoprecipitation (anti-pTyr IP) and IMAC. The resulting peptides were stable isotope labeled using iTRAQ and identified and quantitated using liquid chromatography–tandem MS.

five sites was equivalent in U87-H and U87-SH cells, suggesting a saturation point is reached at about two million copies of EGFRvIII per cell.

Since considerable knowledge on EGFR signaling pathways has been accumulated over the past several years (10), the various phosphorylation sites quantitated in this global study were mapped onto these canonical pathways. While previous studies have shown that wild-type EGFR activation led to a rapid (i.e., 5 min) increase in the active form of Erk1, Erk2, and signal transducer and activator of transcription 3 (STAT3) in human mammary epithelial cells (11), increasing EGFRvIII receptor expression levels in the glioblastoma cells had an insignificant effect on these protein's phosphorylation levels. Increasing EGFRvIII receptor levels caused a more than threefold increase in the phosphorylation levels of phosphatidylinositol 3-kinase (PI3K) and Grb2-associated binding protein 1 (GAB1), its upstream adaptor protein. Considering that it has been implicated in promoting cell proliferation, survival, and migration (12), activation of the PI3K pathway may explain the observed tumorigenic properties of constitutively activated EGFRvIII *in vivo*. These results also indicate that EGFRvIII utilizes different downstream pathways compared to wild-type EGFR, indicating that different treatments may possibly inhibit EGFRvIII-activated signaling without disrupting EGFR signaling.

6.2.2   Validation of the Role of c-Met Signaling

In any experiment designed to capture a "global" view of a specific molecular characteristic, the challenging step is finding valuable information within the large data set that was invariably acquired. For this study, the entire data set was clustered into tyrosine sites whose phosphorylation state changed in a similar fashion as the EGFRvIII levels were modulated. Within the cluster of phosphorylation sites that significantly increased as a function of increasing EGFRvIII expression level were Y1234 of the c-Met receptor tyrosine kinase (sixfold increase) and Y62 on tyrosine phosphatase SHP-2 (10-fold increase). While little was known about the role of this phosphorylation site on SHP-2, it was know that this protein is downstream of the c-Met receptor.

These data suggested that EGFRvIII was constitutively activating the c-Met receptor pathway through increased phosphorylation at Y1234. This initial observation was confirmed when additional phosphorylation sites identified in the global analysis were mapped onto known components downstream of the c-Met receptor. Many of these sites were increased more than threefold as EGFRvIII expression levels rose. Not only were many of these proteins downstream of c-Met, but they are also known to be activated by EGFRvIII as well. It is possible that many of these sites require stimulation of both c-Met and EGFRvIII for activation.

Considering the observed link between EGFRvIII and c-Met signaling (13), the investigators treated U87-H cells with the c-Met kinase inhibitor, SU11274 (14). Analysis of the c-Met receptor Y1234 phosphorylation site using quantitative MS showed a decrease in the receptor's phosphorylation upon inhibitor treatment. In addition, the data also showed that the phosphorylation of the multiple tyrosine sites within EGFRvIII was not affected by treatment with SU11274.

The data also revealed an obvious overlap in proteins that were stimulated by coactivation of EGFRvIII and c-Met receptors. As shown in Figure 6.2, these proteins include SHC, GAB1, PLCγ, p85, and p110. To determine cell synergy between these two receptors, U87-H cells were treated with either AG1478 (EGFRvIII inhibitor) (15) or SU11274. For both treatments, high doses of inhibitors were required to decrease cell viability. When the same cells were treated with a static dose of AG1478 (5 μM) and an increased dose of SU11274, cell viability and cell death were significantly decreased and

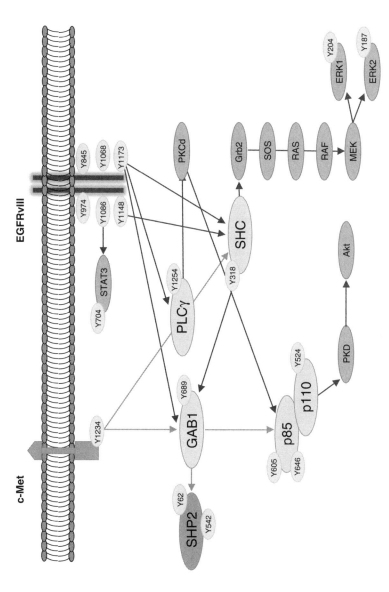

Figure 6.2. Illustration showing selection of tyrosine phosphorylation sites identified in U87MG cells in which EGFRvIII or c-Met were stimulated. The phosphorylation states of the proteins in blue were found to be highly abundant in both EGFRvIII and c-Met-activated cells.

increased, respectively, even at much lower overall doses than when either inhibitor was used solely. These results suggest that the biological responses to the c-Met inhibitors were driven by on-target effects against c-Met itself or, perhaps, c-Met and another kinase, which is a shared off-target molecule.

GBM xenografts that express EGFRvIII are known to be resistant to cisplatin unless this chemotherapy is administered along with AG1478 (16). In addition, activation of the c-Met receptor confers resistance to a wide variety of chemotherapeutics (17). Considering the strong overlap in downstream proteins activated by EGFRvIII/c-Met stimulation shown in this study, the investigators tested if the chemoresistance of EGFRvIII-expressing tumors was related to EGFRvIII-mediated activation of c-Met. U87-H cells were treated with various levels of SU11274 along with a constant 10 μg/mL dose of cisplatin. This combination of chemotherapy and c-Met inhibitor caused a dramatic decrease in cell viability compared to cisplatin alone. The results of this assay shows that the chemoresistance observed with EGFRvIII-positive GBM tumors may be at least partially related to c-Met activation.

This global phosphorylation analysis using MS revealed that c-Met activation is mediated through EGFRvIII phosphorylation. While this study did not quantitate thousands of phosphorylation sites, it did uncover a dense amount of important information centered on EGFRvIII and c-Met stimulation. It would be very challenging to obtain the same level of information using a hypothesis-driven method relying on tyrosine-specific antibodies to each of the proteins identified in this study. Distilling the global results was able to show that treatment of cells expressing high levels of EGFRvIII with c-Met kinase inhibitors and cisplatin or c-Met kinase inhibitors and EGFR kinase inhibitors demonstrated enhanced cytotoxicity. So how important or relevant are these findings? There are presently a number of phase I, II, and III clinical trials being conducted to test the efficacy of c-Met inhibitors on many different tumor types (18). In addition, c-Met expression is currently linked to poor prognosis in a number of cancers ranging from colon to GBM (19).

6.3 APOPTOSIS AND CANCER

Programmed cell death, known as apoptosis, is essential for a number of critical biological functions including embryonic development and cell turnover (20). Apoptotic signaling pathways have been associated with many human diseases including cancer (21). It has also been recognized for some time now that mutations in oncogenes can lead to tumorigenesis, progression, and metastasis through the disruption of apoptosis. In addition, many cytotoxic anticancer agents function through the induction of apoptosis, suggesting that treatment failures may be related to defects in apoptotic pathways. Apoptotic progression is highly regulated by the proteolytic cleavage of various cell survival proteins by caspases (i.e., cysteine-dependent aspartic acid-directed proteases) (22). Not only are caspases important in apoptosis but so is phosphorylation of their target sequences. Phosphorylation of either caspases or their substrates at or near the cleavage site, can lead to inhibition of caspase-mediated proteolysis.

6.3.1   Global Discovery of Overlapping Kinase and Caspase Targets

Using a discovery-driven proteomics approach, a recent study attempted to clarify the global role of phosphorylation in the regulation of caspase signaling (23). This study began by utilizing an increasingly underutilized tool in today's world of science: knowledge. If the reader would allow me a go off on a small tangent for a moment. I have observed that in today's fast-pace world of science where the objective is to collect copious amounts of data and publish quickly; too many studies begin by ignoring what has already been discovered. The net result is poorly designed experiments that do not add much (if anything) to the existing literature. You can decide to agree or disagree.

In this global phosphorylation/caspase study, the investigators first used a peptide matching program to evaluate consensus sequences for phosphorylation by 10 protein kinases, Akt, aurora A, calcium/calmodulin-dependent protein kinase II, Cdk1, casein kinase 2 (CK2), CK1, ERK2, glycogen synthase kinase-3β, Nek6, and adenosine 3′,5′-monophosphate-dependent protein kinase, which are involved in cell survival or tumorigenesis. The investigators looked for the presence of overlapping kinase and caspase-3 recognition motifs, which differ depending on the kinase or caspase being examined. Of all the kinases examined, CK2 was found as the most prominent that had consensus phosphorylation sequences that overlapped with caspase-3 recognition motifs. The consensus motifs for both CK2 and caspase were dominated by acidic residues (24). CK2 is known to be overexpressed in many human cancers, and inhibition of CK2 can result in spontaneous apoptosis of cancer cells (25, 26). Evidence suggesting an antiapoptotic role for CK2 via protection of prosurvival proteins, such as Bid, Max, phosphatase and tensin homolog deleted from chromosome 10, and connexin 45.6, is also increasing (27). As outlined below, the information gathered in this global proteomic study suggests that this kinase might play an even more important role in inhibiting caspase-mediated degradation than was previously appreciated.

The peptide match programs identified over 300 potential CK2 and caspase targets that have a CK2 phosphorylation consensus sequence directly adjacent to the aspartic acid residue that directs caspase cleavage (23). Prior to conducting any empirical experiments, these targets were evaluated against databases containing comprehensive lists of kinases and their targets (e.g., Phosida and Phospho.ELM). To determine whether the putative targets of CK2 and caspases were phosphorylated at the predicted CK2 and caspase sites in cells, a comprehensive evaluation of phospho-databases, including Phosida and Phospho.ELM, was conducted (28, 29). Thirty-seven proteins identified in the peptide match search were found to be phosphorylated at the predicted (Ser or Thr) site in cells (Figure 6.3). Five proteins were found to be cleaved by caspases through searching the CASBAH database (30). A vast majority of the proteins (~90%), however, had not been characterized whereas the evaluation of the caspase substrate database CASBAH revealed five proteins (2%) that are cleaved by a known caspase, although 87% have not been characterized.

Obviously, this putative data gathered via database searching required empirical studies for validation. To determine whether these putative CK2 and caspase target sequences were indeed phosphorylated by CK2, a peptide array target screen was

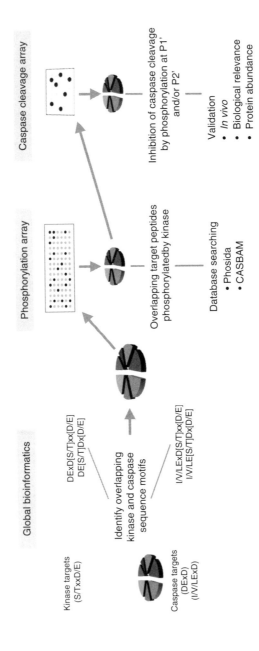

Figure 6.3. Schematic of strategy used to identify global overlapping protein kinase and protease targets (23). In the initial phase a bioinformatics analysis of the human proteome was conducted to identify kinase and caspase targets that contain overlapping phosphorylation and caspase recognition motifs. Customized arrays were generated to measure the phosphorylation of the predicted peptides by CK2. A caspase cleavage array (see Figure 6.4) was generated based on the CK2-specific peptides identified within the phosphorylation array. The reactivity of the peptides within this array was tested against various members of the caspase family.

performed. An array containing 13 amino acid long peptides that contained putative CK2 and caspase target sequences was synthesized. After soaking in ethanol and washing with kinase buffer, the peptides were incubated in the presence of GST-tagged CK2 (GST-CK2α) and [γ-^{32}P]ATP. After quenching the reaction and washing the array, the phosphorylation of the peptides was visualize using a phosphorimager. The relative incorporation of γ-^{32}P incorporated was measured and only those peptides that incorporated γ-^{32}P by a factor of three times higher than observed in a negative control peptide were considered to be phosphorylated by CK2. Each peptide was screened in triplicate and included a number of positive controls to validate the veracity of the array.

The next piece of the puzzle was to determine whether the candidate target peptides of CK2 and caspases phosphorylated by CK2 in the peptide arrays were also cleaved by caspases. To collect these data, an *in vitro* fluorescence peptide cleavage assay, called the caspase substrate identification (CSI) assay, was developed (Figure 6.4). For the CSI assay, N-terminal fluorescein (FL) label and C-terminal biotin groups were incorporated into the synthesized CK2 and caspase target peptides. The labeled peptides were incubated in 96-well plates in the presence of either caspase-3 or active caspase-8. For negative control experiments, the peptides were incubated with inactive caspase mutants or left untreated. After the cleavage assay, peptides were transferred to streptavidin-coated 96-well plates, incubated and washed to remove any unbound peptide. The fluorescence resulting from the FL label was then measured. If the peptide was cleaved, two fragments were produced: an N-terminal FL-labeled and a C-terminal biotinylated peptide. If the peptide was cleaved, the FL label would be washed away and the peptide rendered undetectable. Only uncleaved peptides containing the FL and biotin labels were detected by the CSI assay. The caspase-dependent cleavage of these peptides was quantitated by measuring the fluorescein counts.

Procaspase-3 was identified in both the peptide match screen and phosphopeptide array as a candidate substrate of CK2 and caspases. As well designed discovery-driven experiments often do, this one now led to hypothesis-driven experiments. To specifically determine the effect of phosphorylation on the cleavage of procaspase-3 by caspases, a series of FL- and biotin-labeled procaspase-3 peptides phosphorylated at the P2 (IEpTD) or the P1' (IETDpS) residues (corresponding to residues 174 and 176 in the full-length protein) were incubated in the presence of caspase-8. The phosphorylated FL-biotin–caspase-3 peptides were significantly protected from caspase-mediated cleavage compared to nonphosphorylated versions of the peptides. The results suggested that phosphorylation of either P2 or P1' sites inhibited caspase-mediated cleavage, showing the phosphorylation on either side of the aspartic acid residue was sufficient to prevent caspase-directed cleavage. Full-length procaspase-3 was confirmed as a target of CK2 in an *in vitro* kinase assay showing that the protein was indeed phosphorylated at Thr^{174} and Ser^{176}. The protection of the full-length phosphorylated protein from caspase-8 (and caspase-9) cleavage was also confirmed using an *in vitro* assay. In concordance with other results, these support a role for phosphorylation in the inhibition of caspase-mediated cleavage of proteins. Phosphorylation of procaspase-3 at Thr^{174} and Ser^{176} would likely inhibit activation of the key effector caspase, revealing a novel mechanism in caspase signaling that plays a major role in carrying out apoptosis.

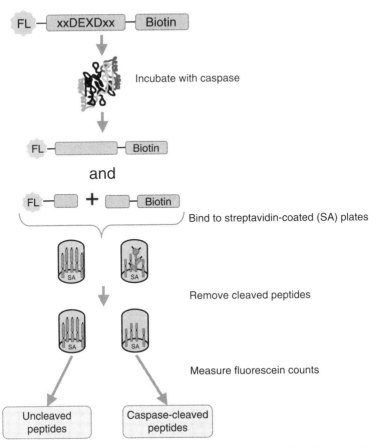

Figure 6.4. Schematic of CSI method to detect caspase cleavage. Candidate peptide targets of CK2 and caspases were synthesized to contain a C-terminal biotin and N-terminal FL. The peptides were coupled to streptavidin-coated wells within a 96-well plate and incubated with active or inactive caspases. After incubation and washing (to remove cleaved FL fragments), the fluorescein counts were measured within each well. Only the peptides cleaved by active caspase have a reduced fluorescent signal.

6.3.2 Functional Relevance of Overlapping CK2 and Caspase 3 Motifs

Obviously finding interesting biology *in vitro* is exciting, but proving its functional relevance requires studying the system's behavior *in vivo*. Phosphorylation of procaspase-3 at Thr[174] and Ser[176] resulted in protection from caspase-8- and caspase-9-mediated cleavage *in vitro*. Accordingly, the regulation of caspase-3 activation by phosphorylation might provide a mechanism by which CK2 could protect cells from apoptosis, as well as facilitate tumorigenesis. The first thing the investigators did was to determine whether CK2 and procaspase physically interact. An immunoprecipitation targeting hemagglutinin-tagged α- and β-subunits of CK2 (CK2α–HA and HA-CK2β) revealed a 32-kDa band corresponding to procaspase-3 as determined using Western blotting. In the reverse experiment, an immunoprecipitation experiment targeting myc-tagged procaspase-3 was able to pull-down 50-kDa and 30-kDa bands corresponding to CK2α-HA and HA-CK2β, respectively. Beyond showing a physical interaction, the investigators were also able to demonstrate that the phosphorylation of procaspase-3 was dependent on CK2 activity *in vivo* using the CK2 inhibitor TBB (4,5,6,7-tetrabromobenzotriazole).

So, ultimately where does this physical and functional interaction between CK2 and procaspase-3 lead. HeLa cells expressing myc-tagged procaspase-3 were treated with TBB or dimethyl sulfoxide (DMSO) for 12 h and the extent of cleavage of the caspase-3 target, PARP1, was measured. TBB-treated cells showed an increase in the abundance of cleaved PARP1 compared to untreated (DMSO) cells, showing that inhibiting CK2 caused an increase in caspase-3 activity not only against PARP1, but in general. In addition, when CK2 activity was knocked-down using RNA interference (RNAi), HeLa cells underwent spontaneous apoptosis and a band corresponding to the cleavage of caspase-3 could be observed by Western blotting. These observations suggest a direct interaction between CK2 and caspase-dependent apoptosis.

The most specific finding within this global phosphorylation/caspase study was the intimate relationship between CK2 and caspase 3. Considering the dynamic relationship between kinases and caspases and cancer, these findings may lead to additional drug targets that can promote apoptosis in tumor cells. While this study highlighted the apoptotic connection, it is important to remember that caspases are involved in a large number of nonapoptotic functions such as cell differentiation. Beyond the CK2/caspase 3 discovery, this study has identified a large number of peptides that are both phosphorylated by CK2 and cleaved by various caspases. This study followed a very logical path in that it took a single important result from a global screen and developed experiments to show its functional relevance *in vivo*. The remaining data identified in the phosphopeptide array or CSI assays can still be pursued in similar studies either by this laboratory or another. Just the database analysis revealed that 54 kinases contained overlapping kinase and caspase consensus sites, and approximately 75 kinases are known targets of caspases. All of the data, both computational and empirical, suggests that the cooperation between kinase and caspase pathways plays a major role in cell survival in both healthy and tumorigenic cells.

6.4 DASATINIB AND PROTEIN SIGNALING

The history of deranged signaling pathways as a hallmark of cancer has been well established. Not surprisingly kinases, which are the key drivers of signaling pathways, are the most important target for drug development within the pharmaceutical industry. Roughly one third of all drugs being developed within the pharmaceutical industry target protein or lipid kinases. With over 500 known kinases within the human proteome and the unique characteristics of each cancer, designing drugs with the necessary specificity is a major issue. While the specificity toward its intended target is critical, it is also important to understand a drug's global affect on cell signaling to better understand its overall mechanism and recognize potentially adverse side effects.

One of the greatest success stories in cancer treatment in the past decade involves chronic myelogenous leukemia (CML) (31). This cancer is characterized by overproliferation of myeloid cells resulting from the fusion of chromosomes 9 and 22 (i.e., the Philadelphia chromosome), which produces the fusion protein breakpoint cluster region-Abelson (BCR–ABL) (32). BCR–ABL is a tyrosine kinase that is constitutively active. In May of 2001, Gleevec (also known as imatinib) was approved by the U.S. Food and Drug Administration for treatment of CML and this drug is now presently approved for the treatment of 10 different cancers. Gleevec functions by inhibiting the kinase activity of BCR–ABL resulting in the selective death of the leukemia cells. In June of 2006 a next generation tyrosine kinase inhibitor, dasatinib (33), was approved for treatment of Philadelphia chromosome-positive acute lymphoblastic leukemia (ALL) (34). Dasatinib, like Gleevec, inhibits BCR–ABL activity; however, it can also inhibit mutant forms of this fusion protein that other inhibitors are ineffective against. In addition, dasatinib is used to treat ALL in cases where treatment with other tyrosine kinase inhibitors may cause adverse hematologic and cytogenetic responses (35).

6.4.1 Global Phosphorylation Analysis of Dasatinib-Treated Leukemia Cells

As mentioned above, it is critical to fully understand how a drug affects a cell, especially in today's era of systems biology and personalized medicine. To assess the global effect of dasatinib on the phosphoproteome, Dr. Matthias Mann's group used triple labeling SILAC (stable isotope labeling by amino acid in cell culture) to compare K562 (human immortalized leukemia cells) treated with vehicle (DMSO) and 5 and 50 nM dasatinib (36). Proteins extracted from each cell culture were digested into peptides and phosphopeptides were crudely separated from nonphosphorylated peptides using strong cation exchange (SCX). Phosphopeptides were then enriched from the SCX fractions using titanium-dioxide (TiO$_2$) chromatography (Figure 6.5). The peptide mixtures were then separated using RPLC prior to being analyzed using an LTQ-Orbitrap mass spectrometer.

The total analysis resulted in the identification and quantitation of almost 5700 phosphopeptides that originated from almost 1900 proteins. In total, cells treated with 5 nm dasatinib showed a twofold decrease in phosphorylation for about 12% (595) of the phosphopeptides, while increasing the treatment to 50 nM resulted in a downregulation of about 17% (842) of the phosphopeptides. Overall 465 phosphopeptides were found to be downregulated in both 5 and 50 nM treatments, while 130 and 377 were identified as being downregulated when cell were treated with 5 and 50 nM dasatinib, respectively. Less than 1% (76) of the phosphopeptides were downregulated when the cells were treated with dasatinib. This disparity between up- and downregulation of phosphorylation makes perfect sense considering that this drug acts as a tyrosine kinase inhibitor. The changes in the degree of phosphorylation were not related to changes in protein abundance as less than 1% of the proteins quantified in the same cells showed a greater than twofold change in abundance.

So, what did this global study specifically discover? The activating autophosphorylation sites of ABL1, SRC, MAPK1, and MAPK3, and hence their kinase activities,

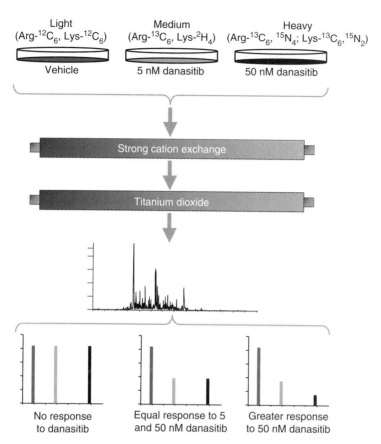

Figure 6.5. Quantitative MS strategy to identify the effects of kinase inhibitors on cell signaling. Three cultures of human immortalized leukemia cells (K562) were stable isotope labeled with light, medium, or heavy forms of the SILAC reagents. After treatment with different levels of danasitib, the extracted proteins were digested and separated using strong cation exchange and titanium dioxide chromatography to extract phosphopeptides. Analysis of the samples by liquid chromatography–tandem MS results in different peptide profiles in MS mode depending on the affect of the treatments on specific protein targets.

were significantly suppressed upon dasatinib treatment. The tyrosine phosphorylation status of BCR and LYN was drastically downregulated as was the Ser^{62} site of c-myc. Phosphorylation of this residue within c-myc is regulated by mitogen stimulation and enables c-myc to regulate functions such as transcription (37).

Several important BCR–ABL signaling pathways that play a role in chronic myeloid transformation have already been identified (38). It is know that BCR–ABL, along with SHC and Grb2, activates both ERK1/2 and JAK-STAT pathways. These activated pathways result in increased cell growth and proliferation, independent of growth factors. The activation of the PI3-AKT-kinase and JAK-STAT pathways by BCR–ABL enhances cell survival; however, activation of proteins involved in focal adhesion components causes a loss in cell adhesion leading to abnormal interactions between tumor and stroma cells.

Some of the BCR–ABL-related phosphorylation sites that were found to be down-regulated by dasatinib treatment are shown in Figure 6.6. As expected, four Ser residues and a single tyrosine residue within BCR and ABL, respectively, were down-regulated along with specific signaling events within major MAPK cascades such as p38α and ERK1/2. Phosphorylated residues within paxillin were also downreg-ulated. This observed affect on paxillin is consistent with the downstream effects observed on the cell's cytoskeleton when treated with dasanitib (39). Another major class of proteins whose phosphorylation states were downregulated by dasatinib are those associated with apoptosis. Studies have shown that suppression of apopto-sis is a major mechanism of BCR–ABL transformation (40, 41). Several phospho-rylation sites on proteins within the BCL family were found to be downregulated when cells were treated with dasatinib. This connection between downregulation of BCR–ABL phosphorylation and stimulation of apoptosis is similar to that shown in the previous study in which inhibition of the kinase CK2 also promoted apop-tosis through various caspases (23). Probably, the most novel effects of BCR–ABL inhibition discovered in this study were the effects on chromatin remodeling and protein splicing.

The ability to quantitatively measure global effects on the phosphoproteome when cell are treated with a drug, enables off-target and secondary effects of the compound to be determined. These unintended effects can be very detrimental to the patient as illustrated by the hypophosphataemia that is exhibited in 30% of CML patients treated with the BCR–ABL inhibitor, imatinib (42). Owing to their sheer number and the homol-ogy of the ATP-binding pockets within kinases, it is almost probable that any kinase inhibitor will bind to unintended targets. Also considering the complicated circuitry of protein pathways and networks, it is anticipated that a kinase inhibitor will affect pathways outside of the major mechanism required to kill the tumor cells. Even though present technology is still incapable of measuring every molecule within the cell, this study represents an effective method for finding off target and secondary effects of kinase inhibitors. Considering the profound effects that kinases have within cells, it is critical to find these unintended consequences during clinical trial and preferably at the molecular level; not when patients are exhibiting adverse reactions.

6.5   DRIVER AND PASSENGER KINASE MUTATIONS

Many cancers are a result of genetic mutations that cause uncontrolled cell growth and proliferation. These mutations may result from replication errors or from exposure to mutagens. Protein kinases represent the most common family of genes that contribute to cancer, emphasizing the importance of studying this family at both the gene and protein level. These mutations can be classified as either "drivers" or "passengers" depending on whether they confer a selective growth advantage to the cell and directly contribute to tumorigenesis and progression (driver) or whether they have no real functional con-sequence to the cell (passenger). Akin to stopping an out-of-control vehicle, locating and eliminating the driver mutations in tumor cells is more critical than removing the passengers.

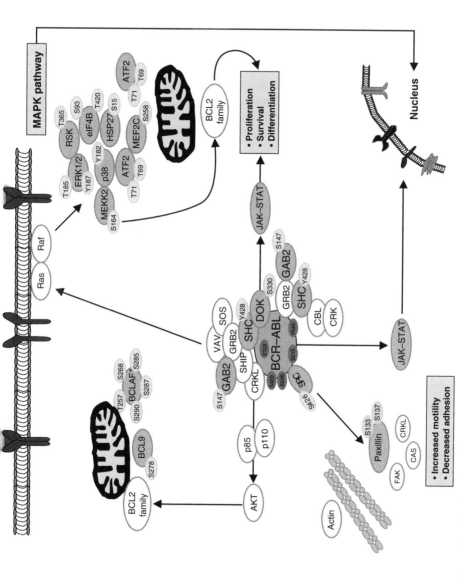

Figure 6.6. Illustration showing the effect of dasatinib on the BCR–ABL signaling pathway in human leukemia cells. All of the phosphorylation sites within specific proteins displayed in this figure were suppressed by dasatinib treatment (both at 5 and 50 nM levels). This wide suppression of phosphorylation affects such cell functions as cell proliferation, survival, differentiation, motility, and adhesion.

Distinguishing driver and passenger mutations has been an important goal of genomics. A common method for doing so involves sequencing genomes from a large cohort of cancer patients and locating genes that are mutated at a high frequency (43). These mutations can then be tested using a specific gene test for population testing. While this strategy has successfully identified a large number of cancer genes, the technology and analysis methods have not identified all the driver mutations in the vast expanse of individuals cancers. The two challenges in distinguishing driver and passenger mutations is the background mutation rate (which quantitates the accumulation of passenger mutations) must be known within a reasonable estimate and the extensive mutational heterogeneity of individual cancers results in mutations from different patients being found in different genes. Since they regularly target genes that encode for kinases, many combinations of driver mutations can aberrantly affect a protein pathway that is important in tumorigenesis and/or cancer progression.

6.5.1 Global Phosphorylation Analysis for the Identification of Driver Kinases

Instead of relying solely on genomics data, a recent study from Eric Haura's laboratory employed a global phosphoproteomics approach to identify driver tyrosine kinase mutations in sarcomas (44). Sarcomas are a rare and heterogeneous group of mesenchymal tumors that too often lead to death (45). While genomic studies have led to new treatment options for sarcomas such as gastrointestinal stromal tumors that harbor activating mutations in the *c-KIT* gene, those without this mutation do not respond as well to the same treatment and patients suffering with metastatic soft tissue sarcomas and osteosarcoma, as well as advanced sarcomas, are still lacking effective treatment options. A number of tyrosine kinase genes have been implicated as drivers in the oncogenesis of sarcoma, including the platelet-derived growth factor and its tyrosine kinase receptors (PDGFR), the insulin-like growth factor receptor (IGF1R), and the aforementioned c-kit (46). While there are a number of potential study design and patient selection issues that may contribute to continued lack of efficacy of tyrosine kinase inhibitors for treating sarcomas, the presence of additional unknown kinases that drive sarcoma progression cannot be ruled out.

Dr. Haura's laboratory began with 10 sarcoma cells lines: MNNG, U2OS, MG63, and Saos2 (osteosarcoma), RD18 and A204 (rhabdomyosarcoma), leiomyosarcoma (SK-LMS1), RD-ES and A673 (Ewing's sarcoma), SK-LMS1 (leiomyosarcoma), and HT1080 (fibrosarcoma) (Figure 6.7) (44). Their hypothesis was that a phosphoproteomics strategy could identify tyrosine kinases and substrate proteins important in the malignant process in sarcoma cells and tumors and identify driver tyrosine kinases responsible for sarcoma cell growth. Proteins were extracted from each cell line and digested into peptides. Tyrosine phosphorylated peptides were immunoprecipitated from these mixtures and analyzed using RPLC coupled directly on-line with an LTQ-Orbitrap mass spectrometer. Each sample was run in duplicate, and the number of spectra observed for each peptide in the LC–MS^2 runs were summed. As opposed to two previous studies described in this chapter (9, 36), this study did not use any type of stable isotope labeling.

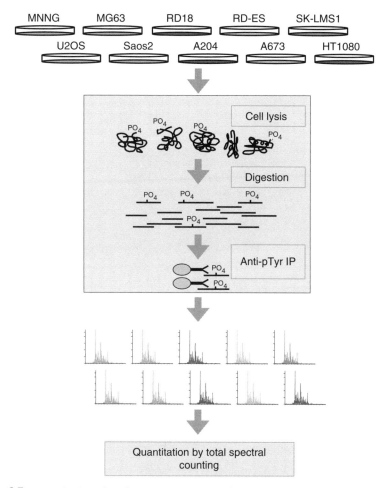

Figure 6.7. Quantitative phosphotyrosine analysis of sarcoma cell lines. The study by Bai et al. extracted phosphotyrosine peptides from 10 individual sarcoma cell lines using immunoprecipitation with a phosphotyrosine-specific antibody (44). Each mixture was analyzed separately using liquid chromatography–tandem MS and the relative abundance of the phosphotyrosine peptides measured using spectral counting.

The relative amount of each tyrosine-phosphorylated peptide within each cell type was based on their total spectral counts.

A total of 1936 unique tyrosine phosphopeptides originating from 844 proteins were identified in these studies. Thirty nine out of a total of 99 known tyrosine kinases were identified (Table 6.1). This coverage is quite an accomplishment considering that phosphorylated tyrosine residues represent only ~0.5% of the total phosphoamino acids within a cell. Saos2 osteosarcoma cells had the highest total spectral count of tyrosine phosphopeptides (264), while RD-ES Ewing's sarcoma cells had the fewest (72).

TABLE 6.1. Tyrosine Kinases Identified by Phosphoproteomics in Ten Sarcoma Cell Lines (44). The Numbers Represent the Number of Times a Tyrosine Phosphopeptide was Identified from Each Protein

	MNNG	U2OS	MG63	Saos2	RD18	A204	RD-ES	A673	SK-LMS1	HT1080
ABL	8	16	2	4	11	6	3	7	5	11
ACK	4	7	15	24	11	15	3	3	8	15
ALK					1			1		
CSK										3
DDR1					12	1		8		
DDR2	1	2	6	2	3	13	2	8		2
EGFR	6	3	9	10	11	3			15	12
EphA2	10	12	13	23	18	34	5	23	11	22
EphA3		4	4	4		1	4	15	3	4
EphA4							8	7		4
EphA5			13	18				6		
EphA7			2		17	4		2		
EphB1			4	7			2	2	3	2
EphB2			16					3		6
EphB3			9	10	4	3		6		
EphB4	5		1	12	5	11	4	15		2
FAK	25	28	28	32	11	19	2	39	7	14
FER		3				2			2	4
FGFR1	4	11	24	8	8	28	3	6	13	
FGFR2		2		3						
FGFR3								1		
FGFR4		4	6		3					
FGR			1			1				
SRC:FYN/HCK	14	17	15	10	17	24	17	21	10	17
IGFR1		8	5	10	9	2	2	5	1	6
INSR							1			
JAK1	2	1	1	5	4	3		7	4	
JAK2		3	1	4	3	11		5	4	4
JAK3			1		1		1			2
KDR	1									
KIT				3			1			
MER	2		4	4	2	6	4	11	3	3
MET	23	7	8	20	14				22	15
MUSK					4					
PDGFRα	10	1	23	10		33			13	
RET								14	1	
TEC	2			1						
TYK2	2	3	2	5	4	4	4	3	3	3
UFO/AXL	5	13	24	19	4	6	3		13	15

TABLE 6.2. Kinase Inhibitor Sensitivity of 10 Sarcoma Cell Lines (44)

	MNNG	U20S	MG63	Saos2	RD18	A204	RD-ES	A673	SK-LMS1	HT1080
Erlotinib	>10	>10	>10	>10	>10	>10	>10	>10	>10	9.63
Dasatinib	>10	>10	0.80	0.39	6.07	0.03	>10	6.75	>10	1.62
Imatinib	>10	>10	>10	>10	>10	0.13	>10	>10	>10	>10
PHA	0.33	8.03	2.50	3.03	2.87	4.20	6.31	3.33	>10	3.10
OSI868	>10	>10	4.83	2.24	>10	9.80	0.10	4.26	>10	4.12
SU5402	>40	>40	34.08	>40	>40	15.24	>40	>40	>40	>40
JAK	3.47	>10	>10	>10	>10	9.23	8.36	>10	>10	0.91
ZD6474	6.37	>10	8.28	>10	9.33	5.77	5.78	5.37	>10	9.04

Phosphopeptides originating from focal adhesion kinase (FAK) were the most commonly identified (316 total spectral counts) and were found in all ten cells lines (316 total spectral counts). C-Src kinase (CSK) produced the fewest phosphotyrosine identifications (one spectral count in HT1080 fibrosarcoma cells). Obviously, not every tyrosine site with a protein has the same function. Some of the identified phosphorylated sites correspond to known autophosphorylation sites that cause kinase activation, while others are related to protein–protein interactions, enzymatic activity, or their function that had not yet been defined. Many receptor tyrosine kinases were identified including EGFR, Ephrin receptors, FGFR receptors, IGF1R, KIT, MET, and PDGFRα (Table 6.1). Several nonreceptor tyrosine kinases were identified including FAK, ACK, FYN, SRC family kinases, JAK members, and ABL. Although not listed in Table 6.1, several tyrosine phosphorylation-dependent signaling proteins (e.g., p130^{CAS}, paxillin, Stat3, and ERK1/2) were also identified. Importantly, many of the results obtained using LC–MS were validated using Western blotting.

The goal of this study was to find driver kinases that regulate tumor cell growth, proliferation, and survival in sarcomas. To identify the driver kinases amongst the 39 tyrosine kinases quantitated in this study, the growth of the sarcoma cell lines was measured after treatment with inhibitors of EGFR (erlotinib), JAK (P6), FGFR (SU5402), Met (PHA665752 or PAH), IGF1R (OSI868), RET (ZD6474), PDGFR (imatinib and dasatinib), and SRC (Table 6.2). MNNG cells were sensitive to Met inhibition (PHA) with IC_{50} = 330 nmol/L while Saos2 cells were sensitive to dasatinib (IC_{50} = 390 nmol/L). A204 cells were sensitive to the PDGFR inhibitors dasatinib and imatinib (IC_{50} = 34 and 135 nmol/L, respectively), while RD-ES cells were sensitive to the IGF1R inhibitor OSI868 (IC_{50} = 96 nmol/L). These data suggested that many of the tyrosine kinases identified in this global phosphorylation analysis may indeed be drivers of sarcoma tumor growth and progression.

6.5.2   Validation of Global Phosphorylation Data

6.5.2.1   PDGFRα Signaling   Based on the quantitative LC–MS data showing that the rhabdomyosarcoma cell line A204 expressed high levels of phosphotyrosine peptides originating from PDGFRα, and that this cell line is sensitive to PDGFRα inhibitors, it made sense that PDGFRα may be a driver kinase and an important target in some rhabdomyosarcoma cells and tumors. To corroborate the PDGFRα results, the effects of dasatinib and imatinib inhibition on downstream signaling events were measured. The AKT pathway was inhibited, while both apoptosis and cell-cycle arrest were induced in dasatinib and imatinib treated A204 cells. siRNA knockdown of PDGFRα caused a significant reduction in cell viability. A204 cells that were engineered to express a PDGFRα mutant that did not bind either inhibitor, were unaffected by dasatinib and imatinib treatment. No evidence of PDGFRα gene amplification or gene mutation was found in the cell line.

As every young scientist is taught; while results in cell lines are encouraging, they do not necessarily represent actual patient samples. Therefore, PDGFRα levels were measured using IHC in 23 rhabdomyosarcoma tumors obtained from patients. Three of the patient tumors were considered positive for PDGFRα with the remaining 19 tumors showing either very weak or negative PDGFRα staining. One of the positive cases was from a patient showing multiple sites of metastasis and of these the brain metastasis was positive while lung and subcutaneous metastasis sites were negative. The other two positive cases were from primary tumors. While collectively, the data show that PDGFRα may be a driver kinase and a potential target for rhabdomyosarcomas, unfortunately, the validation data showed that this receptor is overexpressed only in a small group of tumors.

6.5.2.2 c-Met Signaling

6.5.2.2 c-Met Signaling Phosphotyrosine peptides originating from c-Met were observed with especially high spectral counts in the osteosarcoma cell lines MNNG and Saos2 and the leiomyosarcoma cell line, SK-LMS1. MNNG cells, opposed to all other cell lines, were found to be sensitive to the c-Met inhibitor PHA. Phosphorylated AKT and pERK were also inhibited by PHA treatment in MNNG cells while both apoptosis and cell-cycle arrest were induced. Consistent with these results, siRNA knockdown of c-Met siRNA had a strong negative effect on MNNG cell viability. These results together suggest c-Met is a driver kinase responsible for cell growth and survival in MNNG cells and these cells have been shown to contain a *TPR:MET* gene fusion that leads to constitutive c-Met signaling (47).

6.5.2.3 IGF1R/INSR Signaling

6.5.2.3 IGF1R/INSR Signaling Phosphorylated peptides originating from IGF1R were identified in nine of the 10 cell lines, with the highest levels found in Saos2, RD18, and U20S cell lines. Interestingly though, only RD-ES cells, which only had two spectral counts for IGF1R phosphorylated tyrosine peptides, was sensitive to OSI868 ($IC_{50} = 96$ nmol/L). As with many of the inhibitors targeting PDGFRα and c-Met, the IGF1R inhibitor OSI868 inhibited pAKT, induced cell apoptosis, and cell-cycle arrest, but opposed to the others where ERK status was unaffected, OSI868 treatment also increased phosphorylated ERK levels. RD-ES cells were transfected with siRNA against IGF1R, INSR, and a combination of the two. Tyrosine phosphorylated peptides were observed for INSR in RD-ES cells exclusively in the panel of 10 lines that were tested (Table 6.1). While inhibition of both IGF1R and INSR resulted in decreased RD-ES cell viability, INSR inhibition by itself had no effect. The investigators could not find a siRNA that only affected IGF1R protein levels without affecting those of INSR. Building on these siRNA results, it was found that RD-ES cells were the most sensitive of the 10 sarcoma cell lines to treatment with the dual IGF1R/INSR inhibitor BMS-754807. Interestingly, the effects of OSI868 treatment on RD-ES cells are consistent with other studies in which targeting both IGFR1 and INSR with a dual purpose inhibitor provided superior antitumor efficacy compared with targeting IGF1R alone (48,49).

6.5.3 Global Phosphotyrosine Analysis of Xenograft Models

While the above-mentioned studies were all conducted using cell lines, this group also examined the tyrosine phosphorylation patterns of proteins extracted from xenografts of human liposarcoma, small round blue tumor, and malignant peripheral nerve sheath tumors implanted in mice. Using LC–MS^2, they were able to identify 220 unique phosphotyrosine peptides originating from 170 unique proteins. Amongst tyrosine kinases, the proteins with the highest spectral counts for phosphotyrosine peptides included EphB4, SRC:FYN, FAK, and PDGFRα. While there was insufficient data to determine if these proteins acted as driver kinases in the xenografts, it did demonstrate the ability to partially graduate the technology to tumor models that more closely imitate human tumors *in vivo*.

This study exemplifies much of the promise and frustration of global phosphoproteomic profiling. The goal of identifying tyrosine phosphoproteins present within a panel of sarcoma cell lines was achieved and potential driver kinases were also identified (e.g., c-Met in MNNG cells). The data also illustrate the fact that not every phosphoprotein identified in a discovery data set has functional relevance to the cell model. For example, high levels of both PDGFRα and phosphotyrosine PDGFRα were identified in MG63 cells yet inhibiting PDGFRα had little effect on these cells. The same was observed for phosphorylated EGFR in all the cell lines tested. Thus, in the context of analyzing global phosphoproteomics data, it is important to have integrated strategies with other approaches to identify driver kinases as opposed to bystander (irrelevant) kinases. It is currently difficult to infer the driver kinases based solely on a phosphotyrosine MS data set. While identifying driver kinases is impossible based solely on the LC–MS^2 data sets, it does provide a number of candidates that can be tested using orthogonal techniques as illustrated in this study. Ultimately, that is the goal of discovery science; producing data that can be used to generate specific testable hypotheses.

6.6 CONCLUSIONS

Chapter 4 in this book provides a description of the technology and sample preparation steps needed to characterize phosphopeptides and proteins in simple and complex samples. As I mentioned early in this chapter, I am a big advocate that to fully understand something, it is important to see how it is done. Even more important than seeing how it is done, is doing it. As a gentleman I used to coach hockey with often said "A player needs to score goals to learn to become a goal scorer." This chapter does not provide the hands-on tools needed for a scientist to become an expert in global phosphoproteomic analysis; however, it illustrates examples of what the analysis and data look like. More importantly, the studies described in this chapter illustrate the purpose of conducting discovery-driven studies; using data to generate hypothesis-driven studies.

The studies presented above show that there are many different ways to conduct global phosphorylation studies. Three of the examples in this chapter used quantitative MS while the fourth utilized peptide arrays. Even the studies that used quantitative MS employed different methods for measuring phosphopeptide abundances. One study used metabolic isotope labeling (SILAC), one used chemical isotope labeling (iTRAQ), and the third used label free quantitation (total spectral counting). Each had one thing in common; however, the effort to acquire the discovery data was a small fraction of the entire amount of data within the study. This ratio is a consistent feature of excellent research that captures global proteomic data and is an indication of how discovery studies are gaining prominence in life science research. It was not long ago that studies detailing the global analysis of a cell's phosphoproteome were well received by high impact journals. To generate a truly impactful study today requires validating the global data using targeted approaches to show functional significance of at least one of the findings. Each example presented in this chapter validates at least two important findings within the global data and still contains a number of important observations that can be followed up in additional validation studies.

REFERENCES

1. Beausoleil SA, Jedrychowski M, Schwartz D, Elias JE, Villén J, Li J, Cohn MA, Cantley LC, Gygi, SP. Large-scale characterization of hela cell nuclear phosphoproteins. Proc. Natl. Acad. Sci. U.S.A. 2004;101:12130–12135.

2. Reardon DA, Perry JR, Brandes AA, Jalali R, Wick W. Advances in malignant glioma drug discovery. Expert Opin. Drug Discov. 2011;6:739–753.

3. Farias-Eisner G, Bank AM, Hwang BY, Appelboom G, Piazza MA, Bruce SS, Sander Connolly E. Glioblastoma biomarkers from bench to bedside: Advances and challenges.Br. J. Neurosurg. 2012;26:189–194.

4. Inda MM, Bonavia R, Mukasa A, Narita Y, Sah DW, Vandenberg S, Brennan C, Johns TG, Bachoo R, Hadwiger P, Tan P, Depinho RA, Cavenee W, Furnari F. Tumor heterogeneity is an active process maintained by a mutant egfr-induced cytokine circuit in glioblastoma. Genes Dev. 2010;24:1731–1745.

5. Halatsch ME, Schmidt U, Behnke-Mursch J, Unterberg A, Wirtz CR. Epidermal growth factor receptor inhibition for the treatment of glioblastoma multiforme and other malignant brain tumours. Cancer Treat. Rev. 2006;32:74–89.

6. Cho J, Pastorino S, Zeng Q, Xu X, Johnson W, Vandenberg S, Verhaak R, Cherniack AD, Watanabe H, Dutt A, Kwon J, Chao YS, Onofrio RC, Chiang D, Yuza Y, Kesari S, Meyerson M. Glioblastoma-derived epidermal growth factor receptor carboxyl-terminal deletion mutants are transforming and are sensitive to egfr-directed therapies. Cancer Res. 2011;71:7587–7596.

7. Nishikawa R, Sugiyama T, Narita Y, Furnari F, Cavenee WK, Matsutani M. Immunohistochemical analysis of the mutant epidermal growth factor, deltaEGFR, in glioblastoma. Brain Tumor Pathol. 2004;21:53–56.

8. Shinojima N, Tada K, Shiraishi S, Kamiryo T, Kochi M, Nakamura H, Makino K, Saya H, Hirano H, Kuratsu J, Oka K, Ishimaru Y, Ushio Y. Prognostic value of epidermal growth factor receptor in patients with glioblastoma multiforme. Cancer Res. 2003;63:6962–6970.

9. Huang PH, Mukasa A, Bonavia R, Flynn RA, Brewer ZE, Cavenee WK, Furnari FB, White, FM. Quantitative analysis of EGFRvIII cellular signaling networks reveals a combinatorial therapeutic strategy for glioblastoma. Proc. Natl. Acad. Sci. U.S.A. 2007;104:12867–12872.

10. Han W, Lo HW. Landscape of EGFR signaling network in human cancers: Biology and therapeutic response in relation to receptor subcellular locations. Cancer Lett. 2012;318:124–134.

11. Wolf-Yadlin A, Hautaniemi S, Lauffenburger DA, White FM. Multiple reaction monitoring for robust quantitative proteomic analysis of cellular signaling networks. Proc. Natl. Acad. Sci. U.S.A. 2007;104:5860–5865.

12. Höland K, Salm F, Arcaro A. The phosphoinositide 3-Kinase signaling pathway as a therapeutic target in grade IV brain tumors. Curr. Cancer Drug Targets. 2011;11:894–918.

13. Lal B, Goodwin CR, Sang Y, Foss CA, Cornet K, Muzamil S, Pomper MG, Kim J, Laterra J. EGFRvIII and c-met pathway inhibitors synergize against pten-null/EGFRvIII+ glioblastoma xenografts. Mol. Cancer Ther. 2009;8:1751–1760.

14. Sattler M, Pride YB, Ma P, Gramlich JL, Chu SC, Quinnan LA, Shirazian S, Liang C, Podar K, Christensen JG, Salgia, R. A novel small molecule met inhibitor induces apoptosis in cells transformed by the oncogenic TPR-MET tyrosine kinase. Cancer Res. 2003;63:5462–5469.

15. Nagane M, Narita Y, Mishima K, Levitzki A, Burgess AW, Cavenee WK, Huang HJ. Human glioblastoma xenografts overexpressing a tumor-specific mutant epidermal growth factor receptor sensitized to cisplatin by the AG1478 tyrosine kinase inhibitor. J. Neurosurg. 2001;95:472–479.

16. Johns TG, Luwor RB, Murone C, Walker F, Weinstock J, Vitali AA, Perera RM, Jungbluth AA, Stockert E, Old LJ, Nice EC, Burgess AW, Scott AM. antitumor efficacy of cytotoxic drugs and the monoclonal antibody 806 is enhanced by the EGF receptor inhibitor AG1478. Proc. Natl. Acad. Sci. U.S.A. 2003;100:15871–15876.

17. Engelman JA, Zejnullahu K, Mitsudomi T, Song Y, Hyland C, Park JO, Lindeman N, Gale CM, Zhao, X, Christensen J, Kosaka T, Holmes AJ, Rogers AM, Cappuzzo F, Mok T, Lee C, Johnson BE, Cantley LC, Jänne PA. MET amplification leads to gefitinib resistance in lung cancer by activating ERBB3 signaling. Science 2007;316:1039–1043.

18. Sharma N, Adjei AA. In the clinic: Ongoing clinical trials evaluating c-MET-inhibiting drugs. Ther. Adv. Med. Oncol. 2011;3:S37–S50.

19. Cooke VG, LeBleu VS, Keskin D, Khan Z, O'Connell JT, Teng Y, Duncan MB, Xie L, Maeda G, Vong S, Sugimoto H, Rocha RM, Damascena A, Brentani RR, Kalluri, R. Pericyte depletion results in hypoxia-associated epithelial-to-mesenchymal transition and metastasis mediated by met signaling pathway. Cancer Cell 2012;21:66–81.

20. Raff MC, Barres BA, Burne JF, Coles HS, Ishizaki Y, Jacobson MD. Programmed cell death and the control of cell survival: Lessons from the nervous system. Science 1993;262:695–700.

21. Wu Z, Yu Q. E2F1-mediated apoptosis as a target of cancer therapy. Curr. Mol. Pharmacol. 2009;2:149–160.

22. Vermeulen K, Van Bockstaele DR, Berneman ZN. Apoptosis: Mechanisms and relevance in cancer. Ann. Hematol. 2005;84:627–639.

23. Duncan JS, Turowec JP, Duncan KE, Vilk G, Wu C, Lüscher B, Li SS, Gloor GB, Litchfield DW. A peptide-based target screen implicates the protein kinase ck2 in the global regulation of caspase signaling. Sci. Signal 2011;4:ra30.

24. Turowec JP, Duncan JS, Gloor GB, Litchfield DW. Regulation of caspase pathways by protein kinase CK2: Identification of proteins with overlapping CK2 and caspase consensus motifs. Mol. Cell. Biochem. 2011;356(1-2):159–167.

25. Deshiere A, Duchemin-Pelletier, E, Spreux E, Ciais D, Combes F, Vandenbrouck Y, Couté Y, Mikaelian I, Giusiano S, Charpin C, Cochet C, Filhol O. Unbalanced expression of CK2 kinase subunits is sufficient to drive epithelial-to-mesenchymal transition by Snail1 induction. Oncogene 2013;32:1373–1383.

26. Martins LR, Lúcio P, Silva MC, Gameiro P, Silva MG, Barata JT. On CK2 regulation of chronic lymphocytic leukemia cell viability. Mol. Cell. Biochem. 2011;356(1-2):51–55.

27. Hanif IM, Pervaiz S. Repressing the activity of protein kinase ck2 releases the brakes on mitochondria-mediated apoptosis in cancer cells. Curr. Drug Targets 2011;12:902–908.

28. Gnad F, Gunawardena J, Mann M. PHOSIDA 2011: The posttranslational modification database. Nucleic Acids Res. 2011;39:D253–D260.

29. Dinkel H, Chica C, Via A, Gould CM, Jensen LJ, Gibson TJ, Diella F. Phospho.ELM: A database of phosphorylation sites–update 2011. Nucleic Acids Res. 2011;39:D261–267.

30. Lüthi AU, Martin SJ. The CASBAH: A searchable database of caspase substrates. Cell Death Differ. 2007;14:641–650.

31. Druker BJ, Tamura S, Buchdunger E, Ohno S, Segal GM, Fanning S, Zimmermann J, Lydon NB. Effects of a selective inhibitor of the Abl tyrosine kinase on the growth of Bcr-Abl positive cells. Nat. Med. 1996;2:561–566.

32. Schaefer-Rego K, Arlin Z, Shapiro LG, Mears JG, Leibowitz D. Molecular heterogeneity of adult philadelphia chromosome-positive acute lymphoblastic leukemia. Cancer Res. 1988;48:866–869.

33. Shah NP, Tran C, Lee FY, Chen P, Norris D, Sawyers CL. Overriding imatinib resistance with a novel ABL kinase inhibitor. Science 2004;305:399–401.

34. Brave M, Goodman V, Kaminskas E, Farrell A, Timmer W, Pope S, Harapanhalli R, Saber H, Morse D, Bullock J, Men A, Noory C, Ramchandani R, Kenna L, Booth B, Gobburu J, Jiang X, Sridhara R, Justice R, Pazdur R. Sprycel for chronic myeloid leukemia and philadelphia chromosome-positive acute lymphoblastic leukemia resistant to or intolerant of imatinib mesylate. Clin. Cancer Res. 2008;14(2):352–359.

35. Steinberg M. Dasatinib: A tyrosine kinase inhibitor for the treatment of chronic myelogenous leukemia and philadelphia chromosome-positive acute lymphoblastic leukemia. Clin. Ther. 2007;29:2289–2308.

36. Pan C, Olsen JV, Daub H, Mann M. Global effects of kinase inhibitors on signaling networks revealed by quantitative phosphoproteomics. Mol. Cell. Proteomics 2009;8:2796–2808.

37. Liu J, Martin, HJ, Liao, G, Hayward SD. The kaposi's sarcoma-associated herpesvirus LANA protein stabilizes and activates c-myc. J. Virol. 2007;81:10451–10459.

38. Schenone S, Brullo C, Botta, M. New opportunities to treat the T315I-Bcr-Abl mutant in chronic myeloid leukaemia: Tyrosine kinase inhibitors and molecules that act by alternative mechanisms. Curr. Med. Chem. 2010;17:1220–1245.

39. Serrels A, Macpherson IR, Evans TR, Lee FY, Clark EA, Sansom OJ, Ashton GH, Frame MC, Brunton VG. identification of potential biomarkers for measuring inhibition of src kinase activity in colon cancer cells following treatment with dasatinib. Mol. Cancer Ther. 2006;5:3014–3022.

40. Hickman JA. Apoptosis and tumourigenesis. curr. opin. Genet. Dev. 2002;12:67–72.

41. McGahon AJ, Cotter TG, Green DR. The Abl oncogene family and apoptosis. Cell Death Differ. 1994;1:77–83.

42. O'Sullivan S, Lin JM, Watson M, Callon K, Tong PC, Naot, D, Horne A, Aati O, Porteous F, Gamble G, Cornish J, Browett P, Grey A. The skeletal effects of the tyrosine kinase inhibitor nilotinib. Bone 2011;49:281–289.

43. Vandin F, Upfal E, Raphael BJ. De novo discovery of mutated driver pathways in cancer. Genome Res. 2012;22:375–385.

44. Bai Y, Li J, Fang B, Edwards A, Zhang G, Bui M, Eschrich S, Altiok S, Koomen J, Haura EB. Phosphoproteomics identifies driver tyrosine kinases in sarcoma cell lines and tumors. Cancer Res. 2012;72:2501–2511.

45. Martín Liberal J, Lagares-Tena L, Sáinz-Jaspeado M, Mateo-Lozano S,García Del Muro X, Tirado OM. Targeted therapies in sarcomas: Challenging the challenge. Sarcoma 2012;2012:626094.

46. Wardelmann, E, Schildhaus HU, Merkelbach-Bruse S, Hartmann W, Reichardt P, Hohen-berger P, Büttner R. Soft tissue sarcoma: From molecular diagnosis to selection of treatment. Pathological diagnosis of soft tissue sarcoma amid molecular biology and targeted therapies. Ann. Oncol. 2010;21:vii265–269.

47. Soman NR., Wogan GN, Rhim JS. TPR-MET oncogenic rearrangement: Detection by poly-merase chain reaction amplification of the transcript and expression in human tumor cell lines. Proc. Natl. Acad. Sci. U.S.A. 1990;87:738–742.

48. Buck E, Gokhale PC, Koujak S, Brown E, Eyzaguirre A, Tao N, Rosenfeld-Franklin M, Lerner L, Chiu MI, Wild R, Epstein D, Pachter JA, Miglarese MR. Compensatory insulin receptor (IR) activation on inhibition of insulin-like growth factor-1 receptor (IGF-1R): Rationale for cotargeting IGF-1R and IR in cancer. Mol. Cancer Ther. 2010;9:2652–2664.

49. Zhao H, Desai V, Wang J, Epstein DM, Miglarese M, Buck E. Epithelial-mesenchymal transition predicts sensitivity to the dual IGF-1R/IR inhibitor OSI-906 in hepatocellular carcinoma cell lines. Mol. Cancer Ther. 2012;11:503–513.

7

THE SEARCH FOR BIOMARKERS IN BIOFLUIDS

7.1 INTRODUCTION

There is no grander prize being pursued in omics today than the search for biomarkers. Whether it is genomics, transcriptomics, proteomics, or metabolomics, the goal of finding biomarkers is of utmost importance. The number of lives that would be saved and the impact on public health if more effective biomarkers could be discovered for cancer, traumatic brain injury, viral and bacterial infection, heart disease, etc., would be staggering. If biomarkers for early detection could be discovered, the number of deaths due to cancer would drop dramatically. Biomarkers that could monitor therapeutic effectiveness would enable the correct treatment to be recognized sooner in cases where the best course of treatment is not obvious.

While biomarkers from any source would be helpful, finding protein biomarkers within easily accessible biofluids (e.g., blood, urine, saliva, etc.) would be invaluable. To illustrate this point, consider the case of an individual with pancreatic cancer. Assume that the tumor is in its early stages yet expresses and secretes a highly specific, but as yet undiscovered, biomarker. There are a number of advantages in analyzing the tumor tissue itself as illustrated in Figure 7.1. First, the concentration of the biomarker is highest at the site of the tumor making it easier to measure analytically. Second, adjacent healthy tissue can be used as a negative control. Third, once the biomarker is discovered,

Proteomic Applications in Cancer Detection and Discovery, First Edition. Timothy D. Veenstra.
© 2013 John Wiley & Sons, Inc. Published 2013 by John Wiley & Sons, Inc.

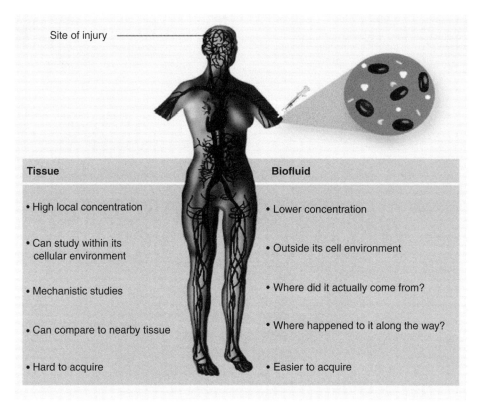

Site of injury

Tissue	Biofluid
• High local concentration	• Lower concentration
• Can study within its cellular environment	• Outside its cell environment
• Mechanistic studies	• Where did it actually come from?
• Can compare to nearby tissue	• Where happened to it along the way?
• Hard to acquire	• Easier to acquire

Figure 7.1. Diagram outlining the various advantages and disadvantages in conducting biomarker discovery projects using tissue versus biofluids.

studies to determine how it functions and where it is localized within the context of the tumor can be performed. The obvious disadvantage is that tissue is not an easily accessible sample.

Compare these advantages to the disadvantages in trying to find a biomarker for this tumor in a blood sample taken from the patient. First, the biomarker's concentration is much lower in blood compared to the tissue making it more difficult to detect. Second, the biomarker may have undergone significant transformation due to the protease activity found in circulation. Third, if a potential biomarker is found, it is virtually impossible to ensure that it originated from the tumor. The major advantage, however, is that it is much easier to obtain a blood sample than tissue.

With the advantages of attempting to discover biomarkers in tissue, why has most of the focus been on analyzing biofluids? The answer is that the advantage of accessibility trumps all others. Biofluids can be acquired during routine physical examinations; tissues are usually only biopsied in cases where there is some obvious indication of present disease. If biomarkers are to be used for early-stage diagnosis, they must be measured in easily acquired samples. Providing a yearly biofluid sample, such as urine or blood, which can be analyzed for a panel of disease-specific biomarkers, would provide the early diagnostic capability that could save countless lives. Yearly tissue biopsies are impractical for obvious reasons. What tissue(s) should be routinely biopsied if there is no indication of disease? Where in the tissue should the biopsy be taken? While its early stage diagnostic utility is limited, measuring a biomarker in tissue is still important for therapeutic monitoring as most patients would be willing to undergone periodic biopsies if it allowed the physician to determine a therapy is working as soon as possible.

Numerous diagnostic tests for diseases ranging from cancer to diabetes are presently conducted using biofluid samples (http://www.health.harvard.edu/diagnostic-tests/urinalysis.htm; http://www.accessmedicine.com/pocketDiagnostic.aspx). The general consensus, however, is that the archive of information within biofluids has only begun to be understood (1–4). Technologies developed over the past decade have increased our knowledge of the molecular content of biofluids immensely. These developments have been made in the areas instrumentation for data collection, sample preparation methods (5–7), and data analysis, respectively (8–10). In this chapter, strategies for biomarker discovery for diseases (with an emphasis on cancer) will be discussed. Many of the analytical strategies that will be discussed are biofluid independent; however, careful consideration needs to be given when designing the biomarker discovery study regardless of the starting material.

7.2   COLLECTION AND STORAGE OF BIOFLUIDS

One issue that is a constant concern in the proteomic analysis of biofluids is sample collection, preparation, and storage (11–14). These three steps have a major impact on the sensitivity, selectivity, and reproducibility of any analysis. While a number of groups have examined various stages of this process (primarily storage), the definitive study to measure these effects on the sample's proteome has not been done. Most studies concerning collection and storage of biofluids have focused on the effect of the number of freeze–thaw cycles. Getting a firm handle on the variability afforded by using different collection, preparation, and storage conditions is an immensely challenging task. A number of study design issues make the task a major challenge. The optimal study would start with a single patient from whom a single, uniform biofluid sample was drawn and processed using multiple different methods (e.g., plasma vs. serum preparation and ethylenediamine tetraacetic acid [EDTA] vs. heparin plasma preparation). Each of these samples would then be aliquoted into a large cohort, which is then evaluated under a number of different preparation conditions (e.g., immunodepletion and filtering) and storage conditions (e.g., temperature, freeze–thaw cycles, and time). It is pretty easy to envision how many possible combinations and permutations could be developed from such a study. On the analytical side, it needs to be determined the type of analytical instrument that will be used for data collection. A simple liquid chromatography (LC) separation with ultraviolet (UV) detection could be used for simplicity; however, this technique provides an average result at each time point and not a direct signal for each species in the sample. Mass spectrometry (MS) would be a good choice since it is so widely used in biomarker discovery; however, there are a number of different types of spectrometers, each having different performance characteristics. If MS is used, how will the results be compared to one another? Will the comparisons be done on direct peak alignment or will only identified species be counted? How will the intensities be normalized to allow quantitation? Probably, the biggest uncertainty is how to define optimized. Even with the most advanced technology available today, we cannot detect every protein in a biofluid (and may never know when we do anyway). Being unable to detect everything prevents optimum from being absolutely established. Without having the ability to establish optimum conditions of biofluid collection, preparation, and storage, the best scenario that can be achieved is consistency. Consistency needs to be applied to every sample within a defined study to minimize analytical variation.

Over the past few years, studies have made specific observations concerning the collection, preparation, and storage of biofluids (particularly blood) for proteomic analysis. Studies have shown that the primary effects on the blood proteome of using commercially available blood collection tubes are the addition or removal of components. Silicones, and polymers such as polyvinylpyrrolidines or polyethylene glycols, are frequently used to coat the internal surface of these tubes (15). A study using an aqueous saline solution showed that almost 65% of the tubes tested shed polymeric compounds in the m/z range of 1000–3000 into the sample (16). These compounds can potentially compromise the collection of data using MS and/or interfere with LC fractionation. Adsorption of blood proteins to the collection tube may also occur. Significant differences in proteomic results have been found when samples collected using "red-top" tubes (glass tubes containing no preservatives or anticoagulants) and "tiger-top" tubes (also known as serum separator tubes). (17, 18) Other factors in serum preparation such as the anticoagulant used, the clotting time allowed, and the length of the time period before centrifugation were also found to have a significant effect on the serum proteome (19–21). As mentioned earlier, in lieu of a universal standard operating procedure for blood proteome analysis, the key is to ensure consistency in the sample collection tubes used for a specific study.

Consistency is also a critical factor in the collection of urine samples as these are often highly variable in volume, protein concentration, and pH. Issues related to collection vials are not as large as with blood; however, time of collection is a major issue. A study that measured the protein concentrations of urines collected over the course of 10 days from a single subject showed that the high variability was not related to average protein concentration (ranging from 26 mg/mL for 24 h, random spot, and second morning urine to 34 mg/mL for first morning urine), but rather the standard deviations in these measurements (22). Standard deviations of protein concentrations were lowest for the 24-h and first morning collections (39% and 41%, respectively), while the second urine samples of the day and spot collection (54% and 61%, respectively) were highly variable.

Trying to determine sample issues related to storage is very challenging as designing a study to measure the effects of say 20–30 years of storage on a standard sample is outside the scope (and interest) of most laboratories. Probably a bigger roadblock to such a study is the fact that analytical technology dramatically changes over such time periods. To simulate the effect of storage, scientists often measure things such as freeze–thaw cycles or storage at different temperatures. The aim is to measure variability in the sample induced by short periods under "harsh" conditions as a mimic of long periods under "soft" conditions. Some studies have shown that repeated freeze–thaw cycles have a minor effect on the proteome of plasma or serum, while others have shown a more pronounced affect (23, 24). The proteomic profile of urine samples is minimally affected after one to four freeze–thaw cycles as measured using surface enhance laser desorption/ionization-time-of-flight (SELDI-TOF)/MS (25). After the fifth freeze–thaw cycle some peaks showed a decrease in intensity while others were no longer detectable. While this result seems logical, another study showed that some peaks increased in intensity while others decreased when urine samples were subjected to four to seven freeze–thaw cycles (26). Storing serum for more than 4 h at room temperature or 24 h at 4 °C was shown to severely affect the sample's quality. Samples stored for 2 weeks at −80 and −20 °C showed no significant difference in their proteomic profile. These studies were conducted using MALDI-TOF/MS and all that was acquired was a single pattern of each sample. Unfortunately, using this technology in this manner limits the detection to highly abundant proteins (27). To gauge the affect on lower abundant proteins would require at minimum analysis of each sample using LC–MS.

Unfortunately, the conclusive study establishing the optimum collection, processing, and storage conditions remains lacking. One overlooked challenge to establishing these conditions is investigators have not established a clear workflow for discovery of biomarkers in these fluids. Consider the case of a field goal kicker. His path is clearly laid out in front of him. If he kicks the football through the uprights, he achieves success. He knows where the uprights are and the distance to them. What if he was blindfolded and the uprights were moved to arbitrary locations on the field? The path he needs to kick the football would be uncertain as would his chance of success (which would probably be very small). This situation reflects biomarker discovery today. As will be discussed later, there are a number of different paths being taken to discovery biomarkers, yet the chance of success remains low. Optimizing sample conditions is almost impossible unless the downstream discovery process is optimized first. Probably the best we can do at this stage is to ensure samples have been collected under standard operating procedures, processed using widely accepted standards and stored at −80 °C with a minimum number of freeze–thaw cycles.

7.3 COMMONLY USED BIOFLUIDS FOR BIOMARKER DISCOVERY

Just about every conceivable clinical sample, including biofluid, has been used in biomarker discovery projects. Many of the studies, however, may simply report technology developments using a specific biofluid and do not describe actual attempts at looking for biomarkers. The proteomes of plasma, serum, cerebrospinal fluid (CSF), urine, saliva, sweat, cyst fluid, bile, nipple aspirate fluid, lavage fluids, tissue interstitial fluids, etc. have all been analyzed. Owing to the amount of effort put into analyzing these fluids, it is appropriate to gain a better understanding of many of these.

7.3.1 Blood

Blood, in the form of serum and plasma, is the most commonly used sample for biomarker discovery. For simplicity, when I refer to blood in this chapter it can refer to either serum

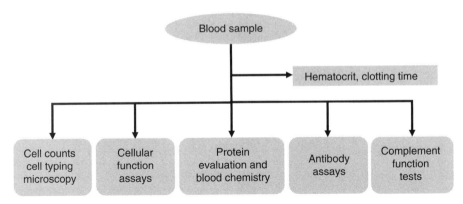

Figure 7.2. Blood is an extremely valuable fluid for monitoring the well being of the patient. A variety of different tests are conducted on a blood sample to assist the physician in diagnosing the patient's physical condition.

or plasma. It must be clarified that blood is not actually a biofluid, but rather a tissue. In the proteomic era, however, it has been most often referred to as a biofluid. Its accessibility and the way it is processed are more similar to other biofluids, such as urine and CSF, than tissues. Five liters of blood circulates through the human body, traversing ~60,000 miles of veins, arteries, and capillaries (http://www.fi.edu/learn/heart/systems/circulation.html). Blood carries oxygen and nutrients to cells and transports carbon dioxide and waste excreted from cells (http://en.wikipedia.org/wiki/Blood; http://health.howstuffworks.com/blood .html). As "The eyes are the windows to the soul"; blood can be thought of as the window to the physical condition of the patient. Every cell in the human body is within four cell units of the circulation system. It is not surprising, therefore, that blood contains information concerning the overall pathophysiology of the patient. Many different characteristics are measured in blood that is acquired at a routine physical to provide a general assessment of the patient's health (Figure 7.2). For example, the basic metabolic panel measures salts, proteins, and metabolites such as sodium, potassium, chloride, calcium, bicarbonate, blood urea nitrogen, creatinine, and glucose levels to determine the state of the patient's renal function (http://www.labtestsonline.org/understanding/analytes/bmp/glance.html). At the cellular level, the concentration of white blood cells, red blood cells, and platelets in the blood is generally measured. Any deviation from a normal range in these cell counts can indicate the presence of a condition such as infection or anemia.

As proteomic investigators are quickly learning, blood is a very complicated at both the cellular and molecular level. Blood consists of many different cell types suspended in plasma, a fluid consisting of metabolites, proteins, gases, nutrients, minerals, and hormones. The major cell components of blood are red (erythrocytes) and white blood cells (including leukocytes and platelets). Plasma is the straw-yellow colored liquid portion of unclotted blood that is left behind after the cells are removed. Plasma is prepared by venipuncture in the presence of an anticoagulant followed by centrifugation of the sample to remove cellular elements. The most commonly used anticoagulants used to prepare plasma include heparin, EDTA, and sodium citrate. Only heparin can potentially

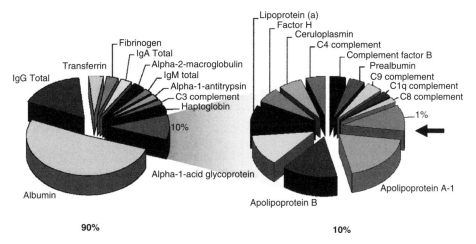

Figure 7.3. The dynamic range of protein concentrations in human serum. The protein concentration of serum is very high, however, only 22 proteins make up 99% of the total protein amount.

interfere with proteomic analysis particularly when using MS. Heparin prevents coagulation by activating antithrombin while EDTA and sodium citrate prevents coagulation by chelating calcium ions. Serum is prepared by collecting blood without any anticoagulant, which allows a fibrin clot to form. This clot is then removed via centrifugation leaving behind serum. While coagulation makes serum and plasma qualitatively different, the difference in their protein concentration is only about 3–4% (28). The major difference in protein content is due to the removal of coagulation factors, which are not of interest in most biomarker discovery studies.

Blood has many positive attributes that make it an ideal clinical sample for proteomic analysis, especially when using MS. Blood is easily acquired and has a protein concentration in the range of tens of mg/mL. Unfortunately, this high protein concentration is not very well distributed. Approximately 99% of the protein content of blood is made up of only 22 proteins (Figure 7.3) (29). The concentration of proteins in circulation is estimated to span 10 orders of magnitude. This large dynamic range is especially problematic for biomarker discovery using MS analysis. The dynamic range of a typical mass spectrometer is only on the order of two to three orders of magnitude and most LC–MS^2 experiments are run in a data-dependent mode, resulting in the most abundant ions being selected for tandem MS. Unfortunately, the prevailing hypothesis is that disease-specific biomarkers will be very low abundant. Using a data-dependent mode of operation means that the most interesting, low abundance proteins will rarely be selected for analysis using MS.

7.3.2 Characterization of the Serum and Plasma Proteome

Although probably not recognized at the time, the dynamic range of protein concentration in serum and plasma was evident in many early analyses. In 1972, a study using an early form of two-dimensional gel electrophoresis (2D-GE), separated serum proteins using a 4.75% gel followed by orthogonal electrophoresis within a 2–30% polyacrylamide linear gradient gel slab (30). One hundred and twelve individual proteins spots were visualized. Most of these proteins were high abundant proteins such as immunoglobulins, haptoglobin, ceruloplasmin, transferrin, and albumin. Even using what we would today consider archaic technology, this group reported one of the first differential proteome findings as several post- and prealbumin components were detected in sera from patients with myeloma, leukemia, and Hodgkin's disease that were not detected in control sera.

This study formed the basis for the fractionation of blood proteins from serum and plasma (as well as other sources) using 2D-polyacrylamide gel electrophoresis (PAGE) in 1975 by Patrick O'Farrell (31). Building on this study and recognizing the utility of 2D-PAGE as developed by Dr. O'Farrell, two pioneers in proteomic analysis of blood, Norman and Leigh Anderson applied this fractionation method for the analysis of human plasma proteins (32). This father and son team was able to resolve and visualize about 300 distinct proteins, which they estimated were composed of 75–100 unique proteins. While this effort almost tripled the number of visualized proteins, it still barely scratched the extent of the plasma proteome.

With the rapid development of mass spectrometers with greater sensitivity and identification speed, a renewed focus on the proteome of blood began at the turn of the century. In 2003, another 2D-PAGE study directed by Dr. Rembert Pieper working with a large number of scientists including Dr. Leigh Anderson produced (in my opinion) seminal results describing the serum proteome (33). This paper was seminal in terms of the proteins it identified, but probably more importantly, in the effectiveness of the technologies it demonstrated. Before the sample was fractionated by 2D-PAGE, the high abundance proteins albumin, haptoglobins, transferrins, transthyretin, α-1-antitrypsin, α-1-acid glycoprotein, hemopexin, and β-2-macroglobulin were depleted using a specially prepared immunoaffinity column. This study was one of (if not the) first to use immunoaffinity to remove high abundance proteins. As described later, high abundance protein removal has become almost standard in the proteomic analysis of blood. Before going directly to 2D-PAGE, the remaining proteins were separated into 66 fractions using sequential anion-exchange and size-exclusion chromatography. Each fraction, including the original and immunodepleted sera and each anion exchange fraction were separated on individual 2D-PAGE gels (Figure 7.4). Coomassie staining of these 74 gels revealed a total of approximately 20,000 individual spots, which corresponded to about 3700 unique spots. Of these 1800 were identified using MS, corresponding to 325 unique proteins. This study shows the extent of fractionation that was going to be required if MS was going to identify a significant portion of the blood proteome. Forty percent of the identified proteins were classified as circulatory proteins, 35% were intracellular proteins, and about 6% were known cell surface proteins. Quite importantly, proteins having known serum concentrations less than 10 ng/mL (e.g., interleukin-6, metallothionein II, cathepsins, and various peptide hormones) were identified in this analysis.

As mentioned, one of the key analytical advances to arise from this study was the value of immunodepletion of the high abundance proteins. Depletion of high abundant proteins from serum and plasma has now become a standard technology prior to using

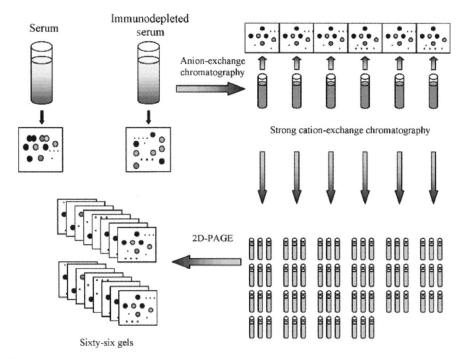

Figure 7.4. Characterization of the serum proteome via immunodepletion/chromatographic/ 2D-PAGE fractionation strategy followed by MS identification. Serum, in which the high abundance proteins had been immunodepleted, was fractionated using anion and strong cation exchange chromatography resulting in a total of 72 fractions that were separated and visualized on 2D-PAGE gels. Raw and immunodepleted serum were directly separated on two other gels. Analysis of the accumulative 20,000 spots resulted in the identification of 350 unique proteins. Data derived from Reference 48.

these biofluids for the discovery of biomarkers. About half of the protein content of blood is albumin Figure 7.3; therefore, removing this protein has an immediate impact on the dynamic range of protein concentration problem. A number of commercially available immunodepletion systems have been developed over the past several years including the Agilent multiple affinity removal system (MARS), which removes six high abundant proteins (i.e., albumin, IgG, IgA, transferrin, haptoglobin, and alpha-1-antitrypsin) (34). As illustrated in Figure 7.5, this column-based product is quite effective in removing these proteins. Immunodepletion columns that remove 20 and 12 proteins from serum or plasma have been introduced by Sigma and GenWay Biotech, respectively (35). These immunoaffinity columns have shown themselves to not only be effective, but robust as well. In a study evaluating the reproducibility of a MARS column, 250 serum samples obtained from prostate cancer patients were immunodepleted and showed a relative standard deviation below 7%, over a 6-week period (34).

One concern with the use of any depletion strategy that removes high abundance proteins is that potentially important biomarkers can also be eliminated due to their

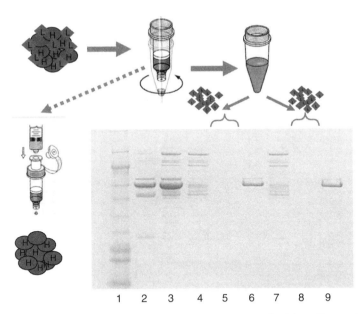

Figure 7.5. Depletion of serum using MARS. Lane 2: serum standard; Lane 3: raw plasma; Lane 4: plasma proteins that flow through MARS immunodepletion column during first washing step; Lane 5: plasma proteins that flow through MARS immunodepletion column during second washing step; Lane 6: elution of high abundant proteins that are retained by the MARS immunodepletion column. Lanes 7, 8, and 9 are replicates of Lanes 4, 5, and 6, respectively, using a second MARS column.

interactions with these high abundant proteins. To evaluate these interactions and possible losses, customized immunoaffinity columns were created using antibodies against albumin, IgM, IgA, transferrin, and apolipoprotein (36). Serum was passed over these columns using conditions that would preserve protein–protein interactions. The proteins that bound to the column were eluted; however, the antibody was not since it was covalently coupled to the column. The eluted proteins were identified using LC–MS^2 resulting in the identification of 210 unique proteins. Approximately two-thirds of the proteins identified were not reported in a manuscript describing the serum proteome that had been characterized up to that date (37). While this study demonstrates that proteins are lost, the benefits of immunodepletion related to the number of proteins that can be identified far outweigh its disadvantages.

Most of the earliest studies of the blood proteome were "technology focused." The initial focus was not on finding disease-specific biomarkers; rather their intent was to develop methods to identify as many proteins in blood as possible. These studies were critical for setting the foundation enabling scientists to even consider doing comparative studies for biomarker discovery. The biggest variations in the technologies used were in the sample preparation and prefractionation strategies. While different mass spectrometers were used, at the time their performance characteristics were not remarkably different. Most studies utilized high abundant protein depletion using techniques such as immunoaffinity (37) or filtration (38). The most popular types of fractionation included 2D-PAGE (33) and strong cation exchange (SCX) of tryptically digested proteins (37, 38). The fractions collected after strong cation exchange were generally analyzed using reversed-phase LC followed by MS^2 analysis. These methods gradually increased the number of blood proteins that could be identified in a single study into the range of 300–500; still a long way from complete proteome coverage.

With the results from four separate global blood proteomic studies in hand, Dr. Leigh Anderson sought to compare the proteins identified within these studies along with those proteins reported in literature as known plasma or serum proteins (39). A total of 1175 unique proteins were identified in the studies. Only about 17% of these, however, were identified in two or more of the studies. Even more disheartening, only about 4% were identified in all four studies. Why the poor overlap? This paper proposed three factors: (1) the methods used were different so that different subsets of proteins were being interrogated; (2) while the samples used were human serum or plasma, the order of medium- and low-abundance protein components may have been different in each; or (3) the identifications provided by bioinformatic analysis of the MS^2 data may have suffered from a high error rate. It is unlikely that factors 1 and 2 are correct as it would suggest that blood samples differ widely and the methods used in these studies are extremely variable. The procedures used in these studies did not vary to a degree that one would expect only 1 out of every 25 proteins to be found in all four studies. While blood samples do vary, the major variation would be found in the low to medium abundance proteins, and none of the studies analyzed possessed the capability of digging deep into this proteome. This argument is strengthened by the fact that low abundance, cytokines and protein hormones were almost completely absent from the experimental data sets. The database of known blood proteins (468 total) contained a larger percentage of proteins that possessed signal sequences indicating they were extracellular proteins. The experimental data sets showed a much greater percentage of cellular proteins. While the results of this study were somewhat disappointing, it provided a "reality check" that significant effort was still required to not only develop technologies to increase proteome coverage but identify the optimum methods for dealing with this important clinical biofluid.

Fortunately, this study only served to increase the verve for characterizing the blood proteome. The Human Proteome Organization (HUPO; http://www.hupo.org/) decided to initiate a global project to increase the knowledge of the blood proteome under the Plasma Proteome Project (PPP; http://www.hupo.org/research/hppp/). Reference specimens of human serum and plasma were sent to 55 participating laboratories worldwide between the years 2003 and 2005 (40). The laboratories were allowed to analyze the samples in which ever manner they desired as long as they reported the procedure that they used. The combined results identified 9504 proteins by at least one peptide per protein. Over 3000 and 1274 proteins were identified by at least two and three peptides, respectively. This study represented a threefold increase in the knowledge of proteins within the blood proteome.

At this time, most laboratories are comfortable with the strategies used for preparing serum and plasma for proteomic characterization. These generally include depletion of high abundance proteins followed by fractionation (ranging from two to four dimensions) and MS^2 analysis. The recent years have seen significant improvements in mass spectrometer technology that continues to increase the number of proteins that can be identified in a single experiment. The focus is now almost solely on comparing proteomes of serum and plasma taken from patients with disease for the discovery of novel biomarkers. I will not discuss quantitative studies as that topic is more suited for Chapter 5 when I describe studies using proteomic technology for the discovery and validation of biomarkers. So what have we learned about the blood proteome besides how to identify lots of proteins? Examination of the proteins that were identified in all of these studies validated the use of serum and plasma for biomarker discovery. As mentioned early in this section, no cell is more than four units removed from circulation. The results show that the blood proteome contains proteins from every portion of the cell. Beyond the expected secreted and membrane proteins, those from the nucleus, cytoplasm, mitochondria, endoplasmic reticulum, Golgi, etc., have been observed in the blood proteome. While some of these proteins are transferred out of the cell, others are likely released by apoptotic or necrotic cells. While these observations may not appear all that important, it is a seminal finding in that if biomarkers for diseases such as cancers that affect internal organs are to be found through analyzing blood samples, proteins from tumors need to be present in blood.

7.4   URINE

After serum and plasma, urine is the third most utilized biofluid for biomarker discovery. Much of the waste that is produced by cells that enters the bloodstream is removed by filtration in the kidneys and, in particular, in the nephrons (41). The action of this filtration produces urine. About 20% of the total blood supply enters the kidneys to be filtered every minute. Much of the fluid within the blood passes through the membranes within the glomeruli and flows into the Bowman's capsule. The filtrate that is produced consists primarily of water, excess salts (primarily Na^+ and K^+), glucose, and the waste product, urea. Before excretion occurs, however, reabsorption needs to occur (42,43). The kidney filters approximately 180 L of water (i.e., the equivalent of 90 two-liter bottles of soda) from the blood every day. Fortunately, $\sim$99% of this total is reabsorbed back into the blood. Glucose is entirely reabsorbed into the bloodstream through active transport in normal, healthy individuals. In diseases, such as diabetes mellitus, the amount of glucose far exceeds the kidneys ability to reabsorb all the material present and high urinary glucose levels are a diagnostic aid indicative of this condition. The kidneys also reabsorb ions (especially sodium) at a rate dependent on the amount of these materials present within an individual's diet.

Blood proteins are filtered and reabsorbed based on their size and charge within the glomeruli (44). After passing through glomeruli, low molecular weight and abundant serum proteins (e.g., albumin, transferrin, immunoglobulin light chain, vitamin D-binding protein, myoglobin, and receptor-associated protein) are reabsorbed, mainly in the proximal renal tubules. Protein concentration in the urine is normally less than 100 mg/L in the urine of a healthy person. The overall protein excretion is normally less than 150 mg/day. Levels higher than this are indicative of proteinuria (i.e., excess presence of serum proteins in the urine).

The protein concentration of urine is normally three orders of magnitude less than that found in serum or plasma. Fortunately what urine lacks in concentration it makes up for in volume. While its protein concentration is significantly lower than that of blood, large amounts of urine can be collected completely noninvasively. While its low concentration may require urine to be concentrated prior to proteomic analysis, clinicians are much more willing to provide large volumes of urine compared to blood.

As discussed with blood, an important aspect of using urine for biomarker discovery was first developing the necessary methods for determining its proteomic content. Based on our present knowledge, urine does not possess the large dynamic range of protein concentrations of blood. Therefore methods, such as immunodepletion, designed to eliminate high abundance proteins are not required when analyzing urine using MS. The characterization of urine followed a similar history to that of blood, where the first objective was to identify as many components as possible. Many of the prefractionation strategies used to analyzed the blood proteome (e.g., 2D-PAGE and MudPIT) were utilized to characterize urine's proteome. In most cases, LC–MS^2 was used for protein identification. The most in-depth characterization of the urinary proteome was published in 2006 (45). In this study, a single urine sample, along with a sample created by pooling urine obtained from nine individuals, were analyzed. After concentrating urinary proteins approximately 50-fold, the sample was split into two aliquots that were fractionated at the intact protein level using SDS-PAGE or reversed-phase liquid chromatography (RPLC). The gel lane containing the separated proteins was pixilated into 14 pieces that were subjected to in-gel tryptic digestion. Albumin was removed from the second aliquot prior to its fractionation into 22 fractions using RPLC. These fractions were also digested into tryptic peptides. The peptides contained within the 36 total fractions were analyzed using LC–MS^2 and LC–MS^3.

An accumulative total of 1543 proteins were identified in all of the fractions that were analyzed. While 337 proteins (26.3%) of the proteins were identified from a single unique peptide, both MS^2 and MS^3 were used to confirm their identity. Of the identified proteins, 335 (22%) and 488 (32%) were classified as extracellular or membrane proteins, respectively. This subset of proteins include kidney-related proteins involved in homeostasis such as transporters for water (e.g., aquaporin 1, 2, and 7), drugs (e.g., multidrug resistance protein 1), sodium, potassium, and chloride (e.g., solute carrier family 12 members 1, 2, and 3). The results showed that the urinary proteome contains a high content of extracellular and membrane proteins; however, the intracellular protein content is low compared to serum or plasma.

While conventional wisdom suggests that urine is useful for discovering biomarkers for diseases related to the kidney, bladder, and urinary tract, the previous studied revealed a potential broader use for this biofluid. Many studies have found promising biomarkers for conditions such as prostate cancer (46), renal cell carcinoma (47), bladder cancer (48), and type 1 diabetes (49). In the study by Adachi et al., a large percentage of lysosomal proteins were identified in urine, suggesting the detection of proteins involved in transporting materials into urine (45). This finding suggests that urine may contain biomarkers for conditions localized outside the kidney, bladder, and urinary tract.

Urine, like blood, is not simply a watery fluid. Urine from healthy individuals contains very small numbers of red and white blood cells, and epithelial cells originating from the bladder. Excessive numbers of these cells typically indicate disease. Normal urine was recently shown to contain tiny vesicles called "exosomes" (50). Exosomes are small (40–80 nm) structures that originate as internal vesicles from multivesicular bodies and are probably formed by most cell types throughout the body. Within the kidney, exosomes are released into urine by fusion of the outer membrane of the multivesicular bodies with the apical plasma membrane. Owing to their origin, exosomes can provide a noninvasive means of examining the physiological state of renal cell. One advantage in working with exosomes is they can be easily isolated by subjecting urine to ultracentrifugation.

In 2004, LC–MS^2 was used to characterize exosomes isolated from urine obtained from a healthy normal subject (50). Almost 300 unique proteins were identified; however, a more recent article claims the ability to detect over 1000 proteins (51). In the 2004 study, proteins originating from every renal tubule epithelial cell type, as well as podocytes and transitional epithelia from the urinary collection system, were found. The exosome proteome was made up of membrane (24%) and cytosolic (45%) proteins, presumably trapped during vesicle formation. Just over 10% of the proteins within the exosome proteome were classified as extracellular. A specific breakdown of the distribution of proteins among various subcellular compartments is shown in Table 7.1. Sixteen proteins identified in exosomes that had been previously implicated in various kidney and systemic abnormalities are listed in Table 7.2. These conditions (and the associated proteins) include hypertension (neprilysin, aminopeptidase A and N, angiotensin converting enzyme isoform-1, etc.), medullary cystic kidney disease 2 (uromodulin), and Liddle syndrome (epithelial sodium channel α, β, and γ). These findings demonstrated that exosomes within urine were also a potentially valuable source of biomarkers. While not as greatly exploited, subsequent studies have shown the presence of exosomes in a

TABLE 7.1. Distribution of Proteins Identified Within Exosomes
Isolated from Urine Based on Their Subcellular Origin

Subcellular Origin	Percentage of Total Proteins
Cytoplasm (endosomal trafficking)	24.7
Cytoplasm (soluble)	22.0
Plasma membrane (integral)	16.3
Extracellular (secreted)	10.5
Plasma membrane (peripheral)	7.8
Unclassified	4.7
Cytoplasm (cytoskeleton motor)	3.4
Lysosome	3.1
Plasma Membrane (GPI linked)	2.7
Nucleus	2.4
Endoplasmic reticulum/Golgi	2.0
Mitochondria	0.3

TABLE 7.2. Proteins Identified in Exosomes Implicated in Various Kidney and Systemic Abnormalities

Protein	Related Kidney Disease
Aquaporin 2	Autosomal recessive nephrogenic diabetes insipidus, type 1
Epithelial sodium channel α, b, g	Autosomal recessive pseudohypoaldosteronism type 1 Liddle syndrome
FXYD domain-containing ion transport	Familial renal hypomagnesemia
Carbonic anhydrase II	Autosomal recessive syndrome of osteopetrosis with renal tubular
Carbonic anhydrase IV	Proximal renal tubular acidosis
Polycystin-1	Autosomal dominant polycystic kidney disease type 1
Uromodulin	Medullary cystic kidney disease 2
Podocin	Autosomal recessive steroid-resistant nephrotic syndrome
Nonmuscle myosin heavy chain IIA	Fechtner and Epstein syndrome
Adenine phosphoribosyltransferase	2,8-dihydroxyadenine urolithiasis
Angiotensin I converting enzyme isoform-1	Hypertension
Aminopeptidase A and P	Hypertension
Neprilysin	Hypertension
Hydroxyprostaglandin dehydrogenase 15	Hypertension
Dimethylarginine dimethylaminohydrolase-1	Hypertension

variety of different body fluids including amniotic fluid, bronchoalveolar lavage fluid, synovial fluid, breast milk, saliva, and blood (52).

7.5 CEREBROSPINAL FLUID

CSF is a clear liquid produced from arterial blood by the choroid plexuses of the lateral and fourth ventricles. Normally a healthy adult produces about 600–700 mL/day while retaining a reservoir of approximately 140 mL. CSF is located between the brain and the skull, providing a mechanical cushion for the brain. This biofluid also circulates throughout the central nervous system facilitating the distribution of neuroendocrine factors as well as the movement of metabolites, toxins, and nutrients to and from cells of the brain and spinal cord. Owing to its direct contact with the brain, CSF is considered a rich source of biomarkers for brain-related disorders. The variety of pathological conditions that may populate CSF with biomarkers is thought to include brain tumors, inflammatory diseases, infection, neurodegenerative diseases, trauma and hydrocephalus. The majority of the CSF proteins are believed to originate from blood that enters at the choroid plexus as well as along this biofluid's flow from the ventricles to the subarachnoid space (53). This entry system results in ventricular and lumbar CSF having different concentrations of blood-derived proteins (54).

(a)   (b)

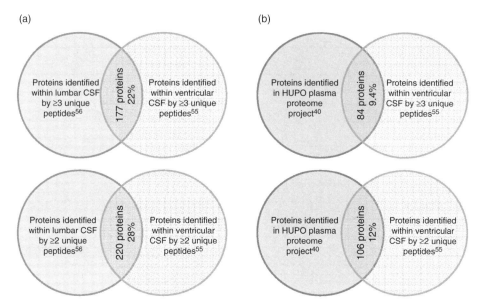

Figure 7.6. (a) Venn diagrams showing overlap in the number of proteins identified in lumbar (Zougman et al., ref. 56) and ventricular CSF (Waybright et al., ref. 55) by at least three (top) and two (bottom) unique peptides. (b) Venn diagrams showing overlap in the number of proteins identified in HUPO PPP (ref 40) and ventricular CSF (ref 55) by at least three (top) and two (bottom) unique peptides.

The protein concentration of CSF from the lumbar region of a typical adult is between 150 and 450 µg/mL (53). While this protein concentration is much less than serum or plasma, CSF is not dominated by a few high abundance proteins; therefore, this concentration is sufficient for global biomarker discovery efforts. Blood contamination, however, is a problem in dealing with CSF. Therefore, albumin depletion is recommended prior to analyzing this biofluid. The proteome of ventricular and lumbar CSF has been characterized using multidimensional fractionation combined with LC–MS^2 protein identification (55, 56). In one of the studies, over 1500 unique proteins were identified, including a number of brain-specific proteins (i.e., amyloid proteins) and numerous low abundance species such as kallikreins, cytokines, and chemokines (55). When the proteins identified in both studies were compared, the overlap was 28% for proteins identified by at least two unique peptides and 22% for those identified by at least three (Figure 7.6a). While much of this level of overlap can be contributed to the nature of peptide identification using data-dependent MS, it also suggests the ventricular and lumbar CSF have unique proteomes. Comparison of proteins identified within ventricular (55) and lumbar (56) CSF to those identified within the HUPO serum proteome initiative (40), showed that the overlap in identified proteins between the two CSF data sets was approximately twice as great as the overlap between the ventricular CSF data set and the plasma proteome (Figure 7.6b). These data show that CSF samples differ significantly from plasma suggesting that CSF may be a more appropriate biofluid for detecting proteins related to brain and neurological disorders.

An obvious disease for which biomarkers are thought to exist in CSF is Alzheimer's disease (AD). Studies have shown that AD pathology begins between 10 and 15 years before the resulting cognitive impairment is obvious enough to warrant medical attention. As with many neurological disorders, finding biomarkers that can detect AD pathology in its early stages and predict dementia onset could possibly direct treatment that post-pones onset. To determine if such markers exist, Craig-Schapiro et al. used a targeted proteomics approach to discover CSF biomarkers that could be used in combination with Aβ42, tau, p-tau181 to augment diagnostic and prognostic accuracy of AD (57). While this book has an obvious focus on MS approaches, this group used a multiplex Luminex platform to measure 190 specific analytes in 333 CSF samples. The 190 ana-lytes measured included cytokines, chemokines, growth factors, hormones, metabolic markers, and other proteins thought to be important in AD. The samples were from patients that were cognitively normal, very mildly demented, and mildly demented. Receiver-operating characteristic curve analyses revealed that small combinations of a subset of the following proteins, cystatin C, VEGF, TRAIL-R3, PAI-1, PP, NT-proBNP, MMP-10, MIF, GRO-α, fibrinogen, FAS, and eotaxin-3, augmented the ability of the tau/Aβ42 ratio to discriminate mildly and very mildly demented patients from normal individuals. Machine learning models used to analyze the data reconfirmed the use of the levels of these proteins as biomarkers for early-stage AD.

7.6 SALIVA

Saliva is produced by the submandibular, sublingual, parotid, and numerous minor salivary glands. Saliva (like most biofluids) is primarily made up of water that contains proteins, peptides, lipids, minerals, and other small compounds (58). Saliva's roles include maintaining homeostasis in the oral cavity, lubricating oral tissues, and assisting in speaking, chewing, and swallowing. Saliva also inhibits bacterial and viral growth within the oral cavity (59).

While not utilized nearly as much as serum or plasma for biomarker discovery, a number of test procedures for measuring different hormones and drugs in saliva have been developed since the late 1980s (60). Saliva collection is noninvasive and easy. The protein concentration of saliva is on the order of a few mg/mL; therefore, it is easy to collect plenty of material to apply MS methods to characterize this easily obtained proteome sample. Unlike serum and plasma, saliva is not dominated by a few high abundant proteins; therefore it can be analyzed without immunodepletion.

The Human Salivary Proteome Project was initiated by the US National Institute of Dental and Craniofacial Research with the goal of cataloging all salivary secretory proteins (http://www.skb.ucla.edu). This interlaboratory, interdiscipline study utilized 2D-PAGE, shot-gun sequencing, free-flow electrophoresis, etc., to produce a human salivary catalogue. The largest portion of this catalogue (1381 unique salivary proteins) was provided using a combination of capillary isoelectric focusing and nano-RPLC coupled on-line with MS^2 analysis (61). The most striking feature was the number of

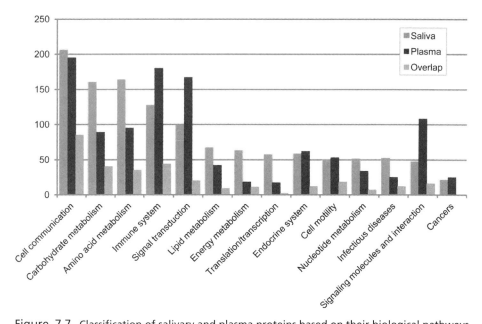

Figure 7.7. Classification of salivary and plasma proteins based on their biological pathways extracted from the KEGG pathway database (63). The number of proteins from each fluid within each KEGG category are represented along the y axis along with the number of proteins found within both proteomes.

proteins with molecular masses less than 20 kDa. A total of 241 (~17.5%) proteins having masses <20 kDa were identified. Forty eight of these proteins are proline-rich proteins, which create a protective film over teeth to promote enamel remineralization (62). Antimicrobial proteins such as FALL-39, β-defensin 125, β-defensin 1, lysozyme, mucins 5B and 2, and six different cystatins were also identified. Proteins originating from 31 different bacterial species were found in the saliva sample. *Streptococcus mutans*, (29 unique proteins) was the most readily detectable species. This bacterium, which causes tooth decay, is ubiquitous within the human population.

A more recent study compared 3020 human proteins identified in plasma with 1939 unique salivary proteins identified using MS (63). Almost 600 (31%) of the 1939 salivary proteins were also identified in plasma. The biological roles of the salivary proteins were compared to those of known plasma proteins by annotating them using the KEGG pathway database (Figure 7.7). Most of the salivary proteins identified were involved in pathway activities such as cell communication, carbohydrate and amino acid metabolism, immune system, and signal transduction. The relative population of these pathways was similar when plasma proteins were examined. An obvious exception was the relatively small number of proteins annotated to signaling and interaction pathways in saliva compared to plasma. While the distribution of proteins based on cellular localization, biological processes, and molecular function were similar in both saliva and plasma, the study showed that the salivary proteome is sufficiently distinct to warrant its study for potential biomarkers for diseases including Sjögren's syndrome and various oral cancers.

An important issue that has not been completely resolved for most biofluids is their stability. A recent study analyzed the effects of different sample collection and handling methods on peptides within saliva (64). The peptide fraction of saliva was isolated using ultrafiltration, and the effects of the various collections and handling methods was evaluated by monitoring the abundance of 41 peptides using MS. The effects of freezing rate, nutritional status of the donors (fed vs. fasted), and room-temperature sample degradation on peptide abundance levels was measured. Increasing the sample freezing rate showed a higher level of peptide abundance, while nutritional status of the donor had no effect on the 41 peptides being monitored. Leaving the peptide samples at room temperature for 5, 10, and 15 min did not result in any detectable degradation of the sample. In addition, the study found that sample processing or instrumental variability was less than 10% the analysis of saliva. The results show that at least the low molecular weight component of saliva is not affected by sample handling or the patient's nutritional status. This result is contrary to what is observed in blood samples, as nutritional status has been shown to play a role on the proteome of this biofluid (65, 66). This fraction of saliva was sensitive to the freezing method used again showing the importance of standardizing sample collection and handling methods when dealing with biofluids.

7.7 OTHER BIOFLUIDS

The above sections have described the major biofluids used for biomarker discovery. Of course, there are a large number of other biofluid types that have been used including pancreatic juice (67), tissue interstitial fluid (68, 69), gastric juice (70), nipple aspirate fluid (71), and ductal lavage (72). Some groups have even attempted to mimic interstitial fluid by incubating surgically resected tissue in buffer with the goal of capturing secreted proteins (73). One of the factors contributing to the sparse use of many alternative biofluids is their accessibility and the limited number of samples available within repositories. The appeal of these biofluids resides in their proximity to the site of disease. These fluids could prove useful for the initial discovery phase followed by a study to determine if an observed biomarker can be detected within an accessible fluid such as serum or urine.

7.8 CONCLUSIONS

The foundation of biomarker discovery rests on the type of biofluid chosen for analysis. Ideally the biofluid closest to the site of disease (e.g., tumor, cyst, etc.) would be available; however, this situation is rare. More commonly, biofluids are chosen based on their accessibility both at the time of the study design and their potential use in the future. This reason is why serum, plasma, and urine have dominated biomarker discovery efforts. While a number of factors need to be considered in analyzing the proteome of a biofluid, the foremost is the protein concentration and dynamic range of this concentration. Plasma and serum represent the extreme in dynamic range of protein concentrations. Being dominated by about 20 high abundance proteins requires these biofluids to be uniquely treated compared to others.

While on the surface the progress in identifying biofluid-based biomarkers appears demoralizing, the knowledge of the proteomes of these samples has grown exponentially over the past decade. In 2002, the ability to identify 400 proteins in serum or plasma was considered monumental; now labs routinely identify thousands of proteins within these biofluids. Has that increased the rate of biomarker discovery? Not yet. Before this rate increases a number of technological advances are required. The research community has yet to establish collection and storage standards for biofluids destined for proteomic analysis. Sample preparation methods for each biofluid type are not standardized, making intercomparison of studies tenuous at best. Employing a commonly cited analogy, proteomic biomarker discovery is akin to finding the differences in a number of different haystacks. Finding what is unique to one group of haystacks would be facilitated if the hay in each stack was as uniform as possible and unexpected variations did not lead us down "rabbit-trails." Standardization is not an easily achieved goal. Many of the samples being used in biomarker discovery were collected decades ago and have been stored over this period of time. Standardizing collection and storage conditions would require an exhaustive study to determine the combination that causes the least variation or degradation over time. The scientific community would then uniformly need to be willing to follow these guidelines. Until then, the chances of finding biomarkers in biofluids should increase as technologies such as MS are able to identify a greater percentage of the proteins within complex proteomes.

REFERENCES

1. Bandhakavi S, Stone MD, Onsongo G, Van Riper SK, Griffin TJ. A dynamic range compression and three-dimensional peptide fractionation analysis platform expands proteome coverage and the diagnostic potential of whole saliva. J. Proteome Res. 2009;8:5590–5600.

2. Carvalho PC, Han X, Xu T, Cociorva D, Carvalho MG, Barbosa VC, Yates JR 3rd. XDIA: Improving on the label-free data-independent analysis. Bioinformatics 2010;26:847–848.

3. Reiter L, Claassen M, Schrimpf SP, Jovanovic M, Schmidt A, Buhmann JM, Hengartner MO, Aebersold R. Protein identification false discovery rates for very large proteomics data sets generated by tandem mass spectrometry. Mol. Cell. Proteomics 2009;8:2405–2417.

4. Xie S, Moya C, Bilgin B, Jayaraman A, Walton SP. Emerging affinity-based techniques in proteomics. Expert Rev. Proteomics 2009;6:573–583.

5. Ahmed FE. Sample preparation and fractionation for proteome analysis and cancer biomarker discovery by mass spectrometry. J. Sep. Sci. 2009;32:771–798.

6. Apweiler R, Aslanidis C, Deufel T, Gerstner A, Hansen J, Hochstrasser D, Kellner R, Kubicek M, Lottspeich F, Maser E, Mewes HW, Meyer HE, Müllner S, Mutter W, Neumaier M, Nollau P, Nothwang HG, Ponten F, Radbruch A, Reinert K, Rothe G, Stockinger H, Tarnok A, Taussig MJ, Thiel A, Thiery J, Ueffing M, Valet G, Vandekerckhove J, Verhuven W, Wagener C, Wagner O, Schmitz G. Approaching clinical proteomics: Current state and future fields of application in fluid proteomics. Clin. Chem. Lab. Med. 2009;47:724–744.

7. Stults JT, Arnott D. Proteomics. Methods Enzymol. 2005;402:245–289.

8. Gucinski AC, Dodds ED, Li W, Wysocki VH. Understanding and exploiting peptide fragment ion intensities using experimental and informatic approaches. Methods Mol. Biol. 2010;604:73–94.

9. Malik R, Dulla K, Nigg EA, Körner R. From proteome lists to biological impact-tools and strategies for the analysis of large MS data sets. Proteomics 2010;10:1270–1283.

10. Tharakan R, Edwards N, Graham DR. Data maximization by multipass analysis of protein mass spectra. Proteomics 2010;10:1160–1171.

11. Ahmad S, Sundaramoorthy E, Arora R, Sen S, Karthikeyan G, Sengupta S. Progressive degradation of serum samples limits proteomic biomarker discovery. Anal. Biochem. 2009;394:237–242.

12. Thongboonkerd V, Saetun P. Bacterial overgrowth affects urinary proteome analysis: Recommendation for centrifugation, temperature, duration, and the use of preservatives during sample collection. J. Proteome Res. 2007;6:4173–4181.

13. Veenstra TD, Conrads TP, Hood BL, Avellino AM, Ellenbogen RG, Morrison RS. Biomarkers: Mining the biofluid proteome. Mol. Cell. Proteomics 2005;4:409–418.

14. Yamamoto T. The 4th Human Kidney and Urine Proteome Project (HKUPP) Workshop. 26 September 2009, Toronto, Canada. Proteomics 2010, Vol. 10, pp. 2069–2070.

15. Weaver R, Riley RJ. Identification and reduction of ion suppression effects on pharmacokinetic parameters by polyethylene glycol 400. Rapid Commun. Mass Spectrom. 2006;20:2559–2564.

16. Drake SK, Bowen RA, Remaley AT, Hortin GL. Potential interferences from blood collection tubes in mass spectrometric analyses of serum polypeptides. Clin. Chem. 2004;50:2398–2401.

17. Luque-Garcia JL, Neubert TA. Sample preparation for serum/plasma profiling and biomarker identification by mass spectrometry. J. Chromatogr. A 2007;1153:259–276.

18. Villanueva J, Philip J, Chaparro CA, Li Y, Toledo-Crow R, DeNoyer L, Fleisher M, Robbins RJ, Tempst P. Correcting common errors in identifying cancer-specific serum peptide signatures. J. Proteome Res. 2005;4:1060–1072.

19. Hsieh SY, Chen RK, Pan YH, Lee HL. Systematical evaluation of the effects of sample collection procedures on low-molecular-weight serum/plasma proteome profiling. Proteomics 2006;6:3189–3198.

20. Kim HJ, Kim MR, So EJ, Kim CW. Comparison of proteomes in various human plasma preparations by two-dimensional gel electrophoresis. J. Biochem. Biophys. Methods 2007;70:619–625.

21. Ostroff R, Foreman T, Keeney TR, Stratford S, Walker JJ, Zichi D. The stability of the circulating human proteome to variations in sample collection and handling procedures measured with an aptamer-based proteomics array. J. Proteomics 2010;73:649–666.

22. Thomas CE, Sexton W, Benson K, Sutphen R, Koomen J. Urine collection and processing for protein biomarker discovery and quantification. Cancer Epidemiol. Biomarkers Prev. 2010;19:953–959.

23. West-Nielsen M, Hogdall EV, Marchiori E, Hogdall CK, Schou C, Heegaard NH. Sample handling for mass spectrometric proteomic investigations of human sera. Anal. Chem. 2005;77:5114.

24. Baumann S, Ceglarek U, Fiedler GM, Lembcke J, Leichtle A, Thiery J. Standardized approach to proteome profiling of human serum based on magnetic bead separation and matrix-assisted laser desorption/ionization time-of-flight mass spectrometry. Clin. Chem. 2005;51:973–980.

25. Schaub S, Wilkins J, Weiler T, Sangster K, Rush D, Nickerson P. Urine protein profiling with surface-enhanced laser-desorption/ionization time-of-flight mass spectrometry. Kidney Int. 2004;65:323–332.

26. Powell T, Taylor TP, Janech MG, Arthur JM. Change in the apparent proteome with repeated freeze–thaw cycles. J. Am. Soc. Nephrol. 2006;17(suppl):436A.

27. Verrills NM. Clinical proteomics: Current and future prospects. Clin. Biochem. Rev. 2006;27:99–116.

28. Lum G, Gambino SR. A comparison of serum versus heparinized plasma for routine chemistry tests. Amer. J. Clin. Pathol. 1974;61:108–113.

29. Anderson NL, Anderson NG. The human plasma proteome: History, character, and diagnostic prospects. Mol. Cell. Proteomics 2002;11:845–867.

30. Wright GL. High resolution two-dimensional polyacrylamide electrophoresis of human serum proteins. Am. J. Clin. Path. 1972;57:173.

31. O'Farrell PH. High-resolution two-dimensional electrophoresis of proteins. J. Biol. Chem. 1975;250:4007–4021.

32. Anderson L, Anderson NG. High resolution two-dimensional electrophoresis of human plasma proteins. Proc. Natl. Acad. Sci. U.S.A. 1977;74:5421–5425.

33. Pieper R, Gatlin CL, Makusky AJ, Russo PS, Schatz CR, Miller SS, Su Q, McGrath AM, Estock MA, Parmar PP, Zhao M, Huang ST, Zhou J, Wang F, Esquer-Blasco R, Anderson NL, Taylor J, Steiner S. The human serum proteome: display of nearly 3700 chromatographically separated protein spots on two-dimensional electrophoresis gels and identification of 325 distinct proteins. Proteomics 2003;3:1345–1364.

34. Darde VM, Barderas MG, Vivanco F. Depletion of high-abundance proteins in plasma by immunoaffinity subtraction for two-dimensional difference gel electrophoresis analysis. Methods Mol. Biol. 2007;357:351–364.

35. Gong Y, Li X, Yang B, Ying W, Li D, Zhang Y, Dai S, Cai Y, Wang J, He F, Qian X. Different immunoaffinity fractionation strategies to characterize the human plasma proteome. J. Proteome Res. 2006;5:1379–1387.

36. Zhou M, Lucas DA, Chan KC, Issaq HJ, Petricoin EF 3rd, Liotta LA, Veenstra TD, Conrads TP. An investigation into the human serum "interactome". Electrophoresis 2004;25:1289.

37. Adkins J.N, Varnum SM, Auberry KJ, Moore RJ, Angell NH, Smith RD, Springer DL, Pounds JG. Toward a human blood serum proteome: Analysis by multidimensional separation coupled with mass spectrometry. Mol. Cell. Proteomics 2002;1:947–955.

38. Tirumalai RS, Chan KC, Prieto DA, Issaq HJ, Conrads TP, Veenstra TD. Characterization of the low molecular weight human serum proteome. Mol. Cell. Proteomics 2003;2:1096–1103.

39. Anderson NL, Polanski M, Pieper R, Gatlin T, Tirumalai RS, Conrads TP, Veenstra TD, Adkins JN, Pounds JG, Fagan R, Lobley A. The human plasma proteome: A nonredundant list developed by combination of four separate sources. Mol. Cell. Proteomics 2004;3:311–326.

40. Omenn GS, States DJ, Adamski M, Blackwell TW, Menon R, Hermjakob H, Apweiler R, Haab BB, Simpson RJ, Eddes JS, Kapp EA, Moritz RL, Chan DW, Rai AJ, Admon A, Aebersold R, Eng J, Hancock WS, Hefta SA, Meyer H, Paik YK, Yoo JS, Ping P, Pounds J, Adkins J, Qian X, Wang R, Wasinger V, Wu CY, Zhao X, Zeng R, Archakov A, Tsugita A, Beer I, Pandey A, Pisano M, Andrews P, Tammen H, Speicher DW, Hanash SM. Overview of the HUPO plasma proteome project: Results from the pilot phase with 35 collaborating

laboratories and multiple analytical groups, generating a core dataset of 3020 proteins and a publicly-available database. Proteomics 2005;5:3226–3245.

41. Veenstra TD Kumar R. 2000. The hormonal regulation of calcium metabolism. In: Seldin D, Giebisch G. eds. *The Kidney: Physiology and Pathophysiology*. Philadelphia, PA: Lippincott-Raven.

42. Brenner B, editor. *The Kidney*. Philadelphia, PA: WB Saunders; 2000.

43. Brunzel NA. *Fundamentals of Urine & Body Fluid Analysis*. Philadelphia, PA: Saunders; 2004.

44. Haraldsson B, Sorensson J. Why do we not all have proteinuria? An update of our current understanding of the glomerular barrier. News Physiol. Sci. 2004;19:7–10.

45. Adachi J, Kumar C, Zhang Y, Olsen JV, Mann M. The human urinary proteome contains more than 1500 proteins, including a large proportion of membrane proteins. Genome Biol. 2006;7:R80.

46. Goo YA, Goodlett DR 2010. Advances in proteomic prostate cancer biomarker discovery. J. Proteomics 73:1839–1850.

47. Kim K, Aronov P, Zakharkin SO, Anderson D, Perroud B, Thompson IM, Weiss RH. Urine metabolomics analysis for kidney cancer detection and biomarker discovery. Mol. Cell. Proteomics 2009;8:558–570.

48. Shirodkar SP Lokeshwar VB. 2009. Potential new urinary markers in the early detection of bladder cancer. Curr. Opin. Urol. 19:488–493.

49. Merchant ML, Perkins BA, Boratyn GM, Ficociello LH, Wilkey DW, Barati MT, Bertram CC, Page GP, Rovin BH, Warram JH, Krolewski AS, Klein JB. Urinary peptidome may predict renal function decline in type 1 diabetes and microalbuminuria. J. Am. Soc. Nephrol. 2009;20:2065–2074.

50. Pisitkun T, Shen RF, Knepper MA. Identification and proteomic profiling of exosomes in Human Urine. Proc. Natl. Acad. Sci. U.S.A. 2004;101:13368–13373.

51. Gonzalez P, Pisitkun T, Knepper MA. Urinary exosomes: Is there a Future? Nephrol. Dial. Transplant. 2008;23:1799–1801.

52. Simpson RJ, Lim JW, Moritz RL, Mathivanan S. Exosomes: Proteomic insights and diagnostic potential. Expert Rev. Proteomics 2009;6:267–283.

53. Huhmer AF, Biringer RG, Amato H, Fonteh AN, Harrington MG. Protein analysis in human cerebrospinal fluid: Physiological aspects, current progress and future challenges. Dis. Markers 2006;22:3–26.

54. Reiber H. Flow rate of cerebrospinal fluid (CSF)—a concept common to normal blood–CSF barrier function and to dysfunction in neurological diseases. J. Neurol. Sci. 1994;122:189–203.

55. Waybright T, Avellino AM, Ellenbogen RG, Hollinger BJ, Veenstra TD, Morrison RS. Characterization of the human ventricular cerebrospinal fluid proteome obtained from hydrocephalic patients. J. Proteomics 2010;73:1156–1162.

56. Zougman A, Pilch B, Podtelejnikov A, Kiehntopf M, Schnabel C, Kumar C, Mann M. Integrated analysis of the cerebrospinal fluid peptidome and proteome. J. Proteome Res. 2008;7:386–399.

57. Craig-Schapiro R, Kuhn M, Xiong C, Pickering EH, Liu J, Misko TP, Perrin RJ, Bales KR, Soares H, Fagan AM, Holtzman DM. Multiplexed immunoassay panel identifies novel csf biomarkers for Alzheimer's disease diagnosis and prognosis. PLoS One 2011;6: e18850.

58. Turner RJ Sugiya H. Understanding salivary fluid and protein secretion. Oral Dis. 2002;8: 3–11.

59. Defabianis P, Re F. The role of saliva in maintaining oral health. Minerva Stomatol. 2003;52:301–308.

60. Lei Z, Hua X, Wong DT. Salivary biomarkers for clinical applications. Mol. Diagn. Ther. 2009;13:245–259.

61. Guo T, Rudnick PA, Wang W, Lee CS, Devoe DL, Bagley BM. Characterization of the human salivary proteome by capillary isoelectric focusing/nanoreversed-phase liquid chromatography coupled with ESI-tandem MS. J. Proteome Res. 2006;5:1469–1478.

62. Van Nieuw Amerongen A, Bolscher JG, Veerman EC 2004. Salivary proteins: Protective and diagnostic value in cariology? Caries Res 38:247–253.

63. Yan W, Apweiler R, Balgley BM, Boontheung P, Bundy JL, Cargile BJ, Cole S, Fang X, Gonzalez-Begne M, Griffin TJ, Hagen F, Hu S, Wolinsky LE, Lee CS, Malamud D, Melvin JE, Menon R, Mueller M, Qiao R, Rhodus NL, Sevinsky JR, States D, Stephenson JL, Than S, Yates JR, Yu W, Xie H, Xie Y, Omenn G .S, Loo JA, Wong DT. Systematic comparison of the human saliva and plasma proteomes. Proteomics Clin. Appl. 2009;3:116–134.

64. deJong EP, van Riper SK, Koopmeiners JS, Carlis JV, Griffin TJ. Sample collection and handling considerations for peptidomic studies in whole saliva; implications for biomarker discovery. Clin. Chim. Acta 2011;412:2284–2288.

65. Corzo A, Kidd MT, Koter MD, Burgess SC. Assessment of dietary amino acid scarcity on growth and blood plasma status of broiler chickens. Poult. Sci. 2005;84:419–425.

66. Mahn AV, Muñoz MC, Zamorano MJ. Discovery of biomarkers that reflect the intake of sodium selenate by nutritional proteomics. J. Chromatogr. Sci. 2009;47:840–843.

67. Gao J, Zhu F, Lv S, Li Z, Ling Z, Gong Y, Jie C, Ma L. Identification of pancreatic juice proteins as biomarkers of pancreatic cancer. Oncol. Rep. 2010;23:1683–1692.

68. Haslene-Hox H, Oveland E, Berg KC, Kolmannskog O, Woie K, Salvesen HB, Tenstad O, Wiig H. A new method for isolation of interstitial fluid from human solid tumors applied to proteomic analysis of ovarian carcinoma tissue. PLoS One 2011;6:e19217.

69. Sun W, Ma J, Wu S, Yang D, Yan Y, Liu K, Wang J, Sun L, Chen N, Wei H, Zhu Y, Xing B, Zhao X, Qian X, Jiang Y, He F. Characterization of the liver tissue interstitial fluid (TIF) proteome indicates potential for application in liver disease biomarker discovery. J. Proteome Res. 2010;9:1020–1031.

70. Hsu PI, Chen CH, Hsieh CS, Chang WC, Lai KH, Lo GH, Hsu PN, Tsay FW, Chen YS, Hsiao M, Chen HC, Lu PJ. Alpha1-Antitrypsin precursor in gastric juice is a novel biomarker for gastric cancer and ulcer. Clin. Cancer Res. 2007;13:876–883.

71. Kuerer HM, Goldknopf IL, Fritsche H, Krishnamurthy S, Sheta EA, Hunt KK. Identification of distinct protein expression patterns in bilateral matched pair breast ductal fluid specimens from women with unilateral invasive breast carcinoma. high-throughput biomarker discovery. Cancer 2002;95:2276–2282.

72. Cazzaniga M, Decensi A, Bonanni B, Luini A, Gentilini O. Biomarkers for risk assessment and prevention of breast cancer. Curr. Cancer Drug Targets 2009;9:482–499.

73. Teng PN, Rungruang BJ, Hood BL, Sun M, Flint MS, Bateman NW, Dhir R, Bhargava R, Richard SD, Edwards RP, Conrads TP. Assessment of buffer systems for harvesting proteins from tissue interstitial fluid for proteomic analysis. J. Proteome Res. 2010;9:4161–4169.

8

PROTEOMIC PATTERNS: A NEW PARADIGM IN DIAGNOSTICS AND THERAPEUTICS?

8.1 INTRODUCTION

As described in the previous chapter, no area has generated more excitement and interest in proteomics than the search for biomarkers. Finding protein biomarkers that can identify cancers at early stages would dramatically increase the success rate of treatments and minimize the burden of this disease on society; a major goal of institutes such as the National Cancer Institute (NCI). Of the top 10 leading causes of death in high-income countries, four are cancer related (Table 8.1). The need for better early-stage cancer detection strategies is still glaring. Not only would biomarkers aid in early detection, they could be used to monitor how well a particular therapeutic regiment is working from taking a simple blood sample. Tools that could routinely assess the condition of the patient using a routinely applicable method would greatly increase the effectiveness of modern medicinal practices. An ideal proteomic test would be simple to conduct, could be carried out quickly (i.e., less than 2 days) on thousands of patients and would utilize a biofluid sample taken at a regularly schedule checkup and determine not only if a patient had a disease (i.e., cancer), but determine the specific type of disease (i.e., stage I ovarian cancer). This vision is one of the goals of proteomic pattern technology,

Proteomic Applications in Cancer Detection and Discovery, First Edition. Timothy D. Veenstra.
© 2013 John Wiley & Sons, Inc. Published 2013 by John Wiley & Sons, Inc.

TABLE 8.1. Top 10 Diseases that Cause the Largest Number of Deaths and Percentage of Deaths in High-Income Countries

Disease	Deaths in Millions	Percentage of Deaths
Coronary heart disease	1.33	16.3
Stroke and cerebrovascular diseases	0.76	9.3
Trachea, bronchus, lung cancers	0.48	5.9
Lower respiratory infections	0.31	3.8
Chronic obstructive pulmonary disease	0.29	3.5
Alzheimer and other dementias	0.28	3.4
Colon and rectum cancers	0.27	3.3
Diabetes mellitus	0.22	2.8
Breast cancer	0.16	2.0
Stomach cancer	0.14	1.8

in which a pattern of signals originating from proteins within a biofluid indicate the current status of the patient and direct the diagnosis or therapeutic strategy.

The potential of developing a simple and reliable proteomic pattern test to diagnose patients with early-stage cancers exploded onto the medical scene in 2002. In this year, Drs. Lance Liotta and Emanuel Petricoin at the NCI and the Food and Drug Administration (FDA), respectively, published a study that was able to correctly diagnose women with various stages of ovarian cancer with almost 100% sensitivity and specificity (1). What was most amazing about these results is that they were obtained using a very simple, inexpensive, and extremely high-throughput mass spectrometer and only required a few microliters of blood. This instrument had the capability of analyzing hundreds of samples a day. George Wright Jr.'s group at Eastern Virginia Medical School concurrently demonstrated the ability to diagnose prostate cancer with exquisite accuracy using the same instrumental platform (2–4). These studies generated a huge excitement in the fields of cancer diagnostics with not only their simplicity, but their high levels of sensitivity and specificity.

8.2   MASS SPECTROMETRY AS A CLINICAL TOOL

There are presently a large number of diagnostic assays that are conducted using MS; however, most of these measure small molecules such as amino acids or lipids. Probably the best known example is the detection of in-born errors of metabolism assay that was devised by Donald Chace (5). This MS-based method analyzes dried blood spots obtained from newborns to screen for over 30 metabolic disorders, including those associated with amino acids (e.g., phenylketonuria) and fatty acids. When people think of MS, however, they typically associate its role on the discovery side of biomarker detection. In a very narrow view, its role is detect and identify proteins that either show an (preferably) increased or decreased abundance in a biologic sample taken from a disease-affected individual. Once identified and validated, the goal would be to design an immunoassay that can analyze samples faster and cheaper. There are a number of issues, both real and perceived, that make using MS as a diagnostic tool impractical. Mass spectrometers are often perceived as difficult instruments to maintain and operate, requiring extremely skilled personnel. While this perception may have been true several years ago, modern day mass spectrometers are user friendly and simple to operate for routine tasks. While the costs associated with purchasing a mass spectrometer can be quite high, considering their technical capabilities they are an excellent long-term investment with a well maintained instrument lasting for 10 years or longer. Probably, the greatest barrier to using MS as a diagnostic tool is throughput. The analysis of a biologic sample using MS is orders of magnitude slower that by immunoassays. The main reason for this lower throughput is that mass spectrometers can only analyze a single sample at a time while immunoassays can run tens to hundreds of samples concurrently. When dealing with a mass spectrometer, the only present method of increasing throughput is to purchase additional instruments. While it was agreed that MS will continue to play a major role in clinical proteomics, it is likely to be focused on the discovery of potential protein biomarkers.

8.3   PROTEOMIC PATTERN TECHNOLOGY

The work presented by Drs. Petricion, Liotta, and Wright in 2002, however, changed how MS was viewed in the scientific and medical communities. The astonishing diagnostic capabilities presented generated an excitement that spanned a number of different scientific disciplines including medicine, analytical chemistry, and instrument technology. These papers created a sensation rarely seen in scientific circles, even leading to a U.S. Congressional resolution supporting the method. Investigators around the world were envisioning a day when a simple blood test could diagnose early stage cancers. Unfortunately, as with many scientific results that generate this much excitement, they also generated an equal amount of controversy.

To understand the excitement surrounding proteomic patterns, it is important to know the technology behind it. The instrumental platform used to conduct these studies was surface-enhanced laser desorption/ionization time-of-flight MS (SELDI-TOF/MS) (6,7). The pioneering MS platform was known as the ProteinChip Reader PBS-II system manufactured Ciphergen in Fremont, CA. The TOF instrument mass spectrometer was an extremely simple instrument (Figure 8.1). It had low resolution and poor mass accuracy compared to most TOF mass spectrometers, but had quite good sensitivity. The instrument had the ability to measure a very wide mass-to-charge (m/z) range (i.e., in excess of 200,000) and could generate a spectrum of intact peptides and proteins. As with most TOF instruments of its day, it did not possess tandem MS (MS^2) capabilities. Peptide and protein ions were ionized and desorbed using a pulsed UV nitrogen laser.

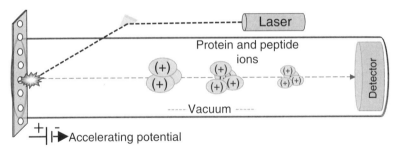

Figure 8.1. Schematic diagram of the SELDI-TOF/MS mass spectrometer. The mass spectrometer is a simple TOF mass spectrometer equipped with an interface that will accommodate protein chips. Biomolecules are ionized and desorbed from the surface of the proteinchip using a UV nitrogen laser.

For all intensive purposes, SELDI was simply a form of matrix-assisted laser desorption (MALDI). The fundamental difference between SELDI and MALDI resides in the sample preparation that is performed prior to MS data acquisition. The SELDI-TOF/MS system was designed to incorporate protein chip arrays that are used to prefractionate complex biologic samples and could also be inserted directly into the mass spectrometer. These protein chip arrays are 2-mm-diameter spots placed onto a 10 mm wide × 80 mm long aluminum surface. Each chip is composed of either eight or 16 spots. Protein chips are available with spots composed of a variety of different surfaces. The spots can be made of chemical surfaces (anionic, cationic, hydrophobic, hydrophilic, metal ion, etc.) designed to bind proteins based on some physicochemical property or they can be activated to couple an affinity reagent of the users' choice (immobilized antibody, receptor, DNA, etc.) to extract a specific protein of interest.

The SELDI protein chips use retentate chromatography instead of typical elution-type chromatography. The protein chips are manufactured with a number of different type of chromatographic surfaces that capture proteins based on a specific physical property or they can be in-laboratory customized, with an antibody for example, to extract a specific protein (Figure 8.2). Biofluids such as plasma, serum, lymph, urine, interstitial fluid, and cerebrospinal fluid are spotted directly onto the protein chip without the need for any cleanup or preseparation. Not only can the protein chips process a wide variety of biological matrices, but only between 1 and 10 μL is required. After the addition of a binding buffer, the chip is incubated allowing proteins to bind to the chip surface. The protein chip is then washed with a series of buffers to remove nonspecifically bound proteins. Similar to what is done in MALDI/MS, a matrix is applied and the sample is allowed to cocrystallize. The protein chip is then placed into the interface of the SELDI-TOF/MS and the ions are formed and desorb via activation of the pulsed UV nitrogen laser. The protein and peptide ions are recorded using TOF/MS. The spectra produced from this type of analysis became popularly known as a proteomic *pattern* since the peaks' identities could not be readily determined from the data provided by the instrument. An example of two proteomic patterns obtained from different cell lines is shown in Figure 8.3. It is pretty evident, that the patterns are quite simple and do

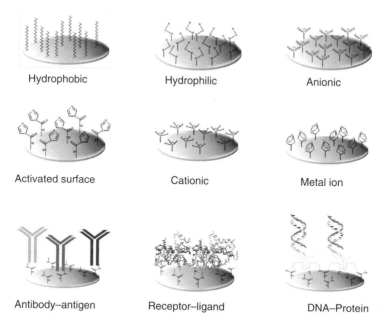

Hydrophobic   Hydrophilic   Anionic

Activated surface   Cationic   Metal ion

Antibody–antigen   Receptor–ligand   DNA–Protein

Figure 8.2. The variety of proteinchip arrays available for preparation of biologic fluids prior to SELDI-TOF/MS analysis. The two upper rows represent chemically modified chromatographic surfaces while the bottom arrays are biochemically modified surfaces. Chemically modified surfaces retain a group of proteins based on a physicochemical property while biochemically modified surfaces are used to isolate a specific protein or functional class of proteins.

not clearly show a huge number of distinct protein signals. This result is in accordance with the dynamic range of proteins in biological systems as well as the performance capabilities of the SELDI-TOF mass spectrometer (i.e., low resolution, mass accuracy, and dynamic range). As described later, the identities of the peaks was not considered necessary in many of the bioinformatic platforms used to analyze the data and ultimately contributed much to the controversy surrounding this technology.

SELDI-TOF/MS system brought one major benefit to clinical diagnostics that other types of mass spectrometers did not possess; throughput. Other MS methods, particularly when it comes to the analysis of biofluids, are generally low throughput. For example, a standard liquid chromatography (LC)–MS^2 analysis of a single blood sample, can take up to a day. The SELDI-TOF/MS platform was designed to concurrently process and analyze tens to hundreds of samples. Robotic liquid handling systems for processing biofluid samples onto the protein chip surfaces were designed and made commercially available. An automated chip loader was available to load protein chips into the interface of the instrument for unattended data acquisition. A single operator could easily acquire the data for over 300 individual biofluid samples per day.

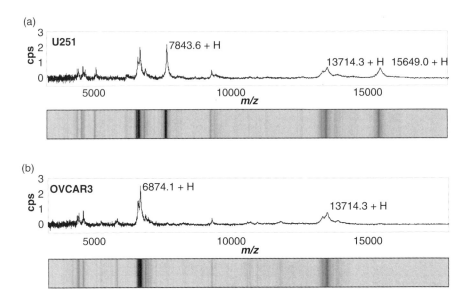

Figure 8.3. Proteomic patterns of lysates obtained from U251 and OVCAR3 cell lines. The patterns are displayed both in the spectral and gel views.

8.4 DIAGNOSIS OF OVARIAN CANCER USING PROTEOMIC PATTERNS

Arguably, the pivotal study that brought proteomic pattern technology to the forefront of proteomics was the one published in Lancet by Drs. Petricoin and Liotta (1). The study focused on the diagnosis of ovarian cancer, the leading cause of gynecological malignancy (8). The challenge with this disease is that almost 80% of women with common ovarian cancer are diagnosed with either stage III or IV disease, where the disease has spread to regional lymph nodes or metastasized (Table 8.2). If the disease is not detected until it has metastasized, the 5-year survival rate is <30%. The 5-year survival rate for patients diagnosed at stage I, where the tumor is still localized, and treated by surgical intervention, however, approaches 95%. These statistics adequately

TABLE 8.2. Stage Distribution and 5-Year Relative Survival Rate by Stage at Diagnosis for all Females at Diagnosis Stage Between the Years 1999 and 2006

Stage at Diagnosis	Distribution (%)	5-Year Relative Survival
Localized (confined to primary site)	15	93.5
Regional (spread to regional lymph nodes)	17	73.4
Distant (cancer has metastasized)	62	27.6
Unknown (unstaged)	7	27.2

illustrate why scientists get incredibly excited about potential biomarkers that can diagnose cancers at early stages. Presently, cancer antigen 125 (CA 125) is the most widely used diagnostic biomarker for ovarian cancer; however, only 50–60% of stage I patients exhibit elevated levels of circulating CA 125 making this marker of limited use for early detection (9). With a positive predictive value (PPV) of only about 10% and combining its measurement with ultrasonography only increasing the PPV to around 20%, shows the limits of using CA 125 measurement as a diagnostic tool (10–12). Compounding this problem is the fact that a significant number of false positives for ovarian cancer are also diagnosed using CA 125 measurements since this protein can be elevated in other nongynecologic and benign conditions (13).

The overall strategy of how this study was conducted is shown in Figure 8.4. Briefly, 5 μL of serum from 100 serum samples from women with ovarian cancer and 116 from healthy women or those with malignant disorders (collectively referred to as control patients) were applied to protein chips made up of a C16 hydrophobic resin (1). After a series of binding and washing steps, the protein chips were placed into the interface of the SELDI-TOF/MS. The acquired proteomic patterns were bioinformatically analyzed to identify changes in *m/z* signals that enabled samples obtained from patients with

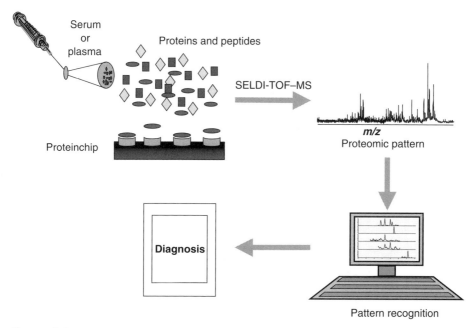

Figure 8.4. Disease diagnostics using proteomic patterns. The sample drawn from the patient is applied to a protein chip, which is made up of a specific chromatographic surface. After several washing steps and the application of an energy-absorbing molecule, the species that are retained on the surface of the chip are analyzed via mass spectrometry. The pattern of peaks within the spectrum is analyzed using sophisticated bioinformatic software to diagnose the source of the biological sample.

ovarian cancer to be distinguished from controls. A high-order self-organizing cluster analysis based on a genetic algorithm was "trained" on SELDI-TOF/MS spectra from serum derived from either control women or women with ovarian cancer. The aim of the algorithm is to identify m/z values that show variability based on their origin (i.e., healthy or diseased patients). The algorithm is initially "trained" on a subset of spectra is used to train the algorithm. During this training phase, the source of the serum (i.e., ovarian cancer or control patients) is made known to the algorithm. The algorithm attempts to recognize m/z values (or features) that enable the sample to be correctly classified. In this study, each spectrum was composed of over 15 000 m/z values along the x-axis with a corresponding amplitude value along the y-axis, representing a very large computational space for the algorithm to find features that provide the correct classification. In subsequent testing and validation phases, the features discovered in the training phase are used to classify the source of the spectra, which is now kept hidden to the algorithm. The algorithm was able to correctly identify all 50 ovarian cancer cases, including all 18 stage I cases, in the blinded validation study. Of the control cases, 63 out of 66 were correctly classified as originating from patients that did not have cancer. These results corresponded to a sensitivity of 100%, a specificity of 95%, and an overall PPV of a remarkable 94%.

Demonstrating that the results presented above for ovarian cancer were not an isolated incident, Dr. George Wright's laboratory demonstrated an equally amazing result for prostate cancer (4). Serum samples from 197 men with prostate cancer, 92 with benign prostatic hyperplasia (BPH), and 96 healthy individuals (total 386) were analyzed by SELDI mass spectrometry. The resulting spectra were randomly divided into training ($n = 326$) and testing ($n = 60$) sets. One hundred and 24 peaks detected by computer analyses were analyzed in the training set using a boosting tree algorithm. The algorithm was able to develop a classifier to separate the prostate cancer samples from the noncancerous (BPH and healthy) groups. The spectra from the test set (30 PCA samples, 15 BPH samples, and 15 samples from healthy men) was then put into the classifier to determine if the algorithm could correctly predict the source of the samples. The classifier was able to correctly predict the prostate cancer and BPH/healthy samples, achieving 100% sensitivity and specificity. Taken together, the astonishing results from these two studies generated a lot of excitement in the fields of oncology, analytical technology, proteomics, and medicine.

8.5   PROTEOMIC PATTERNS SKYROCKET

Subsequent to these studies, there was a huge rush by scientists to determine what other disease conditions could be diagnosed using this technology. The rush was facilitated by the simplicity of the SELDI-TOF/MS analytical platform. It did not require sophisticated expertise to prepare samples and operate the mass spectrometer; therefore, laboratories that were not specialized in MS-based proteomics could conduct these types of studies. Over the next couple of years, studies showing high diagnostic predictability for cancers such as non–small cell lung (14, 15), head and neck squamous cell (16), cervical (17), and bladder (18), to name a few, were published.

Proteomic pattern technology brought a potentially new paradigm to biomarker discovery and disease diagnosis. In direct contrast to conventional LC–MS studies where the goal was to identify a specific protein whose abundance could be directly measured, proteomic pattern technology relied on the signal intensity of a series of unknown peaks within the mass spectrum as the diagnostic determinant. In the time, it took to analyze 10 biofluid samples using LC–MS, the proteomic patterns of hundreds of samples could be measured using SELDI-TOF/MS. In proteomic pattern analysis, no pre-mass spectrometer separation is conducted and the instrument is simply programmed to collect as many peaks as possible. The analysis of the two data sets also differs. For LC–MS data sets, the results are based on identified peptides that are quantitated either by spectral count, differential isotopic labeling, or some other quantitative method. Proteomic patterns are simply analyzed at the peak level, with no knowledge of the identity of the protein or peptide giving rise to the peaks.

Immediately after the publication of the studies by Drs. Petricoin, Liotta, and Wright, SELDI-TOF/MS systems were "selling like hotcakes." As shown in Figure 8.5, a simple PubMed search of the term "SELDI" reveals a rapid increase in the number of publications utilizing this technology. In the year 2002, there were 41 manuscripts that included the word "SELDI." By 2005, this number increased more than fivefold. It seemed that any lab that had a set of biologic samples from healthy and disease affect patients were attempting to have them analyzed using SELDI-TOF/MS. The publicity generated by the perceived uses for this technology was huge. Reports concerning the

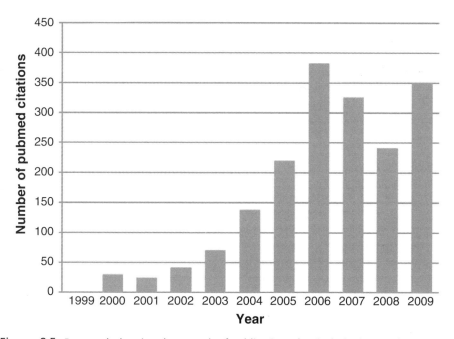

Figure 8.5. Bar graph showing the growth of publications that include the word "SELDI" in their title or abstract as listed in Pubmed.

ability to diagnose cancers from spots of blood were reported by major television networks and media outlets. The NCI funded a laboratory with the purpose of determining the conditions necessary for generating reproducible proteomic patterns that would be capable of providing diagnostic or prognostic information. Paper after paper was published showing the ability to correctly classify the sources of samples based solely on the proteomic pattern they generated. It seemed too good to be true.

8.6   TECHNOLOGY ADVANCES IN PROTEOMIC PATTERNS

As with any recently discovered technology that involves sample preparation and analytical instrumentation, many investigators began developing alternative approaches to acquire proteomic patterns. While the obvious areas of improvement involved simplifying the sample preparation steps and utilizing mass spectrometers with high performance specifications, a critical component was maintaining the throughput afforded using the original PBS-II platform.

8.6.1   The Move to Higher Resolution Mass Spectrometers

In its infancy, proteomic patterns were acquired using the PBS-II system. This mass spectrometer had excellent sensitivity and was very user friendly; however, it had poor resolution. Ciphergen also manufactured a SELDI source that could be coupled directly to the high-resolution hybrid triple quadrupole TOF (QqTOF) mass spectrometer manufactured by Applied Biosystems, Incorporated (ABI). The QqTOF recorded proteomic patterns with a routine resolution >8000 compared to approximately 150 with the PBS-II. A comparison of the mass spectra obtained using a Ciphergen PBS-II and ABI QqTof is shown in Figure 8.6. The spectra obtained from the two instruments were qualitatively similar; however, the increased resolution afforded by the QqTOF was quite evident. A study was conducted to determine if increasing the resolution of the patterns would affect the ability to correctly classify samples based on the condition of the patient. It is important to remember that in any diagnostic test it is important to obtain the highest PPV possible. In a comparative study to determine if increasing the resolution resulted in better PPV, 248 serum samples provided from the National Ovarian Cancer Early Detection Program clinic at Northwestern University Hospital (Chicago, Illinois) were analyzed using the same ProteinChip^{TM} arrays using both PBS-II and QqTOF mass spectrometers (19). The proteomic patterns were acquired on the two different instruments on the same day using the same sample processed on the same proteinchip. The goal of this experimental design was to eliminate all experimental variability other than the use of two different mass spectrometers.

The spectra acquired from the serum samples on the two instrument platforms were divided into three data sets. A training set was used to discover the hidden diagnostics patterns (i.e., m/z values that could discriminate the sources of the serum samples) and a testing and validation set were used to measure the predictive ability of the diagnostic peaks determined in the analysis of the testing set. The algorithm had no prior knowledge of the spectra (i.e., the source of the patient samples) in the testing and validation sets.

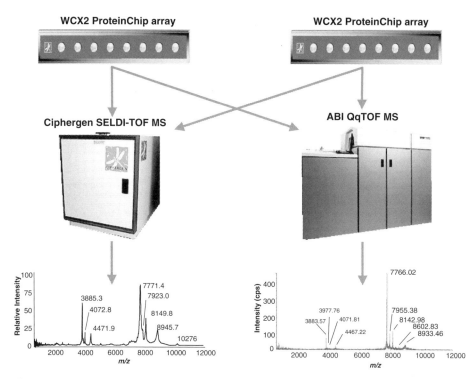

Figure 8.6. Experimental design comparing the effect of resolution on the ability to correctly classify samples based on the condition of the patient. In this study, protein chips were analyzed using both a Ciphergen PBS-II and ABI QqTOF mass spectrometer.

Analysis of the data produced using the high-resolution QqTOF generated four diagnostic models that had 100% sensitivity and specificity in classifying serum samples as acquired from unaffected or ovarian cancer affected women. Each model had a PPV of 100% as no false positives or negatives were found. The PBS-II SELDI-TOF/MS data were unable to provide diagnostic models that were both 100% sensitive and specific. Achieving diagnostic models with 100% sensitivity and specificity cannot be underestimated if this technology is to be used as a screening tool for early-stage cancers. Screening for diseases of relatively low prevalence, such as ovarian cancer, a diagnostic test must exceed 99.6% sensitivity and specificity to minimize false positives, while correctly detecting early stage disease (10). This study showed that movement to higher resolution instruments may enable predictive models possessing the highest possible PPV.

8.6.2 Magnetic Particles and High-Resolution MALDI-TOF

A twist on the original SELDI-TOF/MS platform, was the use of a reversed-phase, magnetic bead format to extract peptides from serum (Figure 8.7). The proteomic patterns

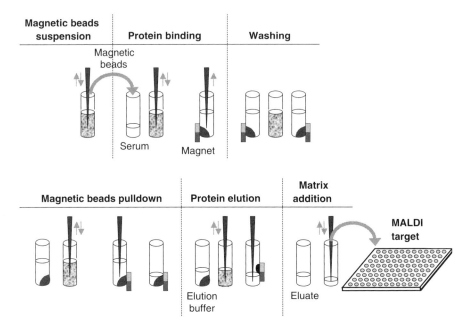

Figure 8.7. Use of magnetic particles to process serum samples for direct MALDI-TOF/MS analysis. A suspension of magnetic beads that contain reversed-phase ligands is created by either multiple pipetting or vortexing. An aliquot of the suspension is transferred to a tube containing serum, which is then thoroughly mixed. A magnet is used to pull the beads to the side of the tube and the supernatant is discarded. The beads are then washed repeatedly. After the final washing step the supernatant is carefully removed and elution buffer is added to remove the bound proteins from the beads. The eulate is transferred to a fresh tube, mixed with MALDI matrix and the sample transferred to a MALDI plate for MS analysis.

of the captured peptides were then acquired using MALDI-TOF/MS (20). Since MALDI-TOF systems are widely available, this sample preparation method was widely useful and compatible with high-resolution mass spectrometers. In this method, a suspension of reversed-phase chemistry coated magnetic beads is mixed with a small volume (i.e., 50 μL) of serum. After this mixture, the beads are extracted through magnetic force. After a series of washing steps, peptides bound to the particles are eluted using a solvent containing 50% (v/v) acetonitrile. This peptide mixture is cocrystallized with matrix and transferred to a MALDI target plate. The plate is inserted into a TOF (or other compatible) mass spectrometer ion source and the proteomic patterns of the various samples acquired. This method has a number of advantages including easy automation of the sample preparation and the ability to generate patterns with high resolution through the use of TOF/MS.

In an application of this technology, proteomic patterns of 34 serum samples obtained from patients with glioblastomas (GBM) and 22 from healthy volunteers were acquired as described above (20). The data were acquired using an Ultraflex MALDI-TOF mass spectrometer, an instrument possessing very high resolution (i.e., 25,000) and mass accuracy. More than 400 distinct peptide peaks were unambiguously detected in a typical serum sample. The two classes (GBM and control) were created by using 55 out of 56 samples as a training set. The 56th sample was used as a test to determine if it would be correctly classified. This training and testing process was repeated 56 times until all the spectra had been used as a test set. Fifty-three out of 55 (96.4%) samples were correctly classified, while two of the control samples were incorrectly classified (3.6%) as GBM samples. The remaining samples could not be classified as either. This study represented the first use of a straight MALDI-based approach to acquire serum proteomic patterns classifying biologic samples.

8.7 TO IDENTIFY OR NOT IDENTIFY: THAT IS THE QUESTION

Diagnosis of samples using proteomic pattern technology did not depend on the identification of any peak within the spectra. The entire result was simply based on the m/z values of the peaks used to construct the classifier. Many scientists protested that the proteins represented by the diagnostic peaks needed to be identified to validate the utility of the method. Identifying the peaks that give rise to the diagnostic model can provide an orthogonal validated test based on measuring only these proteins via an immunoassay, for example, or some mechanistic information regarding the specific condition being studied may be revealed. There was, however, another camp that considered identifying the peaks as irrelevant; all that mattered is that the condition can be diagnosed correctly. In their opinion, the purpose of using proteomic patterns is not to investigate the mechanism of the disease's formation or progression, but simply to provide a quick determination if a patient has some form of disease or not. In addition, knowing the identity of the biomarker does not necessarily provide any insight into the mechanism of the disease [e.g., prostate-specific antigen (PSA) for prostate cancer].

Probably one of the major reasons that peaks within the spectra were not identified is that it is not a trivial task. As mentioned previously, SELDI-TOF/MS does not have direct MS^2 capabilities and the mass measurement accuracy of the instrument makes direct identification based solely on mass impossible. While a SELDI source can be coupled to a QqTOF, which has MS^2 capabilities, direct identification of proteins directly from the protein chips has proven difficult. What the SELDI-TOF/MS spectrum provides is the m/z value of the peak of interest, which can be used to monitor the isolation of the protein during methods designed to isolate it. A series of spin columns packed with a variety of chromatographic resins that can aid in the isolation procedure are available to aid in protein isolation. Once the protein is isolated to a reasonable level of purity, a high-resolution, tandem mass spectrometer can then be used to identify the protein. This overall schema has been used in a number of successful identifications of diagnostic peaks found within SELDI-TOF/MS spectra.

So what did these studies that identified some of the diagnostic peaks find? In many cases, the results were either disappointing or puzzling. Several studies have identified serum amyloid A (SAA) as a biomarker for patients with lung cancer, metastatic renal cancer, and arthritis (21–23). SAA is an acute phase protein and its levels likely rise in all three of the aforementioned conditions; however, it is unlikely that it could be used to accurately diagnose any of the conditions described. This protein has been pervasive of proteomic pattern biomarker discovery: acute phase and inflammatory proteins can be identified; however, lower abundance proteins that could be specific for the conditions being studied are not effectively measured and/or identified. One study examining serum samples from ovarian cancer patients identified the diagnostic peaks as transthyretin, β-hemoglobin, apolipoprotein AI and transferrin (24). Another ovarian cancer study identified SAA1 as a diagnostic peak (25). Two studies identified complement component C3a(desArg) as a biomarker for ductal carcinoma in situ (26, 27), while another identified this protein as a marker for colorectal cancer using serum as well (28). A prostate cancer study identified SAAII as a diagnostic feature in the analysis of serum samples obtained from prostate cancer-affected males (29). It is impossible to list all of the diagnostic peaks identified in SELDI-TOF/MS proteomic pattern analysis; however, a vast majority of the identified peaks originated from highly abundant proteins.

In general, the diagnostic peaks within the SELDI-TOF/MS spectra that were identified do not immediately provide mechanistic information regarding tumor formation or progression. A vast majority of these species are known circulatory proteins and not involved internally within cells as members of signaling pathways. Neither are these proteins directly secreted from deranged cells that comprise the tumor. Is the fact that these proteins do not originate from the tumor a death knell to the technology? An intriguing hypothesis suggests that the proteomic patterns that are unique to serum obtained from cancer-affected individuals result from the action of the tumor not necessarily proteins within the tumor (30). Under this hypothesis, the most important biomarkers for cancers may arise from normal host proteins that are aberrantly clipped or reduced in abundance through the secretion of proteases by the tumor. In essence, the tumor leaves its "footprint" on proteins that are within its environment as an indicator of its presence. This "footprint" can be placed on normal host cell proteins or circulating proteins. Whether the proteins detected as diagnostic within proteomic patterns are somehow acted upon within the microenvironment of the tumor remains to be proven.

There have been cases, however, in which potential disease-specific biomarkers have been identified that appear to originate from the site of the disorder. A 2002 study used SELDI-TOF/MS to discover a biomarker for pancreatic cancer (31), the fourth leading cause of cancer death in both genders in the U.S. and unfortunately current methods for diagnosing pancreatic cancer at early stages are quite ineffective (32). In this study, proteomic profiles of pancreatic juice obtained from 15 patients with pancreatic adenocarcinoma were compared to those obtained from seven patients with other pancreatic diseases. The results revealed a peak with a molecular weight of ~16,572.9 daltons (Da) that showed a higher intensity in the pancreatic juice samples obtained from patients with pancreatic adenocarcinoma. To identify the protein giving rise to this peak, this mass was searched against all of the protein masses listed within the SWISS-PROT and TrEMBL databases. Human pancreatic-associated protein 1 (PAP-1), which is known to originate within the pancreas, had a calculated mass (16,566.5 Da) that matched within the mass measurement error of the SELDI-TOF/MS instrument (i.e., ~1000 ppm). A SELDI-based immunoassay was developed to confirm this result by coupling an antihepatocarcinoma–intestine–pancreas/PAP-I polyclonal antibody to a biochemically activated proteinchip surface. Samples of pancreatic juice obtained from patients with adenocarcinoma analyzed using this immunoassay contained the peak at 16,572.9 Da. Samples that did not display the peak in the original SELDI-TOF/MS screening were negative when analyzed using this immunoassay. Measurement of PAP-I levels by enzyme-linked immuno sorbent assay and immunohistochemistry showed an increase in the abundance of this protein not only in pancreatic juice but also serum of additional adenocarcinoma patients. Unfortunately, I have not been able to find any evidence that this protein graduated into a useful biomarker for pancreatic cancer.

8.8 CRITICISMS

As with any novel scientific claim, the ability of proteomic pattern technology to diagnose disease had its share of critics. One of the most compelling arguments questioning the results provided by SELDI-TOF/MS focused on three prostate cancer studies. A study by Adam et al. reported 83% sensitivity at 97% specificity for prostate cancer detection (2), whereas Petricoin et al. report 95% sensitivity at 78–83% specificity (33) and Qu et al. reported 97–100% sensitivity at 97–100% specificity (4). These results are clearly similar in sensitivity and specificity and clearly superior to those that can be obtained using PSA testing (34). The results generated by two of the groups used different methods resulting in dissimilar distinguishing peaks. Adam et al. and Qu et al. (who were from the same research group) used an IMAC-Cu metal-binding chip for serum adsorption (2, 4), while Petricoin et al. used a hydrophobic C-16 chip (33). The study by Adam et al. found nine peaks (m/z 4475, 5074, 5382, 7024, 7820, 8141, 9149, 9507, and 9656) to sufficiently diagnose prostate cancer, whereas Petricoin et al. used seven peaks (m/z 2092, 2367, 2582, 3080, 4819, 5439, and 18,220). Qu et al. identified 12 major peaks (m/z 9656, 9720, 6542, 6797, 6949, 7024, 8067, 8356, 3963, 4080, 7885, and 6991) for identifying samples of noncancerous from cancer and 9 peaks (m/z 7820, 4580, 7844, 4071, 7054, 5298, 3486, 6099, and 8943) for diagnosing healthy individuals from patients with BPH. A comparison of peak values found in the three studies is provided in Table 8.3. None of the peaks identified by Petricoin et al. were identified in the other two studies as being critical for diagnosing prostate cancer. While this result itself may seem curious, even more surprising is that although they used the same proteinchip and mass spectrometer, Adam et al. and Qu et al. found very different distinguishing peaks. Only the two peaks at m/z 7024 and 9656 were consistent between the studies and a third peak (m/z 7820) was identified by Adam et al. as distinguishing cancer vs. noncancerous and by Qu et al. as distinguishing healthy individuals and BPH patients, but not noncancerous and cancer patients.

How can studies analyzing samples from similar patient pools provide such discordant results? One possible conclusion is that there are thousands proteins in serum capable of distinguishing healthy and cancer patients. The chance of any two groups finding the exact same sets of peaks would have a low probability. This conclusion is unlikely as the resolution and dynamic range of the instruments used to collect the data

TABLE 8.3. List of *m/z* values found in three separate SELDI-TOF/MS studies that discriminated samples taken from men with prostate cancer and healthy controls. Values found in two or more studies are italicized

m/z Values that Indicate Prostate Cancer		
Adam et al. (2)	Qu et al. (4)	Petricoin et al. (30)
4475	3963	2092
5074	4080	2367
5382	6542	2582
7024	6797	3080
7820	6949	4819
8141	6991	5439
9149	*7024*	18,220
9507	7885	
9656	8067	
	8356	
	9656	
	9720	

provide a limited number of peaks for analysis. If this conclusion is accurate, however, it makes it very difficult to see how this technology could ever be standardized as a diagnostic test. The differences in the peaks being selected is more likely due to subtle differences in the methods being used for extracting proteins from serum as the results are sensitive to such parameters as serum storage conditions.

Another study was designed to determine whether proteomic patterns generated by SELDI-TOF/MS could distinguish between serum samples drawn from patients with cancer and those with benign disease among women who possessed suspicious breast abnormalities on mammography or physical examination (35). One hundred and thirty two samples were obtained from three different clinics; 63, 64, and 5 samples were drawn from clinics A, B, and C, respectively. A histopathologic diagnosis of breast cancer was obtained for 96 patients; distributed as ductal carcinoma in situ ($n = 13$), invasive ductal carcinoma ($n = 78$), and lobular, tubulolobular, or mixed carcinoma ($n = 5$). The serum samples were collected in a standardized manner before biopsy, aliquoted, and frozen at $-80\,^\circ$C within 6 h of collection. To acquire the proteomic patterns, IMAC3 chips with Cu(II) as the metal ion from a single manufactured lot were used. The serum samples were prepared and spotted on the protein chips on consecutive 3 days, and SELDI-TOF/MS spectra recorded on consecutive 2 days. The data were analyzed using a variety of different machine-learning and bioinformatic packages; however, none were able to classify patients with breast cancer with any statistical significance. To minimize source-related biases, samples showing reproducible spectra but originating from a single clinic were analyzed. This analysis did not improve disease classification accuracy; however, the machine-learning algorithm could predict the day on which the spectra were acquired and on which day the serum samples were processed on the protein chips. Equally as troubling, distinct spectral features were recognized that enabled the samples to be classified based on the clinics at which they were acquired.

8.9 CONCLUSIONS

A decade ago, proteomic pattern technology brought enormous hope that a simple biofluid-based test would provide accurate diagnosis of various cancers. Some of the earliest reports were published in very high profile journals and produced astonishing results. Many in the field felt that the community was witnessing a revolutionary new method for early cancer detection. It appeared that the method worked for almost every disease and every sample set that was analyzed. Frankly, the fact that it worked so well all the time should have been a bit disconcerting. Very few things in science work all the time on every type of sample. Regardless, the reports continued to come out reporting the use of biomarker discovery using proteomic pattern technology. While there are still a number of publications that utilize the technology, these are generally published in lower impact journals. Presently, the belief that proteomic pattern technology (as described in this chapter) can diagnose cancers is rapidly fleeting or nonexistent amongst scientists in the proteomics community.

The original development of the blood-based test to diagnose ovarian cancer was marketed as OvaCheck^{TM} by Correlogic Systems Inc since this company had been part of the original development through a Cooperative Research Agreement with the NCI. As previously mentioned, OvaCheck^{TM} generated a similar level of frenzy as that surrounding SELDI-TOF/MS as this method was profiled on NBC's Today Show as well as reports describing it in several major US daily newspapers. In 2010, Correlogics announced that it had obtained CE (Conformité Européene or European Conformity) marking from the European Union for the detection of epithelial ovarian cancer using its OvaCheck. While the company continues to seek FDA approval for the test in the United States, the CE marking allows the company to offer OvaCheck commercially in Europe. As of August 2010, OvaCheck had still not been approved by the FDA. This development makes the future of OvaCheck in the United States somewhat bleak.

Looking back and assessing the impact of proteomic patterns overall it would be easy to say it was a waste of time. This view, however, is short-sighted and does not fully grasp the impact this area had on proteomics. Prior to the reports from George Wright, Lance Liotta, and Emanuel Petricoin most of the resources in proteomics were focused on analyzing samples such as yeast or mammalian cells in culture. The diagnostic results published by these scientists made the entire community start to think about using proteomics technology (primarily MS) for clinical applications. Labs that had never before analyzed biofluids using high-resolution, high mass accuracy instruments began characterizing such samples. The number of characterized proteins in serum and plasma rapidly increased and novel technologies, such as high abundant protein depletion, were developed as a result of this new focus. It could be strongly argued that clinical proteomics was born out of proteomic pattern technology. Ultimately, the greatest legacy that proteomics pattern technology may leave behind is not its use as a diagnostic method but rather that it got scientists to start thinking about how they could use proteomics as a clinical tool.

REFERENCES

1. Petricoin EF, Ardekani AM, Hitt BA, Levine PJ, Fusaro VA, Steinberg SM, Mills GB, Simone C, Fishman DA, Kohn EC, Liotta LA. Use of proteomic patterns in serum to identify ovarian cancer. Lancet 2002;359:572–577.

2. Adam BL, Qu Y, Davis JW, Ward MD, Clements MA, Cazares LH, Semmes OJ, Schell-hammer PF, Yasui Y, Feng Z, Wright GL Jr. Serum protein fingerprinting coupled with a pattern-matching algorithm distinguishes prostate cancer from benign prostate hyperplasia and healthy men. Cancer Res. 2002;62:3609–3614.

3. Cazares LH, Adam BL, Ward MD, Nasim S, Schellhammer PF, Semmes OJ, Wright GL Jr. Normal, benign, preneoplastic, and malignant prostate cells have distinct protein expression profiles resolved by surface enhanced laser desorption/ionization mass spectrometry. Clin. Cancer Res. 2002;8:2541–2552.

4. Qu Y, Adam BL, Yasui Y, Ward MD, Cazares LH, Schellhammer PF, Feng Z, Semmes OJ, Wright GL Jr. Boosted decision tree analysis of surface-enhanced laser desorption/ionization mass spectral serum profiles discriminates prostate cancer from noncancer patients. Clin. Chem. 2002;48:1835–1843.

5. Naylor EW, Chace DH. Automated tandem mass spectrometry for mass newborn screen-ing for disorders in fatty acid, organic acid, and amino acid metabolism. J. Child Neurol. 1999;1(Suppl 1):S4–S8.

6. Hutchens TW, Yip TT. New desorption strategies for the mass spectrometric analysis of macromolecules. Rapid Commun. Mass Spectrom. 1993;7:576–580.

7. Issaq HJ, Conrads TP, Prieto DA, Tirumalai R, Veenstra TD SELDI-TOF MS for diagnostic proteomics. Anal. Chem. 2003;75:148A–155A.

8. Menon U, Jacobs I. Recent developments in ovarian cancer screening. Curr. Opin. Obstet. Gynaecol. 2000;12:39–42.

9. Cohen LS, Escobar PF, Scharm C, Glimco B, Fishman DA. Three-dimensional power doppler ultrasound improves the diagnostic accuracy for ovarian cancer prediction. Gynecol. Oncol. 2001;82:40–48.

10. Kainz C. Early detection and preoperative diagnosis of ovarian carcinoma. Wien. Med. Wochenschr. 1996;146:2–7.

11. Hamilton W, Round A, Sharp D. Ovarian cancer not the silent killer. BMJ 2009;339:b2719.

12. Joyner AB, Runowicz DC. Ovarian cancer screening and early detection. Womens Health (Lond. Engl.) 2009;5:693–699.

13. Croswell JM, Kramer BS, Kreimer AR, Prorok PC, Xu JL, Baker SG, Fagerstrom R, Riley TL, Clapp JD, Berg CD, Gohagan JK, Andriole GL, Chia D, Church TR, Crawford ED, Fouad MN, Gelmann EP, Lamerato L, Reding DJ, Schoen RE. Cumulative incidence of false-positive results in repeated, multimodal cancer screening. Ann. Fam. Med. 2009;7: 212–222.

14. Xiao XY, Tang Y, Wei XP, He DC. A preliminary analysis of non-small cell lung cancer biomarkers in serum. Biomed. Environ. Sci. 2003;16:140–148.

15. Yanagisawa K, Shyr Y, Xu BJ, Massion PP, Larsen PH, White BC, Roberts JR, Edgerton M, Gonzalez A, Nadaf S, Moore JH, Caprioli RM, Carbone DP. Proteomic patterns of tumour subsets in non-small-cell lung cancer. Lancet 2003;362:433–439.

16. Soltys SG, Le QT, Shi G, Tibshirani R, Giaccia AJ, Koong AC. The use of plasma surface-enhanced laser desorption/ionization time-of-flight mass spectrometry proteomic patterns for detection of head and neck squamous cell cancers. Clin. Cancer Res. 2004;10:4806–4812.

17. Wong YF, Cheung TH, Lo KW, Wang VW, Chan CS, Ng TB, Chung TK, Mok SC. Protein profiling of cervical cancer by protein-biochips: Proteomic scoring to discriminate cervical cancer from normal cervix. Cancer Lett. 2004;211:227–234.

18. Zhang YF, Wu DL, Guan M, Liu WW, Wu Z, Chen YM, Zhang WZ, Lu Y. Tree analysis of mass spectral urine profiles discriminates transitional cell carcinoma of the bladder from noncancer patients. Clin. Biochem. 2004;37:772–779.

19. Conrads TP, Fusaro VA, Ross S, Johann D, Rajapakse V, Hitt BA, Steinberg SM, Kohn EC, Fishman DA, Whitely G, Barrett JC, Liotta LA, Petricoin EF 3rd Veenstra TD. High-resolution serum proteomic features for ovarian cancer detection. Endocr. Relat. Cancer 2004;11:163–78.

20. Villanueva J, Philip J, Entenberg D, Chaparro CA, Tanwar MK, Holland EC, Tempst P. Serum peptide profiling by magnetic particle-assisted, automated sample processing and MALDI-TOF mass spectrometry. Anal. Chem. 2004;76:1560–70.

21. D Liu L, Liu J, Wang Y, Dai S, Wang X, Wu S, Wang J, Huang L, Xiao X, He D. A combined biomarker pattern improves the discrimination of lung cancer. Biomarkers 2011;16:20–30.

22. Walter M, Heinze C, Steiner T, Pilchowski R, von Eggeling F, Wunderlich H, Junker K. Immunochemotherapy-associated protein patterns in tumour tissue and serum of patients with metastatic renal cell carcinoma. Arch. Physiol. Biochem. 2010;116:197–207.

23. de Seny D, Fillet M, Ribbens C, Marée R, Meuwis MA, Lutteri L, Chapelle JP, Wehenkel L, Louis E, Merville MP, Malaise M. Monomeric calgranulins measured by SELDI-TOF mass spectrometry and calprotectin measured by ELISA as biomarkers for arthritis. Clin. Chem. 2008;54:1066–1075.

24. Kozak KR, Su F, Whitelegge JP, Faull K, Reddy S, Farias-Eisner R. Characterization of serum biomarkers for detection of early stage ovarian cancer. Proteomics 2005;5:4589–4596.

25. Moshkovskii SA, Serebryakova MV, Kuteykin-Teplyakov KB, Tikhonova OV, Goufman EI, Zgoda VG, Taranets IN, Makarov OV, Archakov AI. Ovarian cancer marker of 11.7 kDa detected by proteomics is a serum amyloid A1. Proteomics 2005;5:3790–3797.

26. Li J, Orlandi R, White CN, Rosenzweig J, Zhao J, Seregni E, Morelli D, Yu Y, Meng XY, Zhang Z, Davidson NE, Fung ET, Chan DW. Independent validation of candidate breast cancer serum biomarkers identified by mass spectrometry. Clin. Chem. 2005;51:2229–2235.

27. Solassol J, Rouanet P, Lamy PJ, Allal C, Favre G, Maudelonde T, Mangé A. Serum protein signature may improve detection of ductal carcinoma in situ of the breast. Oncogene 2010;29:550–560.

28. Habermann JK, Roblick UJ, Luke BT, Prieto DA, Finlay WJ, Podust VN, Roman JM, Oevermann E, Schiedeck T, Homann N, Duchrow M, Conrads TP, Veenstra TD, Burt SK, Bruch HP, Auer G, Ried T. Increased serum levels of complement C3a anaphylatoxin indicate the presence of colorectal tumors. Gastroenterology 2006;131:1020–1029.

29. Malik G, Ward MD, Gupta SK, Trosset MW, Grizzle WE, Adam BL, Diaz JI, Semmes OJ. Serum levels of an isoform of apolipoprotein A-II as a potential marker for prostate cancer. Clin Cancer Res 2005;11:1073–1085.

30. Petricoin EF, Liotta LA. SELDI-TOF-based proteomic pattern diagnostics for early detection of cancer. Curr. Opin. Biotechnol. 2004;15:24–30.

31. Rosty C, Christa L, Kuzdzal S, Baldwin WM, Zahurak ML, Carnot F, Chan DW, Canto M, Lillemoe KD, Cameron JL, Yeo CJ, Hruban RH, Goggins M. Identification of hepatocarcinoma-intestine-pancreas/Pancreatitis-associated protein I as a biomarker for pancreatic ductal adenocarcinoma by protein biochip technology. Cancer Res. 2002;62:1868–1875.

32. Greenlee RT, Murray T, Bolden S, Wingo PA. Cancer statistics 2000. CA Cancer J. Clin. 2000;50:7–33.

33. Petricoin EF 3rd, Ornstein DK, Paweletz CP, Ardekani A, Hackett PS, Hitt BA, Velassco A, Trucco C, Wiegand L, Wood K, Simone CB, Levine PJ, Linehan WM, Emmert-Buck MR, Steinberg SM, Kohn EC, Liotta LA. Serum proteomic patterns for detection of prostate cancer. J. Natl. Cancer Inst. 2002;94:1576–1578.

34. Diamandis EP. Proteomic patterns in biological fluids: Do they represent the future of cancer diagnostics?. Clin. Chem. 2003;49:1272–1275.

35. Karsan A, Eigl BJ, Flibotte S, Gelmon K, Switzer P, Hassell P, Harrison D, Law J, Hayes M, Stillwell M, Xiao Z, Conrads TP, Veenstra T. Analytical and preanalytical biases in serum proteomic pattern analysis for breast cancer diagnosis. Clin. Chem. 2005;51:1525–1528.

THE EMERGENCE OF PROTEIN ARRAYS

9.1 INTRODUCTION

The present era of proteomics has been dominated by separations, both gel and liquid chromatography (LC) and mass spectrometry (MS). As discussed within other chapters, combinations of various technologies such as polyacrylamide gel electrophoresis, LC, stable isotope labeling, MS, and data analysis have been very useful in obtaining global views of proteomes. There have always been two major limiting factors that affect MS-based proteomics. The first is the "undersampling effect." Undersampling is a result of the fact that any combination of fractionation and MS cannot analyze all of the material that is present in a complex proteome sample. State-of-the-art mass spectrometers are able to select thousands of species for sequencing per hour; however, during the same period tens of thousands of species will enter the instrument and not be analyzed. Taking into account such variables as alternative gene splicing events, individual coding variants, and posttranslational modifications, the number of unique protein species in humans may possibly exceed 1,000,000 proteins (1). Rough calculations would suggest even today we are only able to characterize less than 10% of the human proteome. The second limiting factor is what I call the "you-get-what-you-get" effect. Since most MS experiments are operated in a data-dependent tandem MS (MS^2) mode, researchers are

Proteomic Applications in Cancer Detection and Discovery, First Edition. Timothy D. Veenstra.
© 2013 John Wiley & Sons, Inc. Published 2013 by John Wiley & Sons, Inc.

at the mercy of the mass spectrometer as far as which proteins are selected for analysis. Many times investigators have scrolled through exhaustive lists of proteins looking for one in particular. If it is not found, a peptide from that protein was simply not selected by the mass spectrometer for identification or a peptide was selected but its signal intensity was low so that the MS^2 spectrum was of poor quality.

Many systems biology studies would like to compare global genomic, transcriptomic, and proteomic results. Unfortunately, this comparison is not that straightforward, as there is a fundamental difference experimentally in how the three data sets are collected. Genomic data collected through shotgun sequencing can provide information covering the entire sequence of a particular genome. Transcript data are collected using probes designed against specific transcripts and, therefore, the number and identity of the species being measured is controlled. In MS-based proteomics the proteins that are measured is dependent on them being selected by the mass spectrometer. In most studies, this selection is based on the abundance of their signal and whether or not a particular species will be selected is generally not known *a priori*. This signal intensity dependence means that the lowest abundance proteins, which are hypothesized to be the most important, are overlooked in a $LC–MS^2$ experiment. In addition, even the fastest mass spectrometers today are unable to characterize anywhere near the number of species that can be measured using microarrays. Fulfilling the promise of systems biology is going to require concordant data sets in which gene, transcript, and protein data can be compared with completely empirically data. Unfortunately, we will not eliminate the undersampling issue in the near future; however, we may be able to minimize the "you-get-what-you-get" problem using other technologies, such as protein arrays.

9.2 PROTEIN ARRAYS

Protein arrays are rapidly being developed as the next possible generation of proteomics tools for characterizing protein interactions as well as measuring their abundance levels and identifying their functions within cells, tissues, and biofluids. Arrays are expected to enable concurrent multiplex screening of thousands of protein abundances, and protein interactions, encompassing protein-antibody, protein-protein, protein-drug, enzyme-substrate screening, and multianalyte diagnostic assays. This technology is poised to become an important proteomics technology for both basic research and clinical diagnostics.

While we generally think of protein arrays as being recently invented, Dr. Roger Ekins initially described microarray enzyme-linked immunosorbant assays (ELISA) in the 1980s (2). In a summary of the field of ligand assays, which he published in 1998, he expounded on how antibody-based microarrays would enable multiple analytes to be simultaneously measured and proved valuable in identifying antigens and antibodies (Abs) (3). Dr. Ekins had already recognized that multiplexed immunoassays would require less samples and save both time and cost to researchers and hospital laboratories. Beyond diagnostics applications, protein arrays have the potential to accelerate the discovery of protein–protein interactions (4, 5) and enable protein abundance profiling (6, 7) across entire proteomes.

Array formats   **Uses**

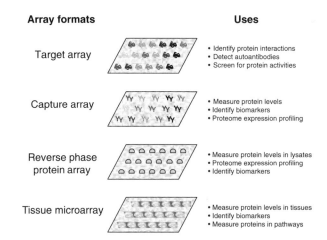

Target array
- Identify protein interactions
- Detect autoantibodies
- Screen for protein activities

Capture array
- Measure protein levels
- Identify biomarkers
- Proteome expression profiling

Reverse phase protein array
- Measure protein levels in lysates
- Proteome expression profiling
- Identify biomarkers

Tissue microarray
- Measure protein levels in tissues
- Identify biomarkers
- Measure proteins in pathways

Figure 9.1. Illustration of four different types of protein arrays commonly used today and some of the uses for each.

9.3 TYPES OF PROTEIN ARRAYS

Protein arrays can be defined as solid-phase ligand binding assay systems in which proteins or affinity probes are immobilized on a surface. The types of surfaces used in protein arrays include membranes, glass, microtiter wells, beads, and other particles (e.g., gold nanoparticles) (8,9). Protein arrays are designed to be multiplexed and small, enabling the collection of a lot of data using very little sample. There are generally considered to be four different types of proteins arrays, target protein arrays, capture arrays, reverse phase protein arrays (RPPAs), and tissue microarrays (TMAs). Each of these formats and some of their intended uses are shown in Figure 9.1. However, be aware that scientific creativity can design additional reagents and combinations to build upon these basic structures providing additional affinity reagents to interrogate proteome samples.

9.4 TARGET PROTEIN ARRAYS

Target protein arrays consist of large numbers of purified proteins immobilized on a solid surface (10–12). These types of arrays can be used to assay protein–protein, protein–DNA, and protein–small molecule interactions as well as monitor enzyme activity. They are also commonly used to detect Abs and assess their specificity for a particular protein. Ideally, a protein array consisting of thousands of individual and functional proteins would permit thousands of proteins to be interrogated in single experiments. Generating this level of density and maintaining each protein's structure and function has been a major challenge.

The first major hurdle in producing a target protein array is the need to produce pure proteins (13). Getting a recombinant protein to express in prokaryotic or eukaryotic

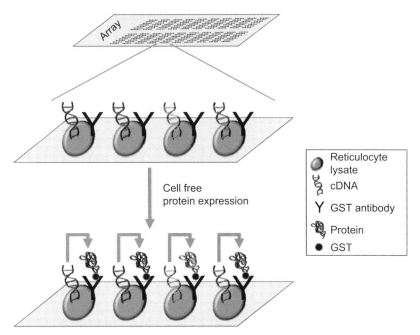

Figure 9.2. A NAPPA is constructed by spotting expression clones that express proteins in a cell-free reticulocyte system. Each protein is expressed with a GST tag that is used to capture the protein at the spot via an antibody directed toward GST.

systems is only the first hurdle. During purification, proteins may not fold properly and fail to mimic their true *in vivo* functional capacity. In other instances, proteins may become unfolded during the purification process. This variation affects experimental reproducibility and makes cross-study comparisons difficult.

To overcome the challenges of protein production and purification, Dr. Joshua LaBaer's laboratory developed a protein microarray method called nucleic acid programmable protein array (NAPPA) (14). This method allows functional proteins to be synthesized *in situ* directly from printed cDNAs as illustrated in Figure 9.2. The cDNAs were combined with a T7-coupled rabbit reticulocyte lysate *in vitro* transcription-translation system to translate the protein as a chimera tagged with glutathione-*S*-transferase (GST). After expression, the proteins are captured on a slide via an antibody targeted to the GST tag. This method eliminates the need for isolating individually expressed proteins, which is incredibly time consuming and can result in contamination by other proteins. In the NAPPA method, proteins are expressed within minutes before the experiment is run, assuring that fresh, functional protein products are used for each study.

Different methods are available for producing a NAPPA. In one approach, a multiple spotting technique prints an *Escherichia* (*E.*) *coli*-based *in vitro* transcription-translation extract directly onto a polymerase chain reaction (PCR) template (15). Another method immobilizes an mRNA-DNA hybrid to which is added a cell-free translation mix to express the specific protein encoded by the mRNA (16). A third approach called DAPA

(DNA array to protein array), translates proteins on a cDNA array, followed by diffusion of the expressed proteins across a cell-free extract-infused membrane to a surface on which the proteins are immobilized (17).

In this initial report, Dr. Labaer's group also developed an improved NAPPA method for producing high-content arrays by preparing fresh protein *in situ* (14). For this improved NAPPA, supercoiled DNA purified using a resin derivatized with diamine chemistry, is printed as the substrate. This DNA purification method allows a single scientist to process approximately 5000 samples per week, resulting in yields that are 5- to10-fold greater than commercially available. The NAPPA was printed using a novel chemistry that relies on the ability of bovine serum albumin (BSA) to substantially improve DNA-binding efficiency. How BSA increases the DNA-binding efficiency is unclear. Regardless, BSA and the capture antibody are coupled to an amine-coated glass surface using an activated ester-terminated homo-bifunctional cross-linker. The process resulted in approximately 64% of the DNA being captured onto the glass surface as determined using fluorescent DNA.

To test the expressed protein yield and reproducibility, an array of 96 cDNAs, a nonexpressing plasmid DNA, and purified recombinant protein were spotted at various concentrations. Ninety-seven percent of the printed samples could be detected using Picogreen staining, which detects double-stranded DNA. A GST-specific antibody confirmed that full-length translation of the protein had occurred for 99% of the printed cDNAs.

While the above validation study was useful, 96 cDNAs does not cover much of the human proteome. The NAPPA system was further tested using 1000 human genes for which isolated, sequence-verified plasmids were available. DNA signal was detected for 99% of the samples with a coefficient of variation of 18%. While a slight variation in protein yield was observed, readily detectable signal was observed for 96% of the genes. Kinases and transcription factors were expressed and captured onto the glass surface with success rates of 98% and 96%, respectively. In addition, 93% of the membrane proteins tested showed decent signal. A small reduction in the protein display success rate was observed as the size of the proteins increased.

Obviously, having a measurable expression level does not mean the protein has retained its function. The function of a protein is intimately related to its structure; therefore, it is critical that the proteins fold correctly on the array. To test if the expressed proteins retained their native functions, an array expressing 647 unique genes in duplicate was printed. The array was tested using a series of established binary interactions, such as Jun-Fos and p53-MDM2. In this test, the query protein was coexpressed with the arrayed protein by addition of the appropriate cDNA to a cell-free expression lysate. After expression and washing, the arrays were interrogated with specific Abs targeted to the query proteins to reveal where they bound within the array. Selective binding of Jun, Fos, and MDM2 was observed with their established binding partners. While correct folding and function requires testing on a protein-by-protein basis, the quality control studies conducted on this NAPPA showed correct behavior for all of the interactions tested.

It is always reassuring in science when the control experiments work properly and you can show that your device works in principle. The true test in this case, however, is expanding the array to address a real scientific problem. To prove the utility of their array, Dr. Labaer's group created a NAPPA containing 4988 expressed proteins (18).

These proteins served as antigens for the detection of autoantibodies (AAb) in sera from patients with breast cancer (Figure 9.3a). AAb have previously been detected in the sera of cancer patients and are produced against changes in protein or glycan expression in tumors (19, 20). Since AAb are stable, highly specific, easily purified from serum, and readily detected with well-validated secondary reagents, they are considered to have advantages over other serum proteins as cancer biomarkers.

The protein arrays were screened using age- and location-matched sera obtained from screening and diagnostic mammography clinics, to control for women undergoing routine screening mammography and women with benign breast disease (18). A three-phase screening strategy (Figure 9.3b) was used to identify AAb from 4988 candidate tumor antigens to limit the false discovery rate inherent to large-scale proteomic screening. In the first phase, 53 cases and 53 controls (Cohort 1) were screened against all 4988 candidate tumor antigens. From this phase, 761 antigens were selected for further testing, while the remaining antigens were considered uninformative. In the second phase, 51 cases and 39 controls from a separate cohort were screened in the second phase against a NAPPA consisting of these 761 antigens. From this phase, 119 potential candidate AAb biomarkers were identified. A protein array made up of these 119 antigens was then screened using a cohort of 51 cases and 38 controls. This study confirmed the validity of 28 potential AAb biomarkers for the early detection of breast cancer.

Most of the top 28 antigen biomarkers identified using the protein array have previously been shown to be involved in breast cancer tumor biology and pathogenesis. Of particular note, the combined analysis of the array containing 761 and 119 antigens revealed ATPase, H+ transporting, lysosomal accessory protein 1 (ATP6AP1) as the most significant individual autoantigens detected. ATP6AP1 was previously identified in the screening of a melanoma cDNA library that was immunoblotted with postvaccination sera from a melanoma patient (21). The ATP6AP1 protein is overexpressed in multiple subtypes of breast cancer and has been detected in ER+ breast cancers (MCF-7 and ZR751 cell lines) as well as basal-like triple-negative breast cancers (BT549 and MDA-436) (22).

As mentioned in other chapters of this book, validation studies are absolutely critical to proving the veracity of any potential biomarker. To confirm the data from the protein microarrays, an independent set of sera ($n = 148$) from women with stage I ($n = 29$), stage II ($n = 70$), and stage III ($n = 49$) breast cancers were tested using an ELISA. Of the cases, 12.8% (19/148) had evidence of ATP6AP1 AAb, with a specificity of 95%. None of the negative controls showed specific AAb binding. While this level of validation is excellent for a research study, clinical validation requires larger, blinded cohorts of sera obtained from multiple institutions. Hopefully, future studies examining ATP6AP1 and the 27 other AAb biomarkers found in this study will yield a useful panel for detecting early stage breast cancers.

9.5   CAPTURE ARRAYS

When individuals think of protein arrays, they most likely envision capture arrays. These form the basis of diagnostic chips for detection of clinically important ligands, such as biomarkers, and arrays for expression profiling. Capture arrays employ high affinity

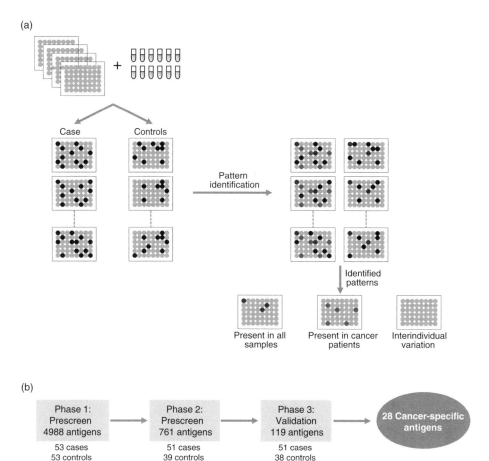

Figure 9.3. (a) Schematic of detection of AAb with NAPPA. Replicate NAPPAs expressing 4988 candidate tumor antigens were probed sera from patients with cancer (cases) and healthy controls. AAb detected in patient and control sera were compared with control sera to identify patterns of immune response. Antigens were classified based on specificity to identify AAb detected only in sera from cancer cases, which are selected for validation. (b) Identification of AAb in breast cancer sera was performed in three phases. The first phase screened case–control sera against a NAPPA expressing 4988 unique full-length cDNAs. In the second phase, a second cohort of sera was screened against a NAPPA expressing only the 761 unique full-length cDNAs that produced a positive result in the first phase. For the final phase, additional case–control sera were screened against a NAPPA of 119 expressed cDNAs. These studies resulted in the identification of 28 antigens that reacted specifically with AAb found within breast cancer sera.

reagent such as Abs (23), peptides (24), glycans (25), and nucleic acids (i.e., aptamers) (26) to bind to and detect ligands (primarily proteins) within biological samples. These types of arrays are also commonly known as forward-phase arrays since the affinity molecule is coupled to the surface of the slide and the sample to be interrogated is applied to the array.

9.5.1   Binding Molecules

Probably, the most popular choice of affinity reagents used in capture arrays are Abs (23, 27, 28). Besides custom designed Ab arrays prepared in labs all around the world, there are a number of commercially available arrays used to detect specific groups of proteins such as kinases. Antibodies for capture arrays are made either by conventional immunization (polyclonal sera and hybridomas) (29), or as recombinant single chain fragments (30), usually expressed in *E. coli*.

The first attempt to mimic DNA microarray analysis for measuring proteins used Abs coated onto microscope slides (31). The proteins that were incubated on these Ab arrays were labeled with amine-reactive forms of fluorescent dyes. Using an Ab array is quite straightforward as it requires incubating the fluorescently labeled protein sample onto the slide, washing to remove the unbound proteins, and detecting the captured proteins using a fluorescence scanning device. The major issue in dealing with any type of Ab-based detection system is the specificity of the Abs for their intended targets (32). Figure 9.4 illustrates three basic methods for detecting proteins in an array format. In the forward phase array, proteins are labeled with a probe and detected after binding to a single Ab that is coupled to a slide. Owing to the concern of cross-reactivity, many Ab arrays use a sandwich format. In a sandwich assay, two Abs that bind to different parts of the same protein are used. Both Abs must bind simultaneously for a signal to be

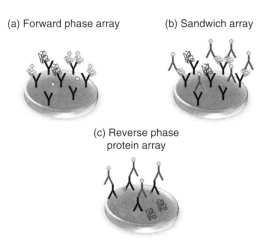

Figure 9.4. The three basic methods for detecting proteins in an array format.

observed. While it is highly probable that both Abs will cross-react with other proteins, the possibility that they cross-react with the same proteins is small. Unfortunately, there are often trade-offs when optimizing an assay. In this case, a sandwich format requires two distinct Abs and can result in increased background signal. In a RPPA (described below), two Abs are used. The primary Ab is for detecting the protein of interest within the spotted lysate, while the secondary Ab that binds to the primary Ab is labeled with a probe enabling its detection.

Antibody arrays have been used to detect protein expression signatures in serum from patients with a wide variety of different cancers (33–35). As with most technologies, when more laboratories become involved the technology improves. The most important improvements in Ab arrays have included the development of better surfaces for Ab immobilization (36), generation of phage-display Abs (37), and improved methods for labeling of proteins in biological samples such as plasma and tissue lysates (38).

The trend over the past few years was to increase the space that technologies survey. Conventional wisdom has been that more is better. Identifying more genes would lead to more single nucleotide polymorphisms that could be associated with cancers. Identifying more proteins would lead to the identification of a greater number of cancer biomarkers. This trend is as true for gene sequencing, proteomic analysis, and metabolomics as it is for protein arrays. However, this trend has not always proven to be efficient. A recent study bucked this trend and designed a targeted protein chip to quantify the levels of only a dozen cytoplasmic and membrane proteins. Not only was the array designed to measure only a few proteins, it made these measurements in single cells. This platform, called the single-cell barcode chip (SCBC), isolates a single (or defined number of cells) within an approximately 2 nL volume microchamber that contains an antibody array (39). The microchamber contains all of the elements necessary for cell lysis, as well as capture and detection of the protein panel.

The SCBC was used to study signal transduction in glioblastoma multiforme (GBM), a primary malignant brain tumor (40). While GBM has been genetically characterized, the key signaling pathways downstream of oncogenic mutations, such as epidermal growth factor receptor activating mutation (EGFRvIII) and phosphatase and tensin homolog (PTEN) tumor suppressor gene loss associated with receptor tyrosine kinase (RTK)/PI3K signaling, are incompletely understood. The 11 proteins quantified were (p indicates phosphorylation) p-Src, p-mammalian target of rapamycin (p-mTOR), p-p70 ribosomal protein S6 kinase (p-p70S6K), p-glycogen synthase kinase-3 (p-GSK-3α/β), p-p38 mitogen-activated protein kinase (p-p38α), p-extracellular-regulated kinase (p-ERK), p-c-Jun N-terminal kinase (p-JNK2), p-platelet-derived growth factor receptor β (p-PDGFRβ), p-vascular endothelial growth factor receptor 2 (p-VEGFR2), tumor protein 53 (P53), and total EGFR. All of these proteins are directly or potentially associated with PI3K signaling (Figure 9.5). The proteins were measured in three isogenic GBM cell lines: U87 (expressing wild-type p53, mutant PTEN, and low levels of wild-type EGFR, no EGFRvIII), U87 EGFRvIII (U87 cells stably expressing EGFRvIII deletion mutant), and U87 EGFRvIII PTEN (U87 cells coexpressing EGFRvIII and PTEN). The cells lines were analyzed under standard conditions, in response to EGF treatment, and after erlotinib treatment followed by EGF stimulation.

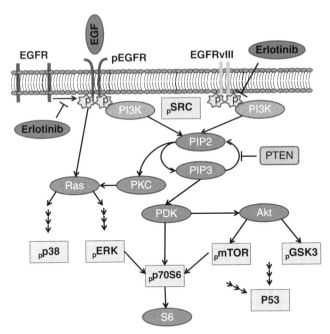

Figure 9.5. PI3K pathway activated by EGF-stimulated EGFR or by the constitutively activated EGFRvIII. All proteins in yellow background were assayed. EGFRvIII and PTEN (light blue) were expressed in the cell lines U87 EGFRvIII or U87 EGFRvIII PTEN. The oval, orange background components are the investigated molecular perturbations.

The SCBC is composed of a two-layer microfluidic network with valves that isolate the array into 120 microchambers (Figure 9.6). The microchambers are used for cell compartmentalization, cell lysis, and protein assays. After loading the cells, the number that enters each microchamber (typically zero to five) is counted through the transparent chip. A buffer diffuses into the microchamber from a neighboring well and lyses the cells. While the SCBC is an Ab array, it begins as 20-µm-wide DNA barcodes. After assembly, the DNA barcodes are converted into Ab barcodes using DNA-encoded Ab libraries. The captured proteins are detected using biotinylated Abs and fluorophore-labeled streptavidin. The antibody pairs were selected to detect only phosphorylated proteins, except in the cases of p53 and EGFR. All 11 proteins were detected in the single-cell experiments, except that p-PDGFRβ was detected only at low levels or not at all. Nine proteins (EGFR, p53, p-VEGFR2, p-ERK, p-p38α, p-GSK3α/β, p-p70S6K, p-mTOR, and p-Src) were detected in single U87 EGFRvIII PTEN cells that were stimulated with EGF.

So what did the actual analyses show? At basal level, U87 cells showed low p-EGFR levels and modest activation of the signaling proteins p-Src, p-mTOR, p-p70S6K, p-GSK3α/β, p-p38α, and p-ERK. U87 EGFRvIII cells showed increased baseline levels of p-Src, p-mTOR, p-p70S6K, p-ERK, and p-JNK2 compared with those expressing wild-type EGFR. In U87 EGFRvIII PTEN cells, coexpression of PTEN diminished

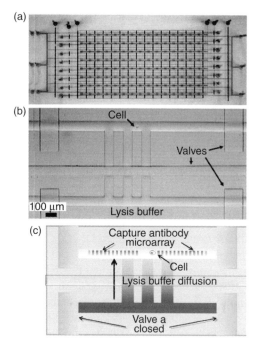

Figure 9.6. The SCBC protein array. (a) A photograph of an SCBC. The flow layer (red) and the control valve layer (blue) are delineated with food dyes. A photograph (b) and a drawing (c) of a single microchamber, with critical parts labeled. A cell is isolated in the cell chamber by the valves. The neighboring chamber contains cell lysis buffer. The duplicate DNA barcode copies are converted into an antibody array prior to cell loading, counting, and lysis. (Reproduced with permission from Reference 39.)

baseline phosphorylation of p-Src, p-mTOR, p-p70S6K, p-ERK, and p-JNK2 compared with U87 EGFRvIII. When the cells were treated with EGF, p-EGFR levels increased resulting in downstream pathway activation in all three cell lines, including activation of p-p70S6K and p-ERK. While EGF-dependent p-GSK3α/β levels were increased in U87 and U87 EGFRvIII cells, they remained unchanged in U87 EGFRvIII PTEN cells.

Erlotinib inhibition followed by EGF stimulation diminished both p-EGFR and p-EGFRvIII levels relative to EGF stimulation. This treatment exhibited little impact on U87 EGFRvIII cells, suggesting that PTEN loss confers resistance to EGFR tyrosine kinase inhibitors. The phosphoprotein expression levels decrease, but are above the unstimulated levels. Representative proteins include p-Src and p-p70S6K. Erlotinib treatment reduced p-ERK, p-p70S6K, p-mTOR, and p-Src levels only in U87 EGFRvIII PTEN cells. Taken together, these results agree with previous studies that have demonstrated that coexpression of EGFRvIII and PTEN protein within GBM cells is consistent with a response to EGFR kinase inhibitor therapy.

The results obtained using this Ab microarray were generally consistent with Western blotting of much larger cell numbers. The SCBC, however, provides certain advantages including the ability to accurately normalize cell numbers. The SCBC also permits cell-to-cell fluctuations to be quickly measured, enabling cellular heterogeneity to be recognized within tumors. In the future, this microarray could be expanded to an even greater number of proteins to provide a more complete mapping of protein interactions involved in GBM. Another obvious direction is to customize the array to measure proteins involved in the many other types of cancers that affect the general public. Hopefully, in the future devices such as these will become commonplace in the clinic bringing the medical community closer to realizing the dream of true translational medicine.

9.5.2 Aptamer Arrays

Aptamers are a popular choice among the nonprotein capture molecules that are used to manufacture arrays. Aptamers are single-stranded DNA or RNA that bind ligands with affinity constants within the range of Abs (41). Aptamers are, however, much smaller in size (e.g., 10 of bases), have greater conformational flexibility, and may be capable of binding a range of ligands far exceeding that of Abs. Aptamers have high selectivity and can be designed to distinguish similar chemical entities and posttranslationally modified proteins. Aptamers are produced using a process known as SELEX (systematic evolution of ligands by exponential enrichment) (42, 43) as shown in Figure 9.7. In this process, an oligonucleotide library is screened against a specific target to find nucleic acid sequences with high affinities. Those RNA or DNA sequences showing high affinity are iteratively selected and amplified via PCR. Since the development of SELEX in 1990, aptamer arrays have been developed for applications such as drug candidate validation, and therapeutic and diagnostic determination.

A recent study used an aptamer array to identify clinical protein biomarkers in serum obtained from lung cancer patients (44). As most people know, lung cancer is the leading cause of cancer deaths worldwide and unfortunately most cases are diagnosed at a stage where curative surgery is not possible. In this study, an aptamer array capable of measuring 813 protein targets starting with 15 μL of serum was used. A total of 1326 sera samples acquired from patients diagnosed with pathologic or clinical stage I-III non–small cell lung carcinoma (NSCLC) and a high-risk control population with a history of long-term tobacco were used. The control populations were selected randomly to represent the patient population at risk for lung cancer.

This multicenter case–control study used archived serum samples from four independent studies of NSCLC in long-term tobacco-exposed populations. A standard operating procedure was used to collect and process the sera to minimize variation from the different centers. Sera were collected from 291 NSCLC patients within 8 weeks of the first biopsy-proven lung cancer and prior to tumor removal by surgery. Control sera were collected from 1035 asymptomatic study participants who had smoked for at least 10 years. Quantitative analysis of 813 proteins using the aptamer array found 44 potential protein biomarkers that could distinguish NSCLC and asymptomatic patients. From these candidates, a 12-protein panel (Table 9.1) that discriminates NSCLC from controls was recognized. This panel correctly diagnosed NSCLC with 91% sensitivity

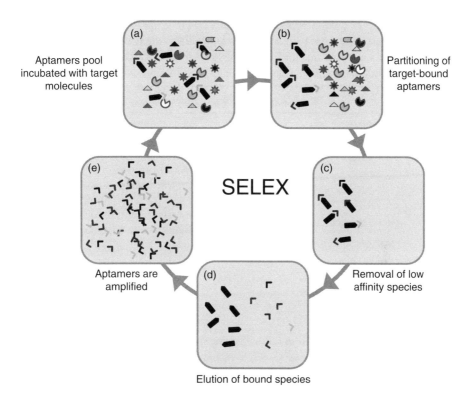

Aptamers pool incubated with target molecules

Partitioning of target-bound aptamers

SELEX

Aptamers are amplified

Removal of low affinity species

Elution of bound species

Figure 9.7. Creation of high affinity aptamers to target proteins using Systematic Evolution of Ligands by Exponential (SELEX) enrichment. (a) In the first step a pool of aptamers is incubated with target molecules. (b) Aptamers that bind to the target are separated and (c) low affinity species are removed from the mixture. (d) The bound aptamers are eluted from target protein. (e) These high affinity aptamers are then amplified. This pool of high affinity aptamers is then used as the starting point in subsequent SELEX cycles. Typically, 10 to 20 cycles of SELEX are conducted before the affinity parameters of the aptamer are fully characterized.

and 84% specificity in cross-validated training and 89% sensitivity and 83% specificity in a separate verification set. Interestingly, the panel had similar levels of sensitivity and specificity for both early and late stage NSCLC.

The 12 biomarkers identified in this study are involved in a variety of cellular functions including cell movement, inflammation, and immune monitoring. Most of these proteins have been associated generally with cancer biology. Some of these proteins have been identified as potential biomarkers (cadherin-1, endostatin, HSP90, and pleiotrophin); however, none have been validated as being clinically useful for diagnosing lung cancer (45, 46). The remaining eight proteins (CD30 ligand, LRIG3, MIP-4, PRKCI, RGM-C, SCF-sR, sL-Selectin, and YES) have not been previously suggested to be potential lung cancer biomarkers.

The authors pointed out some limitations within their study, both of which can be generally associated with most published biomarker discovery studies utilizing serum

TABLE 9.1. List of 12 Potential Protein Biomarkers that Discriminated Sera Obtained from NSCLC Patients and Control Sera Obtained from Individuals Who had Smoked for at Least 10 Years but were Asymptomatic for NSCLC. The Proteins were Identified Using an Aptamer Array that Measured the Abundance Level of 813 Proteins (44)

Potential Biomarker	Expression Level	Function
MIP-4	Up	Monokine
CD30 ligand	Up	Cytokine
Edostatin	Up	Inhibition of angiogenesis
HSP 90a	Up	Chaperone
Pleiotrophin	Up	Growth factor
PRKCI	Up	Serine/threonine kinase, oncogene
YES	Up	Tyrosine kinase, oncogene
LRIG3	Down	Tumor suppressor, protein binding
RGM-3	Down	Iron metabolism
sL-selectin	Down	Cell adhesion
Cadherin-1	Down	Cell adhesion, transcription regulation
SCF sR	Down	Decoy receptor

or plasma (44). Firstly, they did not show organ-specificity for the identified markers. Many of the biomarkers have been shown to be elevated in other cancers, and their data do not pinpoint the source of the proteins they measured. Making a connection between a biomarker found in a biofluid such as serum, plasma, and urine and its point of origin is nearly impossible in these types of discovery studies. Not only does the potential biomarker have tens of thousands of kilometers of veins, arteries, and capillaries to traverse prior to collection, but the biomarker is likely to be expressed within multiple sites in the body. Secondly, their findings were not validated in an independent set of clinical samples. While a multicenter study was used in this study to minimize the effects of potential preanalytical variability, it still does not eliminate all sources of variability.

9.6   REVERSE-PHASE PROTEIN MICROARRAYS

The second type of protein microarray that utilizes Abs for capture and/or detection are RPPAs (47). The name "reverse-phase" is used because cell lysates are immobilized onto the solid surface instead of affinity molecules such as Abs, aptamers, and lectins. After the lysate is immobilized, the arrays are interrogated using Abs targeting specific antigens (i.e., protein) within the lysate. The array format allows multiple replicates of the same lysate to be spotted permitting, numerous Abs to be tested. Conversely, a large number of different lysates can be spotted and tested against a single Ab. The amount of material required for printing (i.e., each spot is approximately 2 nL in volume) an entire array is so small that only a few thousand cells will provide enough lysate to create an entire RPPA. Therefore, laser-capture microdissection is often combined with RRPAs to focus the analysis strictly on cancer cells. The low sample requirement enables hundreds of patient samples to be printed on a single array.

9.6.1 Applications of Reverse-Phase Protein Arrays to Clinical Specimens

The most common use of RPPAs has been to target specific proteins in cancer tumors using a small number of Abs. Proteins that regulate cell growth, differentiation, and apoptosis are all targets of interest in the interrogation of human cancers. In a demonstration of the throughput and effectiveness of RPPAs, Paweletz et al. compared the proteomic signatures of prostate cancer samples obtained from 10 patients (48). Microdissected samples of histologically normal prostate epithelium, prostate intraepithelial neoplasia, and invasive prostate cancer were arrayed and interrogated with Abs directed against phosphorylated forms of Akt (pAkt) and extracellular regulated kinase (pERK). Akt and ERK, along with their phosphorylated forms, are of keen interest in cancer research owing to their associations with cell growth and differentiation (49). Both pAkt and pERK activate cell survival pathways and effect cell proliferation. Monitoring the phosphorylation status of tumor cells obtained from cancer patients pre- and posttreatment, may enable personalized therapeutic responses to be measured at early time points allowing the optimal treatment to be ascertained quickly.

The RPPA results showed a statistically significant increase in phosphorylation of Akt and a decrease in pERK levels in premalignant and invasive prostate cancer compared to the other samples (48). These results pointed to activation of prosurvival pathways within cases of invasive cancer. In concordance with this conclusion, the same study revealed that cleaved and noncleaved caspase-7 and poly-adenosine diphosphate ribose polymerase (downstream components of the apoptosis pathway) were shifted toward prosurvival functions during cancer progression. The cell survival proteins Akt, Erk1/2, and GSK3β, have also been profiled in ovarian tissue using RPPAs (50). One of the most striking features of this study was the heterogeneity that was observed amongst patients. The levels of pERK varied widely in tumors of the same histologic type; however, significant differences between histologic types were not observed. The data from this study show the potential application of RRPAs to personalized medicine as they suggested that cell-signaling pathway activation may be patient-specific rather than type-specific or tumor stage specific.

While RPPAs, along with many other types of proteomic technology, focus primarily on the discovery of diagnostic markers, finding biomarkers that predict response to therapy are also invaluable. A recent study used a RPPA to identify patterns of protein expression in patients with advanced high-grade serous ovarian cancer to determine if any correlations with Ca-125 normalization could be found (51). Normalization of Ca-125 levels after chemotherapy treatment is a significant predictor of recurrence-free and overall survival in patients with high-risk, early-stage epithelial ovarian cancer (52). The earlier a physician can predict therapeutic outcome, the better. Unfortunately, there are no clinically useful biomarkers that predict response to first-line chemotherapy and facilitate the use of investigational drugs earlier in treatment.

In this study, 5–25 mg of frozen tissue was homogenized, lysed, and normalized to a final total protein concentration of 1 μg/μL (51). The protein samples were boiled in denaturing buffer containing 1% SDS. Samples were serially diluted and spotted on nitrocellulose-coated glass slides using a robotic printer. To assess printing and staining quality, negative and positive controls were printed on each slide. Each arrayed slide was probed with a specific primary Ab followed by a secondary Ab as a means of amplifying the signal. The printed arrays were interrogated with a total of 80 different Abs.

Protein expression levels were not associated with stage, residuum, or age, with the exception of YKL-40, which correlated with stage. Proteins showing a significant association with Ca-125 normalization were EGFR, JNK, JNKp183_185 (JNK phosphorylated at residues 183 and 185), plasminogen activator inhibitor 1 (PAI-1), Smad3, TAZ, and YKL-40. All of these except Smad3, had higher levels associated with failure of Ca-125 normalization. Strangely, higher Smad3 expression was associated *with* Ca-125 normalization. When analyzed using a multivariate logistic regression analysis, the data revealed that only EGFR, JNK, and Smad3 were independent determinants of Ca-125 normalization. Assessment of the expression levels of these proteins had a sensitivity of 80%, with a specificity of 75%, and an overall accuracy of 78% for predicting Ca-125 normalization.

While the main objective was to find relationships between protein expression and Ca-125 normalization, the same data were analyzed to see whether any correlations exist between these proteins' levels and progression-free survival. Patients who fail to normalize Ca-125 by the third chemotherapy cycle have a much greater rate of progression. A univariate Cox regression analyses showed that higher JNK levels were significantly associated with a poorer progression-free survival, while increased EGFR and TAZ expression were both of borderline significance for adverse outcomes.

This study provided a number of significant steps forward both in technology and cancer treatment. It demonstrated that RPPAs can be used to develop predictors of chemoresistance in ovarian cancer. As well-designed and executed hypothesis-driven studies often do, this study also provided direction for follow-up studies. The data show that valuable prognostic information relating to Ca-125 normalization can be captured by measuring the levels of only three proteins at the time of diagnosis: Smad3, EGFR, and JNK. Ultimately, validating proteomic markers that can predict treatment outcome will allow alternative approaches to be attempted sooner rather than later. Since RPPAs require minimal amounts of samples and can concurrently test hundreds of patient samples, they represent an excellent platform for assessing therapeutic efficacy on a personal basis.

9.7 TISSUE MICROARRAYS

Recently, investigators have combined automation, Abs, tissue sections, and bioinformatics to create TMAs (53, 54). TMAs are essentially RPPAs without the cell lysis and are capable of conducting hundreds of immunohistochemical (IHC) experiments concurrently. In the construction of a TMA, either fresh frozen or formalin-fixed paraffin

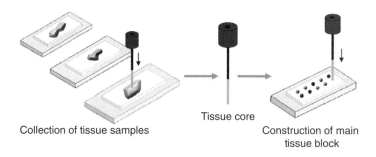

Collection of tissue samples

Tissue core

Construction of main
tissue block

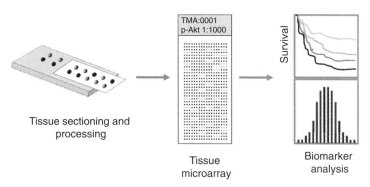

Tissue sectioning and
processing

TMA:0001
p-Akt 1:1000

Survival

Tissue
microarray

Biomarker
analysis

Figure 9.8. Construction of TMAs. Cores are taken from a series of tissue samples and deposited onto a single main tissue block. Thin sections are then cut from this block and after processing, attached to a glass slide. The tissue cores are subjected to IHC analysis using selected Abs. The staining results for these biomarkers can be compared to other clinical parameters such as survival data.

embedded tissues, can be used (Figure 9.8). Specific regions of interest are cored from each tissue block with the diameters of the cores typically being 0.6–2.0 mm. The core is sectioned into thin slices, which are transferred onto a glass slide for automated IHC analysis using specific Abs of interest. The results of the IHC analysis can be assessed either manually or automated and linked to other available clinical data, such as prediction of patient outcome, type of therapy provided, and stage of diagnosis. The obvious advantage of TMAs is the high degree of precision and throughput it provides.

9.7.1   Many Tissues, One Antibody

There are two popular modes for using TMAs; both of which are directly a consequence of their high throughput. In one mode, a single Ab is used to interrogate a very large number of tissues. For example, one study measured the levels of Ets-related gene (ERG) in 11,483 cancer tumors and 72 different normal tissue types using a TMA format (55). The abundance of ERG was found to be high in the nucleus of 36.7% of prostate carcinomas as well as Kaposi sarcomas (91.7%), angiosarcomas (100%) and hemangiomas (90.9%). Moderate to strong nuclear ERG immunostaining was observed in thymoma (6.1%), while only low levels of ERG levels were detected in skin squamous cell carcinomas, lung squamous carcinomas, malignant mesotheliomas, uterine carcinosarcomas, gastrointestinal stromal tumors, hepatocellular carcinomas, testicular teratomas, thyroid anaplastic carcinomas, tendon sheath giant cell tumors, and benign skin fibrous histiocytomas. None of the other 8886 samples from 132 other tumor types and subtypes showed ERG overexpression. Within normal tissues, ERG overexpression was restricted to endothelial cells and subsets of lymphocytes. Having the ability to conduct a large number of hypothesis-driven studies on so many different tissue types enabled cancer researchers to begin to pinpoint a role for ERG in a narrow group of tumors.

9.7.2 Many Tissues, Many Antibodies

Another popular use of TMAs is to screen large numbers of tissues for multiple antigens. This strategy is exemplified by a study that examined the levels of the eukaryotic translation initiation factor complex 4E (eIF4E) and its relationship with the PTEN/Akt and Ras/Mek/ERK pathways in NSCLC (56). More than 75–85% of lung cancers are NSCLC at diagnosis and lung cancer remains the leading cause of cancer-related deaths worldwide. In this study, p-eIF4E, p-Akt, PTEN, phosphorylated tuberin (p-TSC2), phosphorylated mTOR (p-mTOR), phosphorylated S6 (p-S6), and phosphorylated Erk1/2 (p-Erk1/2) were measured in 300 NSCLCs via IHC using a TMA.

The Akt pathway regulates many diverse biological functions and p-Akt results in gain or loss of function of its downstream proteins whose deregulation can promote tumorigenesis in many cancers such as brain, breast, prostate, and endometrium (57–59). Many of these p-Akt responsive proteins are attractive targets for therapy, and several inhibitors are being developed and clinically tested (60,61). Similar to the Akt pathway, the RAS/MEK/ERK pathway, specifically Erk1/2, regulates cell proliferation and cell survival (62). Although its associated with advanced disease, the relationship between p-Erk1/2 abundance and the downstream mediators of the Akt pathway in NSCLC have not been reported (63). Besides p-Erk1/2, Akt's relationship with mTOR status has also not been well documented. mTOR is a protein kinase that controls cell growth, proliferation, and survival and there is great interest in using rapamycin or other homologues to inhibit mTOR's ability to control cell growth and proliferation in cancers (64).

This TMA study analyzed 300 surgically resected primary lung cancer tumor samples from patients that had not received neoadjuvant therapy (56). These 300 tumor samples included 150 adenocarcinomas and 150 squamous cell carcinomas, which were compared with 100 adjacent normal pulmonary tissues in the same tissue block. Representative tumor areas were selected based on the matched hematoxylin and eosin (H&E)-stained slides. A 0.6 mm in diameter core was retrieved from the tissue sample at the selected region and extruded directly into coordinated wells within the recipient block. A series of 4-μm sections were cut with a microtome from the tissue block and eventually transferred to charged glass slides. Prior to IHC staining, adhesive tapes were peeled off the glass slides immediately after immersing the slide in TPC solvent. As a control, every 50th tissue section was H&E stained and examined for histopathology and tissue retention. The tissue samples were stained using Abs specific to the following seven proteins: p-Akt, PTEN, p-TSC2, p-mTOR, p-S6, p-eIF4E, and p-Erk1/2. The tumors were graded histologically according to WHO guidelines and staged according to the International Union against Cancer's tumor-node-metastasis classification (65). Survival time and outcome data were available for 84% of the patients with a median follow-up of 3.4 years.

The staining results showed that a positive p-Akt phenotype was found in 78.8% of NSCLC tumors and that this activation of Akt correlates with a poor prognosis. No correlation, however, was observed between Akt activation and tumor stage. Positive p-mTOR expression was observed in 46.7% of total NSCLC tumors with a positive phenotype being observed in squamous cell carcinoma when compared with adenocarcinoma. These results suggested that activation of mTOR may contribute to tumor progression in lung squamous cell carcinoma; something that had not been postulated previously.

The Ab-staining results were clustered into phenotypes based on the coexpression of various proteins. For example, expression of p-Akt correlated significantly with expression of p-TSC2, p-S6, and p-eIF4E, but not with expression of p-mTOR expression. Both p-TSC2 and p-eIF4E expression was strongly correlated. Phenotypes characterized by positive p-Akt, p-mTOR, p-S6, and p-eIF4E expression had significantly lower survival rates than the overall survival rate of all cases. Another phenotype classified as showing negative p-Akt, p-TSC2, p-eIF4E, and p-S6 expression but positive p-mTOR expression, had significantly higher survival rates than the overall survival rate. Taken together, these phenotypic classes suggest that a pathway could exist in which Akt regulates TSC2 and eIF4E, dependent or independent of mTOR.

Erk1/2, a member of the RAS/MEK/ERK pathway, is known to contribute to tumorigenesis in many cancers (66), and this study found overexpression of p-Erk1/2 in 16.6% of NSCLCs. Expression of p-Erk1/2 strongly correlated with expression of p-TSC2 and p-S6, but not with the expression of p-eIF4E. With other studies suggesting that p-Erk suppresses TSC2 activation of S6 kinase resulting in Akt-independent cellular proliferation and transformation, this TMA study indicated that Erk may be regulating S6 through TSC2 in NSCLC. It was interesting to find that the TMA results did not show a correlation between p-Akt expression and p-Erk1/2 expression, considering that p-Akt and p-Erk1/2 strongly correlated with p-TSC2 and p-S6. Overall, the results suggest that Akt and Erk1/2 act independently in NSCLC tumorigenesis and survival through TSC2. A more favorable prognosis was observed in the phenotypic cluster that included cases with loss of expression of p-Akt and p-Erk1/2, than the other groups, suggesting that Akt and Erk1/2 are important survival factors for NSCLC patients.

The results from this multitissue/multi-Ab TMA study showed for the first time that TSC2, eIF4E, and S6 are more likely to be phosphorylated in tumors with p-Akt in NSCLC, yet no correlation between p-mTOR with p-Akt and p-TSC2 was observed. In addition, it showed that eIF4E overexpression is an independent prognostic factor in NSCLC. These findings indicate that in NSCLC, TSC2, eIF4E, and S6 are likely regulated through Akt activation that is independent of mTOR. Obviously, dysregulation of the Akt pathway plays an important role in NSCLC development but ultimately selecting patients for treatment with mTOR inhibitors based on activation of Akt signaling pathways may not be an optimum selection criteria. While most of the interesting findings were based on comparing expression levels of multiple proteins, it was also found that p-eIF4E expression levels could differentiate stage III or IV NSCLC cases from stages I and II. Overexpression of eIF4E was strongly associated with stages III and IV NSCLC and hence worse prognoses for these patients. While other studies have shown that eIF4E expression is correlated with clinical outcomes and histologic malignancy in a number of cancers (e.g., breast, bladder, colorectum, uterine cervix, and non-Hodgkin lymphoma) (67,68), this TMA study was the first to show its potential role in diagnosis and prognosis of NSCLC.

9.8 CONCLUSIONS

In any field of science, no single tool will provide all of the answers. While proteomics is dominated by MS, other technologies such as protein arrays play a critical role in furthering the application of proteomics to cancer research. The major advantage that protein arrays have over MS methods is their ability to target specific proteins. Over the past decade, there has been a focus on characterizing more and more proteins within the proteome. Unfortunately, our ability to discern information has not kept up with our ability to generate data. The next era in cancer proteomics will likely focus on measuring a select group of proteins considered important to tumorigenesis and cancer progression. At the 2012 American Association for Cancer Research meeting in Washington, D.C., the esteemed cancer researcher Dr. Bert Vogelstein presented his thoughts on continuing massive sequencing projects to identify genes behind cancers (69). Dr. Vogelstein described how his group searched databases for all genetic mutations found in solid cancers over the past 20 years. They found 130,072 mutations in 3142 genes across 353 cancer subtypes. To identify which of these actually contribute to cancer, they assumed that the mutations in suppressor genes had to truncate the gene's protein. They also assumed that a mutation within an oncogene had to be seen at least twice in tumors for it to be involved in cancer. The net result was that only 296 tumor suppressors and 33 oncogenes were concluded to be actually involved in cancer. Classification of these 329 genes showed that almost all of them can be grouped into 12 signaling pathways. Dr. Vogelstein also hypothesized that even as thousands more tumor samples are sequenced it is unlikely that these numbers—329 genes and 12 pathways— are going to appreciably change. The logical conclusion is to target drugs on the proteins within these pathways argued Dr. Vogelstein. Beyond drug design, it also makes sense to focus our analytical methods for specifically measuring features (i.e., abundance, posttranslational state, and interacting partners) associated with these proteins. With the ability to spot specific affinity reagents or interrogate samples with targeted Abs, designer protein microarrays can play a major role in analyzing select proteins in proteomic samples. One could envision customized arrays designed to measure the proteins that comprise each of the 12 pathways suggested by Dr. Vogelstein as critical in cancer.

Another important goal is to bring proteomics technology to the clinic. While many investigators are focused in developing MS-based methods, it is more likely that variations of protein arrays will be more amenable to clinical applications. While multiple reaction monitoring-MS methods are being developed for proteins, this technique is proving to be challenging and may require significant sample preparation prior to MS measurement (70). In addition, a single mass spectrometer can only measure one sample at a time, resulting in a severely limited throughput. Protein arrays, on the other hand, require little sample preparation and can measure thousands of samples concurrently. As described for a number of projects in this chapter, very little biological sample is required when using arrays making them particularly attractive for analyzing tissue biopsies. Protein arrays have a number of technical hurdles still to overcome; however, concurrent developments in protein expression and purification as well as optimization of affinity reagents such as Abs and aptamers will contribute to continuing improvement in this important proteomics technology.

REFERENCES

1. Jensen ON. Modification-specific proteomics: Characterization of post-translational modifications by mass spectrometry. Curr. Opin. Chem. Biol. 2004;8:33–41.

2. Ekins RP. Multi-analyte immunoassay. J. Pharm. Biomed. Anal. 1989;7:155–168.

3. Ekins RP. Ligand assays: From electrophoresis to miniaturized microarrays. Clin. Chem. 1998;44:2015–2030.

4. Katz C, Levy-Beladev L, Rotem-Bamberger S, Rito T, Rüdiger SG, Friedler A. Studying protein-protein interactions using peptide arrays. Chem. Soc. Rev. 2011;40:2131–2145.

5. Tomizaki KY, Usui K, Mihara H. Protein-protein interactions and selection: Array-based techniques for screening disease-associated biomarkers in predictive/early diagnosis. FEBS J. 2010;277:1996–2005.

6. Chandra H, Reddy PJ, Srivastava S. Protein microarrays and novel detection platforms. Expert. Rev. Proteomics. 2011;8:61–79.

7. Horak CE, Snyder M. Global analysis of gene expression in yeast. Funct. Integr. Genomics. 2002;2:171–180.

8. Gong P, Grainger DW. Nonfouling surfaces: A review of principles and applications for microarray capture assay designs. Methods Mol. Biol. 2007;381:59–92.

9. Rusmini F, Zhong Z, Feijen J. Protein immobilization strategies for protein biochips. Biomacromolecules 2007;8:1775–1789.

10. Ayoglu B, Häggmark A, Neiman M, Igel U, Uhlén M, Schwenk JM, Nilsson P. Systematic antibody and antigen-based proteomic profiling with microarrays. Expert Rev. Mol. Diagn. 2011;11:219–234.

11. Berrade L, Garcia AE, Camarero JA. Protein microarrays: Novel developments and applications. Pharm. Res. 2011;28:1480–1499.

12. Stoll D, Templin MF, Schrenk M, Traub PC, Vöhringer CF, Joos TO. Protein microarray technology. Front. Biosci. 2002;7:c13–c32

13. He M, Stoevesandt O, Taussig MJ. In situ synthesis of protein arrays. Curr. Opin. Biotechnol. 2008;19:4–9.

14. Ramachandran N, Raphael JV, Hainsworth E, Demirkan G, Fuentes MG, Rolfs A, Hu Y, LaBaer J. Next-generation high-density self-assembling functional protein arrays. Nat. Methods 2008;5:535–538.

15. Angenendt P, Kreutzberger J, Glokler J, Hoheisel JD. Generation of high density protein microarrays by cell-free in situ expression of unpurified PCR products. Mol. Cell. Proteomics 2006;5:1658–1666.

16. Tao SC, Zhu H. Protein chip fabrication by capture of nascent polypeptides. Nat. Biotechnol. 2006;24:1253–1254.

17. He M, Stoevesandt O, Palmer EA, Khan F, Ericsson O, Taussig MJ. Printing protein arrays from DNA arrays. Nat. Methods 2008;5:175–177.

18. Anderson KS, Sibani S, Wallstrom G, Qiu J, Mendoza EA, Raphael J, Hainsworth E, Montor WR, Wong J, Park JG, Lokko N, Logvinenko T, Ramachandran N, Godwin AK, Marks J, Engstrom P, Labaer J. Protein microarray signature of autoantibody biomarkers for the early detection of breast cancer. J. Proteome Res. 2011;10:85–96.

19. Järås K, Anderson K. Autoantibodies in cancer: Prognostic biomarkers and immune activation. Expert Rev. Proteomics 2011;8:577–589.

20. Naour FL, Brichory F, Beretta L, Hanash SM. Identification of tumor-associated antigens using proteomics. Technol. Cancer Res. Treat. 2002;1:257–262.

21. Hodi FS, Schmollinger JC, Soiffer RJ, Salgia R, Lynch T, Ritz J, Alyea EP, Yang J, Neuberg D, Mihm M, Dranoff G. ATP6S1 elicits potent humoral responses associated with immune-mediated tumor destruction. Proc. Natl. Acad. Sci. U.S.A. 2002;99:6919–6924.

22. Zhao H, Langerod A, Ji Y, Nowels KW, Nesland JM, Tibshirani R, Bukholm IK, Karesen R, Botstein D, Borresen-Dale AL, Jeffrey SS. Different gene expression patterns in invasive lobular and ductal carcinomas of the breast. Mol. Biol. Cell 2004;15:2523–2236.

23. Hause RJ, Kim HD, Leung KK, Jones RB. Targeted protein-omic methods are bridging the gap between proteomic and hypothesis-driven protein analysis approaches. Expert Rev. Proteomics 2011;8:565–575.

24. Katz C, Levy-Beladev L, Rotem-Bamberger S, Rito T, Rüdiger SG, Friedler A. Studying protein-protein interactions using peptide arrays. Chem. Soc. Rev. 2011;40:2131–2145.

25. Rillahan CD, Paulson JC. Glycan microarrays for decoding the glycome. Annu. Rev. Biochem. 2011;80:797–823.

26. Rowe W, Platt M, Day PJ. Advances and perspectives in aptamer arrays. Integr. Biol. (Camb). 2009;1:53–58.

27. Borrebaeck CA, Wingren C. Recombinant antibodies for the generation of antibody arrays. Methods Mol. Biol. 2011;785:247–262.

28. Ayoglu B, Häggmark A, Neiman M, Igel U, Uhlén M, Schwenk JM, Nilsson P. Systematic antibody and antigen-based proteomic profiling with microarrays. Expert Rev. Mol. Diagn. 2011;11:219–234.

29. Charlermroj R, Oplatowska M, Kumpoosiri M, Himananto O, Gajanandana O, Elliott CT, Karoonuthaisiri N. Comparison of techniques to screen and characterize bacteria-specific hybridomas for high-quality monoclonal antibodies selection. Anal. Biochem. 2012;421: 26–36.

30. Ramirez AB, Loch CM, Zhang Y, Liu Y, Wang X, Wayner EA, Sargent JE, Sibani S, Hainsworth E, Mendoza EA, Eugene R, Labaer J, Urban ND, McIntosh MW, Lampe PD. Use of A single-chain antibody library for ovarian cancer biomarker discovery. Mol. Cell. Proteomics 2010;9:1449–1460.

31. Haab BB, Dunham MJ, Brown PO. Protein microarrays for highly parallel detection and quantitation of specific proteins and antibodies in complex solutions. Genome Biol. 2001; 2:1–13.

32. Schwenk JM, Nilsson P. Assessment of antibody specificity using suspension bead arrays. Methods Mol. Biol. 2011;785:183–189.

33. Gallotta A, Orzes E, Fassina G. Biomarkers quantification with antibody arrays in cancer early detection. Clin. Lab. Med. 2012;32(1):33–45.

34. Sanchez-Carbayo M. Antibody microarrays as tools for biomarker discovery. Methods Mol. Biol. 2011;785:159–182.

35. Kopf E, Zharhary D. Antibody arrays–an emerging tool in cancer proteomics. Int. J. Biochem. Cell Biol. 2007;39:1305–1317.

36. Sung D, Park S, Jon S. Facile immobilization of biomolecules onto various surfaces using epoxide-containing antibiofouling polymers. Langmuir 2012;28:4507–4514.

37. Lin M, McRae H, Dan H, Tangorra E, Laverdiere A, Pasick J. High-resolution epitope mapping for monoclonal antibodies to the structural protein ERNS of classical swine fever virus using peptide array and random peptide phage display approaches. J. Gen. Virol. 2010;91:2928–2940.

38. Schröder C, Alhamdani MS, Fellenberg K, Bauer A, Jacob A, Hoheisel JD. Robust protein profiling with complex antibody microarrays in a dual-colour mode. Methods Mol. Biol. 2011;785:203–221.

39. Shi Q, Qin L, Wei W, Geng F, Fan R, Shin YS, Guo D, Hood L, Mischel PS, Heath JR. Single-cell proteomic chip for profiling intracellular signaling pathways in single tumor cells. Proc. Natl. Acad. Sci. U.S.A. 2012;109:419–424.

40. Kesari S. Understanding glioblastoma tumor biology: The potential to improve current diagnosis and treatments. Semin. Oncol. 2011;38:S2–S10.

41. Tuerk C, Gold L. Systematic evolution of ligands by exponential enrichment: RNA ligands to bacteriophage T4 DNA polymerase. Science 1990;249:505–510.

42. Brody EN, Gold L. Aptamers as therapeutic and diagnostic agents. J. Biotechnol. 2000;74: 5–13.

43. Gold L, Brown D, He Y, Shtatland T, Singer BS, Wu Y. From oligonucleotide shapes to genomic SELEX: Novel biological regulatory loops. Proc. Natl. Acad. Sci. U.S.A. 1997;94: 59–64.

44. Ostroff RM, Bigbee WL, Franklin W, Gold L, Mehan M, Miller YE, Pass HI, Rom WN, Siegfried JM, Stewart A, Walker JJ, Weissfeld JL, Williams S, Zichi D, Brody EN. Unlocking biomarker discovery: Large scale application of aptamer proteomic technology for early detection of lung cancer. PLoS One 2010;5:e15003.

45. Greenberg AK, Lee MS. Biomarkers for lung cancer: clinical uses. Curr. Opin. Pulm. Med. 2007;13:249–255.

46. Sung HJ, Cho JY. Biomarkers for the lung cancer diagnosis and their advances in proteomics. BMB Rep. 2008;41:615–625.

47. Speer R, Wulfkuhle J, Espina V, Aurajo R, Edmiston KH, Liotta LA, Petricoin EF 3rd. Development of reverse phase protein microarrays for clinical applications and patient-tailored therapy. Cancer Genomics Proteomics 2007;4:157–164.

48. Paweletz CP, Charboneau L, Bichsel VE, Simone NL, Chen T, Gillespie JW, Emmert-Buck MR, Roth MJ, Petricoin EF 3rd, Liotta LA. Reverse phase protein microarrays which capture

disease progression show activation of pro-survival pathways at the cancer invasion front. Oncogene 2001;20:1981–1989.

49. Gan Y, Shi C, Inge L, Hibner M, Balducci J, Huang Y. Differential roles of ERK and Akt pathways in regulation of EGFR-mediated signaling and motility in prostate cancer cells. Oncogene 2010;29:4947–4958.

50. Wulfkuhle JD, Aquino JA, Calvert VS, Fishman DA, Coukos G, Liotta LA, Petricoin EF 3rd. Signal pathway profiling of ovarian cancer from human tissue specimens using reverse-phase protein microarrays. Proteomics 2003;3:2085–2090.

51. Carey MS, Agarwal R, Gilks B, Swenerton K, Kalloger S, Santos J, Ju Z, Lu Y, Zhang F, Coombes KR, Miller D, Huntsman D, Mills GB, Hennessy BT. Functional proteomic analysis of advanced serous ovarian cancer using reverse phase protein array: TGF-beta pathway signaling indicates response to primary chemotherapy. Clin. Cancer Res. 2010;16:2852–2860.

52. Skaznik-Wikiel ME, Sukumvanich P, Beriwal S, Zorn KK, Kelley JL, Richard SD, Krivak TC. Possible use of CA-125 level normalization after the third chemotherapy cycle in deciding on chemotherapy regimen in patients with epithelial ovarian cancer: Brief report. Int. J. Gynecol. Cancer 2011;21:1013–1017.

53. Hewitt SM. Tissue microarrays as a tool in the discovery and validation of tumor markers. Methods Mol. Biol. 2009;520:151–161.

54. Takikita M, Chung JY, Hewitt SM. Tissue microarrays enabling high-throughput molecular pathology. Curr. Opin. Biotechnol. 2007;18:318–325.

55. Minner S, Luebke AM, Kluth M, Bokemeyer C, Jänicke F, Izbicki J, Schlomm T, Sauter G, Wilczak W. High level of Ets-related gene expression has high specificity for prostate cancer: a tissue microarray study of 11483 cancers. Histopathology 2012;61:445–453.

56. Yoshizawa A, Fukuoka J, Shimizu S, Shilo K, Franks TJ, Hewitt SM, Fujii T, Cordon-Cardo C, Jen J, Travis WD. Overexpression of phospho-Eif4e is associated with survival through Akt pathway in non-small cell lung cancer. Clin. Cancer Res. 2010;16:240–248.

57. Hennessy BT, Smith DL, Ram PT, Lu Y, Mills GB. Exploiting the PI3K/Akt pathway for cancer drug discovery. Nat. Rev. Drug Discov. 2005;4:988–1004.

58. Riemenschneider MJ, Betensky RA, Pasedag SM, Louis DN. Akt activation in human glioblastomas enhances proliferation via TSC2 and S6 kinase signaling. Cancer Res. 2006;66:5618–5623.

59. Thompson JE, Thompson CB. Putting the rap on Akt. J. Clin. Oncol. 2004;22:4217–4226.

60. Huang WC, Chen YJ, Li LY, Wei YL, Hsu SC, Tsai SL, Chiu PC, Huang WP, Wang YN, Chen CH, Chang WC, Chang WC, Chen AJ, Tsai CH, Hung MC. Nuclear translocation of epidermal growth factor receptor by Akt-dependent phosphorylation enhances breast cancer-resistant protein expression in gefitinib-resistant cells. J. Biol. Chem. 2011;286:20558–20568.

61. Gills JJ, Dennis PA. Perifosine: Update on a novel Akt inhibitor. Curr. Oncol. Rep. 2009;11:102–110.

62. Alvarez RH, Valero V, Hortobagyi GN. Emerging targeted therapies for breast cancer. J. Clin. Oncol. 2010;28:3366–3379.

63. Han SW, Kim TY, Hwang PG, Jeong S, Kim J, Choi IS, Oh DY, Kim JH, Kim DW, Chung DH, Im SA, Kim YT, Lee JS, Heo DS, Bang YJ, Kim NK. Predictive and prognostic impact of epidermal growth factor receptor mutation in non-small-cell lung cancer patients treated with gefitinib. J. Clin. Oncol. 2005;23:2493–2501.

64. Fasolo A, Sessa C. Targeting mTOR pathways in human malignancies. Curr. Pharm. Des. 2012;18:2766–2777.

65. UICC, International Union Against Cancer. *TNM Classification of Malignant Tumours*. 6th ed. New York: Wiley and Sons; 2002.

66. Sosa MS, Avivar-Valderas A, Bragado P, Wen HC, Aguirre-Ghiso JA. ERK1/2 and P38α/β signaling in tumor cell quiescence: Opportunities to control dormant residual disease. Clin. Cancer Res. 2011;17:5850–5857.

67. Choi CH, Lee JS, Kim SR, Lee YY, Kim CJ, Lee JW, Kim TJ, Lee JH, Kim BG, Bae DS. Direct inhibition of Eif4e reduced cell growth in endometrial adenocarcinoma. J. Cancer Res. Clin. Oncol. 2011;137:463–469.

68. Inamdar KV, Romaguera JE, Drakos E, Knoblock RJ, Garcia M, Leventaki V, Medeiros LJ, Rassidakis GZ. Expression of eukaryotic initiation factor 4E predicts clinical outcome in patients with mantle cell lymphoma treated with hyper-CVAD and rituximab, alternating with rituximab, high-dose methotrexate, and cytarabine. Cancer 2009;115:4727–4736.

69. http://news.sciencemag.org/scienceinsider/2010/04/a-skeptic-questions-cancer-genom.html.

70. Issaq HJ, Veenstra TD. Would you prefer multiple reaction monitoring or antibodies with your biomarker validation? Expert Rev. Proteomics 2008;5:761–763.

10

THE ROLE OF PROTEOMICS IN PERSONALIZED MEDICINE

10.1 INTRODUCTION

This era of discovery-driven science has spawned a number of novel terms. It seems that the suffix "omics" has been added to almost every class of biological molecule (e.g., genomics, proteomics, lipidomics, glycomics, etc.) as well as many branches of study (e.g., psychogenomics, nutrigenomics, pharmacoproteomics, etc.). Discovery science has also resulted in a number of terminologies describing novel ways of practicing medicine. Two of the more well known phrases are "bench-to-bedside" and "personalized medicine." According to the U.S. National Cancer Institute, bench-to-bedside is "the process by which the results of research done in the laboratory are directly used to develop new ways to treat patients," while personalize medicine is "a form of medicine that uses information about a person's genes, proteins, and environment to prevent, diagnose, and treat disease" (1). Personalized medicine and bench-to-bedside are essentially two sides of the same coin with bench-to-bedside being the action that leads to personalized medicine.

While the term personalized medicine has become popular within the last decade, its origins go back thousands of years. Hippocrates, the Father of Western Medicine, practiced personalized medicine (2, 3). In his works, Hippocrates wrote *"there are great natural diversities"* and *"a great difference in the constitution of individuals."* These

Proteomic Applications in Cancer Detection and Discovery, First Edition. Timothy D. Veenstra.
© 2013 John Wiley & Sons, Inc. Published 2013 by John Wiley & Sons, Inc.

insights led Hippocrates to recognize that "*different [drugs] to different patients, for the sweet ones do not benefit everyone, nor do the astringent ones, nor are all the patients able to drink the same things.*" Although most of us do not think about it, doctors today practice personalized medicine. When we enter a doctor's office for an initial visit, the first thing we are requested to do is fill out a long form asking us questions about our lifestyle, physical condition, and family health history. The doctor uses this information to determine his treatment or provide us advice on lifestyle choices.

The major thrust today is to take personalized medicine to a much deeper molecular level. During his time, Hippocrates believed that disease was a result of four fluids (or humors) that made up the body: blood, phlegm, yellow bile, and black bile (4). These humors were matched to autumn (black bile), winter (phlegm), spring (blood), and summer (yellow bile), and also associated with the four universal elements of earth, air, water, and fire. Obviously, we know a lot more about the physical makeup of the human body today and the information gained through the Human Genome Project has been the major impetus driving the movement toward advanced personalized medicine. Genomic information, in the form of DNA bases, does not (under normal circumstances) change over a person's lifetime potentially allowing disease susceptibility and risk to be assessed at an early age. A simple example is Huntington's Disease (HD). Persons affected by this disease possess an expanded CAG trinucleotide repeat on HD chromosomes that can be detected through a genetic test (5). This genetic test is highly sensitive and very specific and indicates whether or not the individual will develop this condition. The CAG repeat size ranges between 36 and 121 in affected persons and its length is inversely correlated with the age of HD onset (6). There are a number of other genetic tests (e.g., Down's syndrome and mutations in BRCA1 or BRCA2) that can be considered as a form of personalized medicine as they accurately predict a person's susceptibility to a specific condition.

Beyond using it to determine a patient's susceptibility to or diagnose a disease, a major aim of personalized medicine is to predict drug efficacy at the individual level. The general consensus amongst people is that drugs designed to treat a specific condition work for everyone. Unfortunately, this scenario is not reality. A drug that is efficacious in some patients may fail to have the desired effect in others or cause a variety of unpredictable side effects. While this variable response has been recognized for a very long time, understanding its origin has proven elusive. The effort to find ways to accurately predict a drug's response on the individual is of paramount importance in modern medicine and would have a major positive impact on clinical use. Consider a drug that is efficacious for 10% of individuals with stage III lung cancer but causes heart disease in 90% of those also taking the drug. Unfortunately, the reason why it is effective in 10% of the population, but harms the remaining 90%, is unknown. This drug would never gain Food and Drug Administration (FDA) approval for use unless the origin of its success in 10% of the population could be accurately determined and the drug prescribed only to this group. The unpredictability of the side effect would prevent 10% of patients with stage III lung cancer from having access to a treatment that could save their lives. This example is just one of the scenarios that illustrates why personalized medicine is so important.

Variations in drug response are complex and likely involve a combination of different factors. These factors include other drugs the patient is taking, foods or beverages the patient is consuming, and another disease the patient may be suffering from. A person's genetic makeup can affect what happens to a drug when taken by the patient. Some people may metabolize a drug slowly causing it to accumulate in their body resulting in toxicity. Others may metabolize the drug so quickly that the normal dose does not provide enough of the drug to ever make it effective. Although the completion of the human genome project was a major step forward in personalized medicine, genetic variation was already known to be a critical factor in drug response in late 1950s (7, 8). These studies recognized that treated patients that had very high or low concentrations of drugs in their plasma or urine had specific phenotypes. They also found that the biochemical traits related to the drug response were inherited. Another observation that demonstrates the genetic basis of drug response is that the individual variation to a drug treatment within a population is larger than within a single individual who takes a specific drug at different times. Simply put, if aspirin successfully treats your headache today, it will successfully treat your headache tomorrow. However, aspirin may not help your neighbor and would not help your neighbor tomorrow either.

As mentioned, development of sequencing tools to characterize the human genome has enabled the recognition of how alterations in drug target proteins, drug-metabolizing enzymes, and drug transporters can alter drug efficacy and/or side effects resulting in variation in individual drug responses (9–11). This area, known as pharmacogenomics, is increasingly influencing the fields of clinical medicine, drug development, drug regulation, pharmacology, and toxicology. As I mentioned in an earlier chapter, I am a big advocate that to understand something, it is important to see how it is applied. An example of how pharmacogenomics has impacted drug use is illustrated by simvastatin, a HMG-CoA reductase inhibitor (12). Simvastatin is given to patients to lower their blood cholesterol level. Unfortunately, the efficacy and safety of this drug shows wide individual variation. In one study, 156 subjects with low-density lipoprotein (LDL) cholesterol levels of >160 mg/dL, were treated with simvastatin for 6 weeks at doses of 40, 80, and 160 mg/day. The result was a median reduction of LDL cholesterol at the three doses of 41%, 47%, and 53%, respectively. Plasma triglyceride levels were also lowered in these groups by 21%, 23%, and 33%, respectively. While simvastatin treatment was highly effective in reducing LDL cholesterol in many patients, approximately 5% of patients had no significant reduction in LDL cholesterol levels and many demonstrated side effects including diarrhea, constipation, vomiting, myalgia, and back pain. A few subjects exhibited elevated plasma alanine aminotransferase activities at high simvastatin doses indicating liver damage. Pharmacogenomic studies have suggested that the variation in simvastatin's efficacy and side effects can results from a single nucleotide polymorphism in the gene *SLCO1B1*, which encodes the organic ion transporter OATP1B1 (13). This drug transporter regulates hepatic uptake or efflux of statins and statin metabolites. Other pharmacogenomic data have shown that the CYP2C9∗3 polymorphism predisposes individuals for a pharmacologic interaction between simvastatin and warfarin (14). Both of these studies illustrate how our genomic footprint can potentially be used to predict how patients respond to drug treatments. Gene sequencing technologies continue to improve while the costs associated with complete genome sequencing decreases. This combination makes it a real possibility that every individual born in the near future (i.e., 50 years) will have their entire genome sequenced allowing their susceptibility to specific diseases and response to drug treatments be determined shortly after birth.

10.2   THE NEED FOR PROTEOMICS IN PERSONALIZED MEDICINE

As proteomic investigators will enthusiastically remind you, however, the genome does not tell you everything about an individual. The proteome comprises the molecules that carry out the functions encoded by genes within the genome. Unlike genomics, one cannot simply have their proteome characterized at birth and assume it will change little over their lifetime. An individual's proteome is in a constant state of flux both spatially and temporally. So then why do not we just forget about proteomics and personalized medicine and simply wait until genomes can be cheaply sequenced? As shown in Figure 10.1, the cost to sequence a genome has plummeted in the past decade from approximately $80,000,000 to $10,000. This cost should continue to drop into the hundreds of dollars per genome within the next decade. The reason that we cannot simply rely on the individual's genome sequence for personalized medicine is that it tells us what can happen but does not tell us when or where it is going to happen. For example, many thought the genome wide association studies would confidently predict the development of diseases. While this technology has provided enormous insight into

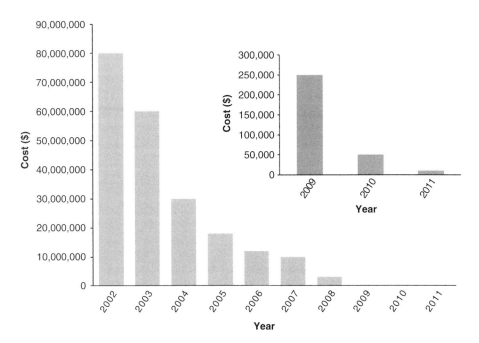

Figure 10.1. The rapid decrease in the cost of sequencing a human genome.

the potential cause and diagnosis of diseases, it has not provided the complete answer for conditions such as type 2 diabetes, obesity, and osteoporosis (15, 16). Other factors such as an individual's environment and lifestyle, which ultimately affect the content of the proteome, also play a role in disease progression and the efficacy of drug treatments (17).

The changing dynamic reflected within the proteome is also another reason why proteomics is going to be critical to personalized medicine. The proteome reflects the physical condition of the patient at a specific point in time and this information is critical in the decision on how to treat the patient. For example, anesthesiologists carefully consider a number of factors before determining the correct anesthesia and dose to give the patient prior to surgery. The major factors are the patient's medical history, current smoking habit, weight, current medications the patient may be taking, and whether the patient followed the rules about eating and drinking prior to surgery. A goal of proteomics in personalized medicine is to provide the physician a molecular view at the protein level that reflects the current physiological state of the patient and indicates the proper diagnosis or directs the course of treatment at a specific time point. It is important to remember that cancers are very specifically classified and the proper therapy may change during the course of treating a single patient. The future goal is that proteomics plays a role in classifying an individual's cancer type and directing the proper treatment course.

For many applications in personalized medicine, proteomics will need to provide accurate results in a short time frame. The need for accuracy is something that everyone understands. The method must have high sensitivity and specificity. But what is the required time frame? It honestly depends on the situation. The weeks to months required for a cancer biomarker discovery project is absolutely unacceptable. For diagnosing and classifying cancer, the time frame can range from hours to several days. For determining the correct therapy, the time frame could also range from hours to several days. For treating cancer via surgery, it could be as short as a few minutes.

10.3   PROTEIN PANEL DISCOVERY

The obvious area where proteomics can impact personalized medicine is diagnostics. This area is not to be confused with biomarker discovery. In general, personalized diagnostics uses the information found in discovery studies and applies it to individuals. Much as sequencing the genome was a key step in pharmacogenomics, the first step in applying proteomics in personalized medicine is discovering what needs to be measured.

In a recent report Cima et al. conducted a proteomics study for the discovery and subsequent application of serum biomarkers for prostate cancer in which the tumor-suppressor phosphatase and tensin homolog (*Pten*) gene is inactivated (18). This gene is inactivated in a number of human cancers including endometrial carcinoma, glioblastoma, breast cancer, and prostate cancer (19). The first stage was the use of a label-free quantitative mass spectrometry (MS) approach to identify 775 N-linked glycoproteins in serum and prostate tissue from wild-type and Pten-null mice. A total of 111 proteins showed a significant difference between in the tissue of *Pten* cKO mice tissue and their healthy age-matched controls. Only 12 proteins, however, were significantly different in the matched serum samples. The candidate biomarker list was then filtered down to 126 proteins based on each marker's *Pten* dependency, prostate specificity, and ability to be detected in serum.

While the trend is to utilize panels of proteins for diagnosing cancers, a panel of 126 proteins would be too analytically and computationally challenging to measure and would likely be prone to error owing to the large dimensionality of the feature space. Therefore, the next obvious step required in this study was to reduce this number of proteins and determine if any of them translate from the mouse model to human patients. Serum samples from patients with *Pten*-inactivated prostate cancer were analyzed via selected reaction monitoring (SRM) and stable-isotope labeled reference peptides to quantitate 57 N-glycosites within 49 specific candidate biomarkers. Thirty-six of the peptides, originating from 33 different proteins, were consistently quantified in 80–105 patients. An additional six proteins were quantified using an enzyme-linked immuno sorbent assay (ELISA). A random forest (RF) classifier algorithm (20) was used to rank and select the candidate biomarkers that would make the best predictive model for discrimination between normal and aberrant *PTEN* status.

Twenty top-ranked variables selected from 100 RF analyses were used to predict Pten-loss through combinations containing up to five serum proteins. This screening resulted in 21,699 different models, which were validated by 100-fold bootstrapping. For each of the resultant 21,699 models, the median area under the receiver operating characteristic curve (AUC) was calculated to identify those that could best predict significantly aberrant *PTEN* status from a data set comprising 54 patients derived from the genetic analysis of 82 patients along with the SRM and ELISA data from 105 patients. A four panel signature consisting of thrombospondin-1, metalloproteinase inhibitor 1 (TIMP-1), complement factor H, and prolow-density lipoprotein receptor-related protein 1 correctly predicted the *Pten* status of 78% of the patients (sensitivity of 79.2% and specificity of 76.7%).

While being able to predict the *Pten* status is interesting, this information can be obtained using genetics based methods that are more reliable. Therefore, the data were also analyzed to determine if a serum-based protein profile could indicate the Gleason grade of the prostate cancer (an indication of its progression and aggressiveness). A five-protein signature comprising polypeptide GalNAc transferase-like protein 4, fibronectin, zinc-α-2-glycoprotein, biglycan, and extracellular matrix protein 1 was identified that predicted patients having tumors with a Gleason score <7 or ≥ 7 with an AUC of 0.788 (sensitivity of 60.9% and specificity of 67.8%). Not only do these results suggest that Pten status is linked to tumor grade, it makes a big step toward personalized medicine in that the biomarker panel can stratify between clinically significant or insignificant prostate cancer and determine whether the patient should undergo immediate therapy or simply be monitored without any intervention. In the last phase of this study, 143 sera from 77 PCa patients and 66 controls was analyzed and a four-protein signature composed of hypoxia upregulated protein 1, asporin, cathepsin D, and olfactomedin-4 that discriminated benign prostatic hyperplasia from prostate cancer was identified (AUC of 0.726, sensitivity of 85%, specificity of 57%). While these results were similar to that afforded by prostate-specific antigen (PSA) testing, when the four-panel biomarker set was combined with PSA testing the result was an AUC of 0.840 (sensitivity of 85%, specificity of 79%). Collectively, these four proteins were as effective in diagnosing an independent set of serum samples as they were when applied to the testing set above.

Using PSA as the standard biomarker for early detection of prostate cancer remains controversial, owing to its lack of sensitivity and specificity (21). The resulting over diagnosis and overtreatment creates a burden to the health care system while putting many men through unnecessary anxiety. While the protein panels identified in the above study are not ready to be applied in a personalized medicine workflow, they do illustrate the potential of proteomics in this area. The study flows from a discovery mode to the focused analysis of a small number that are ultimately filtered to groups of four or five proteins that correctly predict a number of characteristics of each individual's prostate cancer. While one panel can determine the *Pten* status of the cancer, the other two can differentiate prostate cancer from BPH while also staging the tumor's grade. Another critical factor that makes the results from this study potentially useful in personalized medicine is that the proteins are measured in serum. The dynamic nature of an individual's proteome requires repeated measurement at different times, particularly in monitoring the efficacy of treatments. If tissue samples are required every time, the physical burden to the patient will be too extreme in most cases.

10.4 GUIDING THE SURGEON THROUGH IMAGING

Surgery is a critical step, and one of the most personalized treatments of cancer. An important factor in surgery is the complete removal of the tumor. Local tumor recurrence after surgery remains a problem in many cancers. One of the foremost concerns in clinical oncology and pathology is ensuring complete tumor removal (22,23). To ensure complete tumor removal, the surgeon must recognize the margin that exists between the tumor and healthy tissue. Tumor margin status is currently determined through histopathological assessment of the resected tumor, which is normally performed postoperatively but in some cases can be done intraoperatively (24). Since recurrence is a known problem, there must be characteristics within the resected tissue that are not being detected using IHC. Much like disease diagnosis, this situation suggests the need for additional approaches that can increase the accuracy in determining tumor margins during and after surgery. Being an antibody-based hypothesis approach, IHC will not detect anything outside of its intended target.

Before I sound too negative toward IHC, it is fair to say that this technology has many of the aspects of personalized medicine that advanced proteomics techniques are trying to emulate. Not only is it routinely used to diagnose and stage tumors, it has very high throughput and can be done using commonly adopted standard operating procedures making IHC results from different labs comparable. The strengths of IHC are the challenges that proteomics is trying to overcome on the path to personalized medicine.

Is there a role for proteomics in defining margins to ensure that tumor resection during surgery is complete? We could use laser-capture microdissection to isolate tumor and healthy cells and quantitatively compare their proteomes using LC-MS^2. Unfortunately, all we would be doing is finding new IHC markers to complement existing ones. In addition, development of an assay for any new markers would be time consuming and suffer from all of the problems associated with antibody detection. Even if the findings were used to create a MS-based assay, it would lack the throughput and simplicity needed to make it appealing to the clinical community. A possible solution has been proposed by Dr. Richard Caprioli, the pioneer in MS imaging (also known as imaging MS) (25). Developments in his laboratory not only provide me the opportunity to illustrate a novel use of proteomics in personalized medicine, but also briefly discuss MS imaging. While I have not been able to include MS imaging in other chapters of this book, it definitely deserves mentioning as a viable proteomic technology with the potential to impact personalized medicine.

To conduct imaging MS (Figure 10.2), thin (i.e., 5–10-μm-thick) sections are cut from tissue blocks using a microtome and placed onto conductive matrix-assisted laser desorption ionization (MALDI) targets (25–27). Imaging MS can be performed on either frozen or formalin-fixed paraffin-embedded tissues. As with conventional MALDI-MS, a matrix compound is applied to the tissue's surface. The tissue can be optionally treated with trypsin prior to MS analysis to collapse intact proteins into peptides. A laser is fired onto the tissue in an ordered pattern and individual mass spectra are collected at each laser position. Thousands of positions can be interrogated across the entire tissue section or specific regions can be targeted. The individual ablated spots are analogous to a pixel on a computer monitor; however, instead of containing a visual image of the tissue, the spots contain mass spectra containing signals originating from analytes within the tissue. The analyte peaks present within the individual mass spectra can be displayed based on their relative intensity and position within the tissue section. The net result is hundreds (or potentially thousands) of images from molecules within a single tissue section.

One factor that makes imaging MS attractive as a personalized medicine tool is its throughput. Because of their small size, 20 tissue sections can be placed onto a single MALDI plate and analyzed in the same experiment. The high repetition rates of modern lasers (1 kHz or greater), allows the entire set of samples to be analyzed in a few minutes to hours. A recent study showed the complete analysis of a 185 mm^2 section of rat brain in less than 10 min (28).

In an application of MS imaging, Dr. Caprioli's group studied a series of clear cell renal cell carcinoma (ccRCC) tumors that had been removed from cancer patients being treated with surgery (29). Approximately 80% of RCC diagnosed in the United States cases are classified as ccRCC, which owes its name to the yellow appearance and clear cytoplasm resulting from a high lipid and cholesterol content (30). Unfortunately, ccRCC has the poorest prognosis of all common epithelial tumors of the kidney (31).

For cancers such as ccRCC, It is critical to study the molecular characteristics of both the tumor and surrounding tumor margin. Such studies would not only facilitate a better understanding of tumor invasion, but also provide potential markers to aid in histological assessments that help ensure complete tumor removal. The standard treatment for ccRCC is radical nephrectomy (removal of the entire kidney, its fascia, adjacent adrenal gland, and all regional lymph nodes) since individuals do well with a single functional kidney (32). Unfortunately for ccRCC patients with a single functioning kidney or those suffering from bilateral ccRCC radical nephrectomy is not an attractive option. In these cases, as well as for patients with small tumors and those whose other

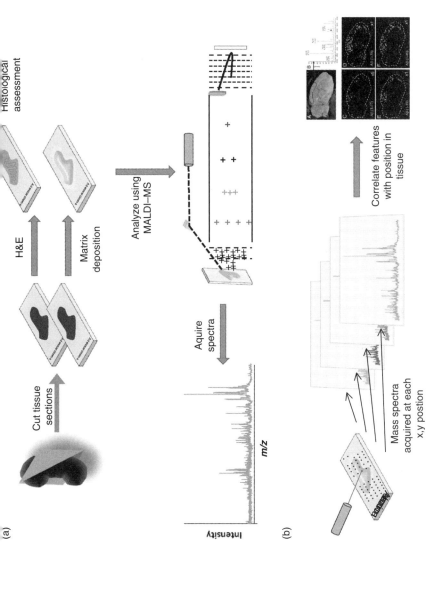

Figure 10.2. Principles of MS imaging. (a) In MS imaging, thin tissue sections are placed onto slides. At least one tissue section is sent for histological assessment after staining with H&E. A matrix compound is applied to tissue sections that are then sent for analysis by MS imaging. (b) During data acquisition, the laser rasters across a defined area of the tissue and mass spectra are acquired at each of these points. The data can be visualized to display the various points within the tissue that specific ions were detected. (Portions of figure reproduced with permission from Reference 25.)

249

kidney may be susceptible to damage via diabetes mellitus, hypertension or a familial history of kidney cancer, partial nephrectomy is more desirable. The effectiveness of partial nephrectomies, however, is in dispute as there are no firm guidelines on how large of a surgical margin is needed to ensure complete tumor removal (33). Bottom line is that surgeons do not accurately know how far outside the observable tumor to cut to ensure the ccRCC does not reoccur.

In this study, 34 kidney samples were obtained from patients undergoing nephrectomies. Each sample contained ccRCC and attached adjacent normal renal tissue (29). These samples were stained with hematoxylin and eosin (H&E) and reviewed by a board-certified pathologies that marked tumor, normal, and histological border regions of the tissue. One goal of this study was to determine how the features in the MS imaging spectra change near the tumor margins and also measure any correlation between protein expression patterns and tumor aggressiveness. Mass spectra were acquired across the tissue samples containing both tumor and attached normal tissue. The distances between the spots interrogated by MS imaging and the tumor border were measured using a software program that calculated the distance between the spot and the tumor margin as assessed by a pathologist examining an H&E-stained tissue section.

To interpret molecular trends contained within the collection of MS imaging spectra, 40 significant features were plotted for 25 patients whose tissue sections provided enough tumor and normal tissue for reasonable assessment. These features included m/z values that were either under- or overexpressed in the ccRCC tumors as well as aberrant molecular features present outside of the tumor margin. The amplitude (i.e., signal intensity) of each m/z value was plotted against its distance from the margin using a locally weighted-regression scatter-plot smoothing (LOWESS) curve (34). By taking the first derivative of the LOWESS trend lines, the maximum rate of change of the intensity of each feature was determined. The result was the identification of numerous m/z values (4888, 5351, 6717, 8577, 9368, 10611, and 12274), or features, whose levels suggested they were underexpressed in the tumor and the margin-normal regions compared to the normal region of the tissue. These m/z values were subsequently identified as originating from the proteins cytochrome c oxidase polypeptides VIII2 (m/z 4888), cytochrome c oxidase polypeptides VIIC (m/z 5351), cytochrome c oxidase polypeptides VIIA2 (m/z 6717), cytochrome c oxidase polypeptides VIC (m/z 8577), NADH-ubiquinone oxidoreductase MLRQ subunit (m/z 9368), cytochrome c oxidase polypeptides VB (m/z 10611), and cytochrome C (m/z 12274). An additional four proteins were identified whose levels were upregulated in the tumor compared to the normal tissue (i.e., calpactin I, calgizzarin, thymosin β10, and macrophage inhibitory factor). The distance from margin-dependent changes in the levels of these proteins were compared to tumor aggressiveness; however, no correlation could be found.

This study provided the first solid evidence suggesting that MS imaging could contribute to more accurately defining tumor margins. While there are a plethora of proteomic studies that have detailed protein abundance differences between tumor and normal tissues from the same organ, the results of such studies are not particularly novel as personalized medicine tools as they simply provide additional targets for IHC assays. While IHC will continue as an invaluable tool in pathology and diagnostic laboratories, it still requires hours to complete an assay that targets only a single protein. While it is still not ready for "prime-time," MS imaging has the potential for automation and throughput to make it a viable personalized medicine technology in the future. In this scenario, a surgeon would resect a tumor from a patient and hand it directly to a technician who will quickly snap freeze the tissue and cut a thin section onto a MALDI plate. A matrix is automatically deposited onto the tissue section, which is then placed into a simple MALDI mass spectrometer for data acquisition. After a few minutes of acquisition, the data are analyzed to identify trends in feature intensity across the tissue. If no significant changes are observed across the tissue, the surgeon can be advised to cut additional tissue to ensure complete resection of the tumor. It is reasonable to assume that this entire process could be completed within 30 min if the necessary instrumentation is properly positioned near the surgical unit.

10.5 PERSONALIZED MEDICINE BY SHRINKING THE PROTEOMIC UNIVERSE

One challenge in personalized medicine is the sheer size of the human proteome and the amount of data that is accumulated in typical proteomic experiments. While sequencing an individual's genome may provide future health benefits, personalized medicine will not require complete knowledge of an individual's proteome. Since it is dynamic, we can never completely know an individual's proteome anyway. While there are a large number of proteins whose abundance or posttranslational modification state may be altered in tumor cells, many more will not show any significant difference when compared with healthy cells. The goal of proteomic personalized medicine should be to discover and then develop assays that only measure those proteins that are directly related to cancer.

10.5.1　Targeted Mass Spectrometry

What each of the above examples described in the previous sections lacked was the ability to target specific proteins. Developing assays that target specific proteins would be the logical step in each case as it makes no sense to continue to assay an entire proteome once the key determinants have been discovered. While antibody-based assays could be developed, a MS approach that combined the specificity along with an ability to measure the molecule's absolute abundance would be optimal. Fortunately, researchers have been working on developing such methods for a number of specific diseases. These methods, known as SRM or multiple reaction monitoring (MRM) are well known in the MS field, having been applied for a number of years for analyzing metabolites, including the well known assay for measuring in-born errors of metabolism (35, 36). For example, Dr. Xia Xu in my laboratory developed an SRM method for measuring 15 estrogen metabolites in a urine or serum sample (Figure 10.3) (37). Many people we work with who are unfamiliar with MS methods do not readily understand that there are a lot more metabolites injected into the instrument than we detect. We need to educate them that we instruct the mass spectrometer to only detect specific peaks at predetermined times during the analysis.

While these methods have been employed in metabolite analysis for several years, their translation into proteomics has not been as straightforward as hoped. The major

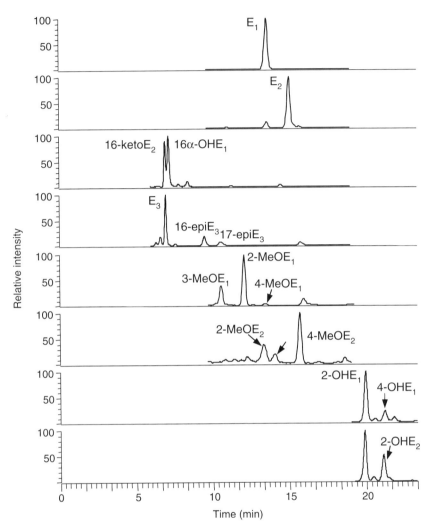

Figure 10.3. SRM-MS chromatographic profiles of 15 estrogen metabolites in a human serum sample. The mass spectrometer method is designed to measure specific precursor and fragment ions at various times during the chromatographic separation.

difference is that proteomics is still dominated by analysis of peptides, not intact proteins. Therefore, the initial step in developing a method to target a protein is to determine which peptide acts as its best surrogate. There are a number of factors that need to be considered: (1) the size of the peptide, (2) the potential for spurious modifications (e.g., oxidation) to residues within the peptide, and (3) its ionization efficiency (38). Currently, the ideal peptide would be about 6–18 nonoxidizable residues, contain no missed tryptic cleavage sites, and provide a very intense ESI signal. The peptide would also have produce about three fragment ions whose combination was unique. There have been a number of papers that have described computational approaches to identify the best surrogate peptide within a protein (39). Peptides suitable for SRM and MRM-MS assays, however, have to be evaluated empirically to identify those that exhibit the best performance. A good place to start is examining the peptides that were detected for a specific protein during the discovery stages of an experiment. Be aware, however, that these peptides may ultimately not be the best candidates for targeted assays.

The process of using SRM- or MRM-MS for developing an assay to measure the absolute quantity of specific targets in samples is shown in Figure 10.4. In MRM-MS, which is usually conducted using a triple quadrupole mass spectrometer, a precursor ion is isolated in the first quadrupole (Q1). This ion is fragmented within Q2, and specific fragment ions (one for SRM and >1 for MRM) are isolated within Q3 and detected. The combination of isolating a specific precursor ion in Q1, the detection of specific ions that are isolated within Q3, and the coelution of a peptide with its stable isotope labeled counterpart provide the certainty that the intended peptide is being measured in these experiments. Absolute quantitation of the targeted peptide is calculated by comparing the peak area ratios between the endogenous peptide and those obtained from spiking a known amount of a stable isotope-labeled internal standard into the sample. Peptides and subsequent transitions suitable for MRM-MS assays have to be evaluated and optimized to achieve best performance with least interference. In fact, peptides seen in discovery stages are not necessarily the best candidates to give best performance in MRM-MS assays. Additional optimization MRM-MS experiments using both best performing peptides seen at discovery stage as well as predicted high-responding peptides of the same protein are still required to narrow down the best peptides and/or the best transitions to monitor in the sample to achieve the best sensitivity and reproducibility.

While the above paragraph and associated figure make the development of these targeted assays seem simple, in reality they are still challenging. It is always worth reflecting on what these types of assays are being asked to do; quantitatively measure a specific peptide(s) within a mixture potentially containing 100,000+ peptides. Achieving this level of filtering requires optimal sample preparation and MS performance. While the performance of modern day mass spectrometers has made the MS part of the equation straightforward, the sample preparation part is generally where success or failure for these assays is determined. Ideally, no more sample preparation than normally done for global LC-MS^2 studies would be required; however, enriching for the targeted protein (or depletion of unwanted proteins, depending on your perspective) is often recommended prior to MS analysis. This enrichment is especially helpful when analyzing serum or plasma samples. Our laboratory developed an MRM assay for measuring different isoforms of prolactin in serum. We found that a simple ultrafiltration step to eliminate the high molecular weight fraction of serum (i.e., >30 kDa) resulted in prolactin peaks with much greater signal-to-noise and higher reproducibility for the assay overall. Other groups have used techniques such as strong cation exchange chromatography or gel electrophoresis to enriched their protein of interest (40, 41).

One of the pioneers of proteomics, Dr. Leigh Anderson, has probably invested more time and energy than anyone in developing targeted proteomic assays for use in personalized medicine. While Dr. Anderson definitely understands the MS requirements, his major contribution has been in the development of Stable Isotope Standards and Capture

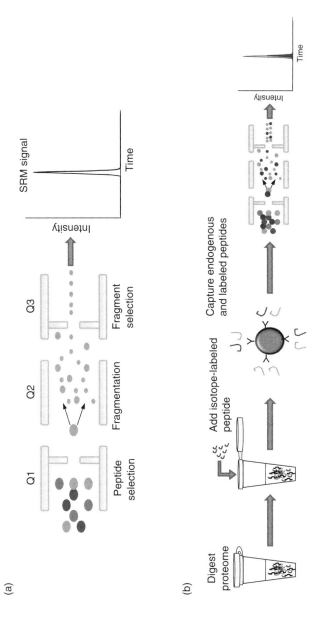

Figure 10.4. (a) Cartoon showing principle of SRM-MS. In MRM-MS, multiple fragment ions would pass through Q3 onto the detector. (b) In SISCAPA, isotope-labeled peptides corresponding to the targeted peptide are added to a proteome sample that has been enzymatically digested. An antibody is used to capture the targeted peptide (and its isotopically labeled counterpart). The peptides are removed from the antibody and the endogenous and isotope-labeled target peptides detected using SRM- or MRM-MS.

by Antipeptide Antibodies (SISCAPA) (42). Stable Isotope Standards and Capture by Antipeptide Antibodies is a sample preparation step that utilizes antibodies to capture specific peptides from a complex mixture (e.g., blood, urine, and tissue lysate) prior to MRM-MS analysis. After the proteome sample of interest is digested into peptides, a known amount of a stable-isotope (usually ^{13}C) labeled version of the targeted peptide is added to the mixture. Antibodies against the targeted peptide are used to capture both forms of the peptide. Once these peptides are extracted from the mixture, they are analyzed using MRM-MS, as described above. SISCAPA assays have shown more than a 1000-fold enrichment for peptides in plasma digests and can result in the detection in the low nanograms per milliliter range with coefficients of variation less than 20% (43).

A recent study illustrates the general use of SISCAPA to measure a peptide originating from the tissue inhibitor of TIMP-1 (44). This protein is known to be aberrantly glycosylated in patients with colorectal cancer (CRC) and has been identified within the serum of humans suffering from this condition (45). An interesting twist to this study, however, was that phytohemagglutinin-L_4 lectin, which binds the β-1,6-N-acetylglucosamine moiety of N-linked glycans on proteins, was initially used to enrich for protein glycoforms from both noncancerous control and CRC serum. Within this glycoprotein fraction were aberrant glycoforms of TIMP-1. These captured glycopeptides were tryptically digested and a $^{13}C/^{15}N$ stable-isotope labeled version of the targeted TIMP-1 peptide GFQALGDAADIR was added to the mixture. The target peptide and its labeled counterpart were enriched using SISCAPA and analyzed using a 15 T MALDI-FTICR mass spectrometer. The MS data showed that the targeted peptide was approximately 5 times more abundant in serum from CRC patients than serum from healthy controls. Since the peptide is to act as a surrogate for the intact protein, this result suggests that glycoforms of TIMP-1 are substantially upregulated in CRC patients and may act as a viable circulating biomarker for this condition. In my opinion, however, it would definitely be more satisfying if the assay was directly measuring the peptide containing the glycan group.

For SRM-MS or MRM-MS approaches to make a wide spread impact in personalized medicine, assays that measure more than a single analyte per assay will be required. The efficiency of using these methods for a single analyte is just too low in terms of time and cost. Since conventional mass spectrometers have only one inlet, multiplexing assays are impossible. The only option is to use the available resolution space available through chromatographic fractionation and within the mass spectrometer to increase the number of proteins measured per unit time. An excellent example of this is illustrated by the measurement of cardiovascular biomarkers in plasma (46).

Cardiovascular disease remains a leading killer of individuals worldwide and represents a major challenge in disease monitoring and treatment (47). All major hospitals that have clinical chemistry units devote a lot of resources to running assays measuring the levels of common biomarkers indicative of cardiovascular health. Recently, an SRM assay was developed to quantitate 10 commonly assessed cardiovascular biomarkers, namely, apolipoprotein (Apo) A1, Apo A4, Apo B, Apo CI, Apo CII, Apo CIII, Apo D, Apo E, Apo H, and C-reactive protein. This assay was designed to measure these markers within a 30 min time period on the mass spectrometer. Obviously, it can be argued that the MS time is only one component of the assay and it is unfair to not include sample preparation time as part of the overall time requirement. Since the sample preparation steps can be multiplexed, however, this makes the MS time the rate limiting step that dictates the time required per patient sample.

In the first step of this study, seven highly abundant proteins were immunodepleted from blood plasma. The proteome sample was digested into peptides and a mixture containing isotope-labeled versions of the targeted peptides was added to the sample. The SRM-MS assay was then performed using LC-MS^2. While this work has not progressed to a point where it has been applied to clinical studies, it demonstrates many of the characteristics that are going to be required for these targeted assays to have an impact in personalized medicine.

10.5.2   Targeted Arrays

Honestly, owing to my background, I have had to constantly remind myself during the course of writing this book that proteomics is more than MS. There are definitely non-MS methods that currently look like they will have a real impact on personalized medicine in the future. Drs. Lance Liotta and Emmanuel "Chip" Petricoin are two scientists at the forefront of developing targeted tissue arrays to measure cancer-related proteins in clinical samples. I have personally known Lance and Chip for several years and have been amazed at how two eminent scientists have worked so well together all these years while putting their egos aside. They each have incredible knowledge and vision, while possessing individual skills that complement each other. Using their knowledge of pathology and cell signaling, they (along with others of course) recognized that cancer is a disease of the signaling pathway network (48). These signaling pathways are controlled by the coordinated activation/deactivation of proteins via kinases and phosphatases. In cancers, this coordination goes awry along with the activation of these signaling pathways. While the complete complement of signaling networks within the human cell is very complex, recent data suggest that many cancers can be categorized based on the activation of specific functional pathways. For example, many human cancers can be characterized by dysregulated AKT/mTOR, Ras-Raf-ERK, growth factor receptor-mediated signal pathways, rather than their location or microenvironment (49).

Recognizing this fact, Drs. Liotta and Petricoin have put great effort into developing and applying reverse-phase protein arrays (RPPA), which are discussed in greater detail in Chapter 9. Briefly as a reminder, RPPA are antibody-based arrays that can be used to measure large numbers of proteins within many different tissues in an automated, high-throughput fashion. In a study of head and neck squamous cell carcinoma (HNSCC), a RPPA was set up to measure the levels of 60 proteins (predominantly phosphoproteins) in matched tumor and nonmalignant biopsy specimens acquired from 23 patients (50). Eighteen of the 60 proteins were found to be more abundant in tumors compared to healthy tissue, while 17 of 60 were of lower abundance in tumors. The most significantly elevated phosphoproteins (and the elevated phosphorylation sites) in the tumors included checkpoint kinase 1 (S345) and 2 (S33/35), eukaryotic translation initiation factor 4E-binding protein 1 (S65), protein kinase C ζ/ι (T410/T412), LKB1 (S334), inhibitor of κB-α (S32), eukaryotic translation initiation factor 4E (S209), Smad2 (S465/67), insulin receptor substrate 1 (S612), and mitogen-activated ERK kinase 1/2 (S217/221). Total PKC ι levels were also shown for the first time to be elevated in HNSCC.

While its present use is primarily as a research tool, technologies such as RPPA offer many of the features required for personalized medicine. Obviously, the required throughput is present as is the automation. Since it shares so many foundational steps with IHC, many of the technologies (tissue preparation, antibody detection, and visualization, etc.) required to make RPPA clinically applicable are already well developed. What is going to be required to take these arrays to the next level from a research tool to a widely used diagnostic (or prognostic) tool is absolute recognition of the protein targets. This recognition would allow the development of kits that contain specific reagents (i.e., antibodies) to test against tumor tissues. Using these antibody panels to recognize which signaling pathway is active, will enable the oncologist to better determine the course of treatment through the use of therapies that shut off the aberrant kinase(s).

10.6 INSTRUMENTATION REQUIREMENTS

The term personalized medicine can make us lose sight of how many patients samples will need to be analyzed. When considering personalized medicine, we often invoke an image of a sample being drawn from a single patient and sent for analysis. In reality, the "personalized" part of personalized medicine refers to using the results of the analysis to tailor the treatment to the individual. Personalized medicine will entail the daily analysis of hundreds of thousands of samples from patients all around the world.

Considering the analytical requirements of personalized medicine, it is prudent to take a step back and examine the available technologies and determine which of these are most likely to provide the capabilities required to make personalized medicine a reality. In proteomics, the available tools include mass spectrometers or affinity-based devices (e.g., protein and tissue arrays, IHC, and ELISAs). Unfortunately, determining which of these technologies has the greatest chance of being developed into a useful personalized medicine device is not quite as easy as picking one over the other based on their present attributes. For example, MS definitely has an advantage since it provides a discrete signal for each protein, whereas affinity-based methods can suffer from cross reactivity. In addition, MS can provide greater detailed information concerning a single protein than antibody-based methods. Mass spectrometry can also analyze a single protein much faster than antibody-based methods. This comparison would make one think that MS is definitely the best option for developing a personalized medicine device.

Unfortunately, MS has several major deficiencies that make it not particularly well suited for personalized medicine. The first deficiency is throughput. While MS can measure several analytes in a single analysis, it can only analyze one sample at a time. Tens to hundreds of samples can be concurrently measured using ELISA or IHC. Unfortunately, the primary method of increasing capacity and/or throughput using MS is to purchase an additional instrument. Mass spectrometers, however, are not cheap. They presently cost in the range of $200,000–600,000 if ancillary chromatography and data analysis requirements are included. Secondly, operating a mass spectrometer requires significant skill. While they are more common than ever, most mass spectrometers still reside within specialty laboratories where they are operated by individuals who have spent their scientific careers in this area. They have definitely not achieved the position where virtually ever chemistry or biochemistry graduate can operate a mass spectrometer like they can a UV spectrophotometer. Wide-spread expertise in MS operation will continue to languish amongst these graduates as the cost of a mass spectrometer prevents many academic institutes from providing the open access required to train large numbers of undergraduate students. Finally, most MS-based assays require significant sample preparation prior to injection into the instrument. These steps can range from high abundance protein depletion (i.e., serum or plasma samples) to affinity chromatography (i.e., phosphoproteins). Many of these sample preparation steps have yet to be routinely automated. The most desirable situation would be akin to an ELISA in which raw blood or urine samples can be injected directly into the mass spectrometer. While MS technologies such as surface-enhanced laser desorption/ionization offers this simplicity, the data provided by this technology (as discussed in Chapter 8) lacks the transparency and certainty required for accurate diagnosis.

Can these deficiencies be overcome in the future? Absolutely! The throughput issue can be resolved through both instrument development and experimental design. The duty cycles of mass spectrometers will continue to increase in speed enabling greater number of analytes to be detected in a given time period. This improvement will both increase proteomic coverage and shorten assay time (particularly for targeted MRM assays). Improvements in separation technologies, either chromatographic or electrophoretic, will also increase the throughput. The lack of wide-spread skill in MS operation can be addressed in two different ways. The dominance of MS as an analytical tool in chemistry and biochemistry will result in an increasing number of university-level courses focusing on this technology. I remember having lunch at Northeastern University with the preeminent analytical chemist Dr. Ira Krull, where he suggested that (and I paraphrase) "owing to its importance, MS should be the only tool we teach in analytical chemistry." I am sure that he did not mean it to that exact extreme, but his opinion is reflective of the growing importance of MS in analytical chemistry. It is likely that within a decade, almost every chemistry and biochemistry undergraduate student will have at least a working knowledge of MS and the percentage of those with "hands-on" experience will exponentially increase. The lack of existing wide-spread expertise will also be addressed by having proteomic clinical assays performed in centralized laboratories specifically designed for these purposes. This situation is akin to what presently exists at contract research organizations and clinical laboratories such as Quest Diagnostics or LabCorp for metabolite analysis. Having centralized laboratories conduct the analysis not only provides the concentration of expertise required, but facilitates the development of standard operating procedures necessary to make these assays reproducible and reliable.

Ultimately, a simplified MS system that allows clinical samples such as blood and urine to be directly injected would solve most of these problems. The idea of a MS clinical analyzer has been discussed for several years. At an Asilomar meeting in the early part of this century, Dr. Leroy Hood described his vision of a portable mass spectrometer that could monitor an individual's blood proteome. While applying personalized medicine will not have to wait for such a device, an instrument that performs a specific analysis chosen from a menu at the simple push of a button will definitely help. This clinical analyzer will have to incorporate on-line sample processing as well as automated data analysis so that the output that the operator receives is a simple list of protein concentrations (and not raw MS spectra).

While all of these challenges to making MS a major technology in personalized medicine may be overcome, antibody-based methods will continue to drive protein diagnostics for the near future (and possibly beyond). The throughput of ELISAs and IHC is presently too much for MS to compete with. The obvious argument against antibody-based tests is the lack of suitable antibodies for all human proteins and the lack of specificity observed for many antibodies in present use. These arguments are definitely valid, however, it needs to be remembered that there exists a number of large programs (both academic and industrial) focusing on developing highly specific antibodies for all of the proteins in the human proteome. While MS will undoubtedly advance in its capabilities in the future, so will antibody-based assays.

10.7 CONCLUSIONS

While there is still a lot of room for growth, it can be argued that proteomics already impacts personalized medicine. Every time protein levels in someone's urine or albumin and globulin levels in blood is measured proteomics is impacting personalized medicine. Doctors and scientists have a much bigger vision for proteomics in personalized medicine. In this vision, a sample is taken from a patient and the concentrations of specific proteins are measured. These protein levels indicate exactly what type and stage of cancer the patient has and also dictates the correct course of required therapy. Getting to the level of diagnostic and therapeutic knowledge is not going to happen through the measurement of a single protein. Much like we cannot accurately describe an individual's appearance through a single characteristic (e.g., hair color and height), it is going to require measuring multiple proteins to accurately describe an individual's cancer. The exact number is presently not known and will undoubtedly vary by cancer type. Personalized medicine, like a vast majority of proteomics, is presently within its discovery phase. Groups are presently discovering which proteins are most important to measure and which technologies will be adaptable to meet the needs of personalized medicine in the near future. These technologies could include some variation of MS or antibody-based array, or be an entirely new technology that has yet to be invented.

REFERENCES

1. http://www.cancer.gov/dictionary
2. Sykiotis GP, Kalliolias GD, Papavassiliou AG. Pharmacogenetic principles in the Hippocratic writings. J. Clin. Pharmacol. 2005;45:1218–1220.
3. Sykiotis GP, Kalliolias GD, Papavassiliou AG. Hippocrates and genomic medicine. Arch. Med. Res. 2006;37:181–183.

4. Yapijakis C. Hippocrates of Kos, the father of clinical medicine, and asclepiads of bithynia, the father of molecular medicine. Review. In Vivo 2009;23:507–514.

5. Goldberg YP, Telenius H, Hayden MR. The molecular genetics of Huntington's disease. Curr. Opin. Neurol. 1994;7:325–332.

6. Langbehn DR, Hayden MR, Paulsen JS PREDICT-HD Investigators of the Huntington Study Group. CAG-repeat length and the age of onset in Huntington disease (HD): A review and validation study of statistical approaches. Am. J. Med. Genet. B Neuropsychiatr. Genet. 2010;153B:397–408.

7. Kalow W, Staron N. On distribution and inheritance of atypical forms of human serum cholinesterase, as indicated by dibucaine numbers. Can. J. Biochem. Physiol. 1957;35:1305–1320.

8. Kalow W, Gunn DR. Some statistical data on atypical cholinesterase of human serum. Ann. Hum. Genet. 1959;23:239–250.

9. Chiam K, Centenera MM, Butler LM, Tilley WD, Bianco-Miotto T. GSTP1 DNA methylation and expression status is indicative of 5-Aza-2′-deoxycytidine efficacy in human prostate cancer cells. PLoS One 2011;6:e25634.

10. Ilic N, Utermark T, Widlund HR, Roberts TM. PI3K-targeted therapy can be evaded by gene amplification along the MYC-eukaryotic translation initiation factor 4E (eIF4E) axis. Proc. Natl. Acad. Sci. U.S.A. 2011;108:E699-E708.

11. Ma Q. Lu AY. Pharmacogenetics, pharmacogenomics, and individualized medicine. Pharmacol. Rev. 2011;63:437–459.

12. Davidson MH, Stein EA, Dujovne CA, Hunninghake DB, Weiss SR, Knopp RH, Illingworth DR, Mitchel YB, Melino MR, Zupkis RV, Dobrinska MR, Amin RD, Tobert JA. The efficacy and six-week tolerability of simvastatin 80 and 160 mg/day. Am. J. Cardiol. 1997;79:38–42.

13. Wilke RA, Ramsey LB, Johnson SG, Maxwell WD, McLeod HL, Voora D, Krauss RM, Roden DM, Feng Q, Cooper-Dehoff RM, Gong L, Klein TE, Wadelius M, Niemi M. The clinical pharmacogenomics implementation consortium: CPIC guideline for SLCO1B1 and simvastatin-induced myopathy. Clin. Pharmacol. Ther. 2012;92:112–117.

14. Andersson ML, Eliasson E, Lindh JD. A clinically significant interaction between warfarin and simvastatin is unique to carriers of the CYP2C9∗3 allele. Pharmacogenomics 2012;13:757–762.

15. Stumvoll M, Tschritter O, Fritsche A, Staiger H, Renn W, Weisser M, Machicao F Häring H. Association of the T-G polymorphism in adiponectin (Exon 2) with obesity and insulin sensitivity: Interaction with family history of type 2 diabetes. Diabetes 2002;51:37–41.

16. Ralston SH, Uitterlinden AG. Genetics of osteoporosis. Endocr. Rev. 2010;31:629–662.

17. Karasik D. How pleiotropic genetics of the musculoskeletal system can inform genomics and phenomics of aging. Age (Dordr). 2011;33:49–62.

18. Cima I, Schiess R, Wild P, Kaelin M, Schüffler P, Lange V, Picotti P, Ossola R, Templeton A, Schubert O, Fuchs T, Leippold T, Wyler S, Zehetner J, Jochum W, Buhmann J, Cerny T, Moch H, Gillessen S, Aebersold R, Krek W. Cancer genetics-guided discovery of serum biomarker signatures for diagnosis and prognosis of prostate cancer. Proc. Natl. Acad. Sci. U.S.A. 2011;108:3342–3347.

19. Hollander MC, Blumenthal GM, Dennis PA. PTEN loss in the continuum of common cancers, rare syndromes and mouse models. Nat. Rev. Cancer 2011;11:289–301.

20. Breiman L. Random forests. Mach. Learn. 2001;45:5–32.

21. Gomella LG, Liu XS, Trabulsi EJ, Kelly WK, Myers R, Showalter T, Dicker A, Wender R. Screening for prostate cancer: The current evidence and guidelines controversy. Can. J. Urol. 2011;18:5875–5883.

22. Ollila DW, Caudle AS. Surgical management of distant metastases. Surg. Oncol. Clin. N. Am. 2006;15:385–398.

23. Altendorf-Hofmann A, Scheele J. A critical review of the major indicators of prognosis after resection of hepatic metastases from colorectal carcinoma. Surg. Oncol. Clin. N. Am. 2003;12:165–192.

24. Raziano RM, Clark GS, Cherpelis BS, Sondak VK, Cruse CW, Fenske NA, Glass LF. Staged margin control techniques for surgical excision of Lentigo Maligna. G. Ital. Dermatol. Venereol. 2009;144:259–270.

25. Seeley EH, Caprioli RM. Molecular imaging of proteins in tissues by mass specrometry beyond the microscope. Proc. Natl. Acad. Sci. U.S.A. 2008;105:18126–18131.

26. Chaurand P, Schwartz SA, Caprioli RM. Imaging mass spectrometry: A new tool to investigate the spatial organization of peptides and proteins in mammalian tissue sections. Curr. Opin. Chem. Biol. 2002;6:676–681.

27. Reyzer, ML, Caprioli RM. MALDI-MS-based imaging of small molecules and proteins in tissues. Curr. Opin. Chem. Biol. 2007;11:29–35.

28. Spraggins JM, Caprioli RM. High-speed MALDI-TOF imaging mass spectrometry: Rapid ion image acquisition and considerations for next generation instrumentation. J. Am. Soc. Mass Spectrom. 2011;22:1022–1031.

29. Oppenheimer SR, Mi D, Sanders ME, Caprioli RM. Molecular analysis of tumor margins by MALDI mass spectrometry in renal carcinoma. J. Proteome Res. 2010;9:2182–2190.

30. Rini BI, Campbell SC, Escudier B. Renal cell carcinoma. Lancet 2009;373:1119–1132.

31. Grignon DJ, Che M. Clear cell renal cell carcinoma. Clin. Lab. Med. 2005;25:305–316.

32. Rodriguez-Covarrubias F, Gomez-Alvarado MO, Sotomayor M, Castillejos-Molina R, Mendez-Probst CE, Gabilondo F, Feria-Bernal G. Time to recurrence after nephrectomy as a predictor of cancer-specific survival in localized clear-cell renal cell carcinoma. Urol. Int. 2011;86:47–52.

33. Sutherland SE, Resnick MI, Maclennan GT, Goldman HB. Does the size of the surgical margin in partial nephrectomy for renal cell cancer really matter? J. Urol. 2002;167:61–64.

34. Cleveland, William S. LOWESS: A program for smoothing scatterplots by robust locally weighted regression. Am. Statistic. 1981;35:54.

35. Bueno MJ, Agüera A, Gómez MJ, Hernando MD, García-Reyes JF, Fernandez-Alba AR. Application of liquid chromatography/quadrupole-linear ion trap mass spectrometry and time-of-flight mass spectrometry to the determination of pharmaceuticals and related contaminants in wastewater. Anal. Chem. 2007;79:9372–9384.

36. Naylor EW, Chace DH. Automated tandem mass spectrometry for mass newborn screening for disorders in fatty acid, organic acid, and amino acid metabolism. J. Child Neurol. 1999;1:S4–S8.

37. Xu X, Keefer LK, Ziegler RG, Veenstra TD. A liquid chromatography-mass spectrometry method for the quantitative analysis of urinary endogenous estrogen metabolites. Nat. Protoc. 2007;2:1350–1355.

38. Brusniak MY, Chu CS, Kusebauch U, Sartain MJ, Watts JD, Moritz RL. An assessment of current bioinformatic solutions for analyzing LC-MS data acquired by selected reaction monitoring technology. Proteomics 2012;12:1176–1184.

39. Blonder J, Veenstra TD. Computational prediction of proteotypic peptides. Expert Rev. Proteomics 2007;4:351–354.

40. Yang X, Lazar IM. MRM screening/biomarker discovery with linear ion trap MS: A library of human cancer-specific peptides. BMC Cancer 2009;9:96.

41. Halvey PJ, Ferrone CR, Liebler, DC. GeLC-MRM quantitation of mutant KRAS oncoprotein in complex biological samples. J. Proteome Res. 2012;11:3908–3913.

42. Anderson NL, Anderson NG, Haines LR, Hardie DB, Olafson RW, Pearson TW. Mass spectrometric quantitation of peptides and proteins using stable isotope standards and capture by anti-peptide antibodies (SISCAPA). J. Proteome Res. 2004;3:235–244.

43. Whiteaker JR, Zhao L, Abbatiello SE, Burgess M, Kuhn E, Lin C, Pope ME, Razavi M, Anderson NL, Pearson TW, Carr SA, Paulovich AG. Evaluation of large scale quantitative proteomic assay development using peptide affinity-based mass spectrometry. Mol. Cell. Proteomics 2011;10:M110.00564544.

44. Ahn YH, Kim KH, Shin PM, Ji ES, Kim H, Yoo JS. Identification of low-abundance cancer biomarker candidate TIMP1 from serum with lectin fractionation and peptide affinity enrichment by ultrahigh-resolution mass spectrometry. Anal. Chem. 2012;84:1425–1431.

45. Kim YS, Hwang SY, Kang HY, Sohn H, Oh S, Kim JY, Yoo JS, Kim YH, Kim CH, Jeon JH, Lee JM, Kang HA, Miyoshi E, Taniguchi N, Yoo HS, Ko JH. Functional proteomics study reveals that N-acetylglucosaminyltransferase V reinforces the invasive/metastatic potential of colon cancer through aberrant glycosylation on tissue inhibitor of metalloproteinase-1. Mol. Cell. Proteomics 2008;7:1–14.

46. Rezeli M, Végvári A, Fehniger TE, Laurell T, Marko-Varga G. Moving towards high density clinical signature studies with a human proteome catalogue developing multiplexing mass spectrometry assay panels. J. Clin. Bioinforma. 2011;1:7.

47. Halim SA, Newby LK, Ohman EM. Biomarkers in cardiovascular clinical trials: Past, present, future. Clin. Chem. 2012;58:45–53.

48. Hollywood D. Signal transduction. Br. Med. Bull. 1991;47:99–115.

49. Marinov M, Fischer B, Arcaro A. Targeting mTOR signaling in lung cancer. Crit. Rev. Oncol. Hematol. 2007;63:172–182.

50. Frederick MJ, VanMeter AJ, Gadhikar MA, Henderson YC, Yao H, Pickering CC, Williams MD, El-Naggar AK, Sandulache V, Tarco E, Myers JN, Clayman GL, Liotta LA, Petricoin EF 3rd, Calvert VS, Fodale V, Wang J, Weber RS. Phosphoproteomic analysis of signaling pathways in head and neck squamous cell carcinoma patient samples. Am. J. Pathol. 2011;178:548–571.

11

THE CRITICAL ROLE OF BIOINFORMATICS

11.1 INTRODUCTION

In his poem *The Rime of the Ancient Mariner*, Samuel Taylor Coleridge pens the classic lines "Water, water, every where, Nor any drop to drink" (1). These lines sometimes reflect how proteomics investigators feel when trying to find cancer-specific biomarkers. We are deluged with data, but cannot turn it into useful information. As bioinformatic tools have been critical in developing next-generation genome sequencing technologies, they have also been critical in advancing proteomics technologies. The amount of data being acquired is simply impossible to analyze manually, making the proteomic community entirely reliant on databases and bioinformatic tools for turning raw data into biologically useful information.

I must admit that this chapter was my least favorite to write. Simply put, I am by no means an expert in bioinformatics and simply not blessed with the vision to analyze data with the creativity possessed by scientists such as Trey Ideker, Rolf Apweiler, Adam Arkin and David Fenyo. While thinking about how to prepare this chapter, I remembered back to one of the goals presented in the original prospectus of this book. This goal was to bridge the knowledge gap that exists between scientists that develop and apply proteomics technologies and oncologists who focus on understanding the biological basis behind cancer manifestation and progression. My laboratory collaborates with scientists

Proteomic Applications in Cancer Detection and Discovery, First Edition. Timothy D. Veenstra.
© 2013 John Wiley & Sons, Inc. Published 2013 by John Wiley & Sons, Inc.

who possess a wide range of specialties (e.g., molecular biologists, oncologists, and neurologists). Before beginning any project, it is always important that I understand what questions they are trying to answer and they understand exactly what type of information the proteomic experiments will provide. This "meeting of the minds" requires educating our collaborator about such things as how the raw data are turned into a protein identification and the confidence level of the result.

A large number of scientists report proteomic data; however owing to the centralization of mass spectrometry (MS) laboratories, only a small percentage of investigators that report proteomic data are actually involved in the acquisition and primary analysis of that data. I think this situation is risky especially if you have to present proteomic results at a seminar or conference. It is a very uncomfortable and embarrassing feeling when someone in the crowd asks you a question about how your raw data were analyzed and you do not have a clue. To alleviate this potential stress, one of the aims of this chapter is to help the non-MS user understand how the results that come from the mass spectrometer are turned into biological information.

11.2 PROTEIN IDENTIFICATION

One of the most basic tasks performed using MS is protein identification. Protein identification probably represents >90% of the experiments submitted to MS analysis. Most non-MS users just simply accept the identification provided by the proteomics laboratory without questioning how the result was obtained. What these users often do not realize is that there is more than one way to identify a protein using MS. If you do not work in an MS laboratory, the acronyms MS, MS^2, and MS^3 have no meaning; however, if you work in the MS laboratory, they each have a special designation and immediately indicate how the protein was identified.

11.2.1 Protein Databases

If an investigator desires to have his/her protein sample characterized by a proteomics laboratory, one of the questions they will be asked is "What species was the sample obtained from?" The reason for asking this question is because the raw MS^2 data are analyzed against a database that contains the amino acid sequences of the proteins from specific species. Only in rare instances will a protein sample be identified *de novo*, without comparing the experimental data to a database of known sequences.

Sequence databases have been absolutely critical to the development of proteomics. Most of the software algorithms discussed below utilize a sequence database to identify peptides from raw MS or MS^2 data. These databases can be composed of either amino acid sequence data or nucleotide sequences that are translated in protein sequences. As one could easily imagine, with the speed at which genomes are sequenced today, the number of entries within these databases has grown exponentially over the years.

There are a large number of sequence databases publicly available to the scientific community; however, most experimental MS data are analyzed against three databases: UniProt knowledgebase (UniProtKB), the NCBI nonredundant (NCBI nr) protein database, and the International protein index (IPI) database (2). Each of these databases is updated regularly to provide proteomic investigators access to the latest available protein information.

The UniProtKB (http://www.uniprot.org) is made up of a combination of the Swiss-Prot, PIR, and TrEMBL databases (3). Each UniProtKB entry contains the necessary core information for use in database search algorithms (e.g., Sequest, Mascot, and X!Tandem), such as the amino acid sequence, protein name or description, taxonomic data, and citation information. This database also contains additional metadata (i.e., data about data), such as each protein's sequence length, positions of potential posttranslational modifications, binding sites, biological ontologies, classifications, and cross-references. Approximately 85% of the protein sequences within UniProtKB are translated from nucleotide coding sequences that have been submitted to the EMBL-Bank/GenBank/DDBJ public databases.

The NCBI nr database (http://www.ncbi.nlm.nih.gov/sites/entrez?db = protein) is constructed by the merging of many different sequence databases (UniProtKB/Swiss-Prot, PIR, PRF, PDB, GenBank, and RefSeq) into a single database (4). The term "nonredundant" is used in the naming of this database as is signifies that identical sequences contained within multiple databases are merged into a single entry. While the merging of data from different sources enables the most comprehensive coverage of protein sequences available, it also results in inconsistency in the annotation. The annotation varies from highly documented entries (e.g., UniProtKB/Swiss-Prot) to machine predictions from of protein sequences from genomic sequences (e.g., RefSeq). Similar to UniProtKB, however, each entry within the NCBI nr database consists of a sequence, its accession number and version, and one or more cross-references to source databases making this database more than acceptable for protein identification search algorithms.

The IPI database (http://www.ebi.ac.uk/IPI), like the NCBI nr database, is composed of a combination of other databases (including UniProtKB, Ensembl, RefSeq, H-invDB, and VEGA) (5). The main difference between the IPI and NCBI nr databases is the method used to reduce sequence redundancy (2). Similar to the NCBI nr database, the annotation and sequence reliability in IPI can be highly variable. Fortunately, like the UniProtKB and NCBI nr databases, the IPI provides sequence information, unique protein accession numbers and version, and one or more cross-references to source sequences making this database perfectly acceptable for protein identification search algorithms. Unfortunately, the closure of IPI was announced in September 2011 with the recommendation that investigators utilize UniProtKB.

11.2.2 Peptide Mapping

Identifying a protein via peptide mapping involves digesting (usually with trypsin) an isolated protein or mixture containing no more than two to three proteins and acquiring the molecular weight (MW) of the resulting peptides (6). The experimental masses of the peptides are compared to masses within a list generated by an *in silico* digest of proteins or translated nucleic acid sequences within a database (7). The *in silico* digest is conducted using the cleavage rules of the enzyme used in the experiment. To be successfully identified, several of the experimental masses must agree with those for a

TABLE 11.1. Software Available for Protein Identification by
Peptide Mass Fingerprinting

Software Program	URL (http://)
MultiIdent	web.expasy.org
Mascot	www.matrixscience.com
MS-Fit	prospector.ucsf.edu
PepMAPPER	www.nwsr.manchester.ac.uk
MassSorter	services.cbu.uib.no
MassWiz	masswiz.igib.res.in
Protein Lynx	www.matrixscience.com
ProFound	prowl.rockefeller.edu

specific protein in the database. To optimize the confidence in the protein identification, it is best to acquire the peptide map on a high mass measurement accuracy instrument, such as a matrix-assisted laser desorption ionization-time-of-flight (MALDI-TOF). Increased confidence in the identification also correlates with the number of matches between the experimental and database peptide masses.

There are a number of useful programs, many freely available via the internet (Table 11.1). The minimum input requirements for these programs include an experimental peptide ion list and database to search the peptide ion list against. Some of these programs can incorporate additional information, such as isoelectric point (pI), MW, and source organism, as an aid to identify the protein. Inputs such as pI and MW can be determined based on the protein's migration on a one- or two-dimensional polyacrylamide gel electrophoresis (1D- or 2D-PAGE) gel. There is little objective basis for selecting which of the programs listed in Table 11.1 to use for analyzing peptide mapping data. The primary reasons an investigator selects a specific program include the fact that it is: (1) bundled with a mass spectrometer purchase, (2) easy to use, and (3) integrates with downstream bioinformatic algorithms.

11.2.3   Tandem Mass Spectrometry

While peptide mapping is useful for identifying proteins within simple mixtures, tandem MS (MS^2) is required for complex samples that may contain upward of 100,000 peptide species (8). With the advent of collision-induced dissociation (CID) and the popularity of ion-trap mass spectrometers, identification of proteins (via their surrogate peptides) using MS^2 dominates the field of MS-based proteomics. Obviously, MS^2 identification is required for achieving significant proteomic coverage of complex samples, but in my opinion, it provides higher confident identifications than peptide mapping even for simple mixtures. The reason is quite simple; the most distinctive characteristic that identifies a protein is its amino acid sequence. Tandem MS provides amino acid sequence information.

Tandem MS results in the fragmentation of peptides ions along its backbone. Most commonly used mass spectrometers yield b- and y-ions, which correspond to cleavages across the amide bond with the charge retained at the NH_2 and COOH termini, respectively (9). While other types of fragment ions, including a-, c-, x-, and z-ions, these are generally less intense and are associated with loss of small groups such as water or ammonia. For identification purposes, the MS^2 spectra are compared to a database of protein sequences using known rules for fragmentation of peptides under low energy conditions. A key for the ability to identify peptides based on their fragmentation patterns is the fact that the energy is not sufficient to completely dissociate every amide bond. If this were so, the mass spectrometer would be producing something akin to amino acid analysis. Instead, the fragmentation occurs in such a way that "ladders" of amino acid residues originating from the peptide are produced, much like a DNA sequencing ladder is produced by Sanger sequencing (10). The amino acid ladders can be interpreted to provide the sequence of residues within the peptide.

The first, and still one of the most popular software programs for analyzing MS^2 data in a high-throughput manner, is Sequest. Sequest was invented in 1994 by Jimmy Eng and John Yates (11). This program initially matches the mass of the precursor ion to possible peptides within a database having the same nominal mass. Theoretical MS^2 spectra are generated for these possible matches and then each is compared to the experimental MS^2 spectrum. A cross-correlation algorithm is used to score the quality of the matches and ranked based on their cross correlation (X_{corr}) score. The program also reports the difference between the first and second ranked peptide solution (ΔCn). Sequest can be used to identify posttranslational modifications as long as the user indicates the possibility of the modification on specific residue types. For example, when searching for a phosphopeptides, the user inputs a potential modification of 80 Da on serine, threonine, and tyrosine residues. The addition is indicated as dynamic, which permits Sequest to consider the possibility of each of these residues being either phosphorylated or not. Data input into Sequest must be in a specific file format (.dta); however, scripts are available for converting data from a variety of different mass spectrometer instruments into this format. One thing to be aware of when using Sequest; it will provide Xcorr scores for even poor-quality MS^2 spectra. Therefore, proteomic laboratories often place thresholds on what they consider an acceptable identification. Several years ago Dr. Li-Rong Yu conducted a meticulous study within our lab demonstrating that utilizing X_{corr} thresholds for tryptic peptide identification of 2.1 for $[M+H]^+$ peptides, 2.2 for $[M+2H]^{2+}$ peptides with $M_r < 1,200$ Da, 2.5 for $[M+2H]^{2+}$ peptides with $M_r \geq 1,200$ Da, 2.9 for $[M+3H]^{3+}$ peptides, and a ΔCn cutoff of 0.08 for all peptide charge states, resulted in a 95% confidence in peptide identification (12). While other laboratories may utilize different thresholds, they are all generally within the same approximate values and generally provide false positive identification rates of less than 5%.

Mascot is also a very popular algorithm for the analysis of MS^2 data (13). Mascot conducts a statistical evaluation of matches between the experimental and theoretical MS^2 data, providing a probability-based assignment of each spectrum to those within a selected database. Like Sequest, it allows for the identification of posttranslational modifications. Unlike Sequest, however, Mascot can utilize several different raw data

TABLE 11.2. Software Available for Protein Identification by Analysis of Tandem MS Data

Software Program	URL (http://)
Sequest	www.thermo.com
Mascot	www.matrixscience.com
MS-Tag	prospector.ucsf.edu
Pep-Frag	prowl.rockefeller.edu
OMSSA	pubchem.ncbi.nlm.nih.gov/omssa
Sonar MS/MS	hs2.proteome.ca/prowl/sonar/sonar_cntrl.html
X!Tandem	www.thegpm.org/tandem
Crux	noble.gs.washington.edu/proj/crux

file formats including .dta files. Like Sequest, Mascot owes much of its popularity to the fact that it is bundled as part of the software component of many popular types of mass spectrometers used today. A list of other publicly available software programs for analyzing MS^2 data is provided in Table 11.2.

So of the possible search engines for MS^2 data analysis, which one is best? Recently Dr. Cheng Lee's group analyzed the same 155,973 MS^2 spectra acquired on a single linear ion-trap mass spectrometer using four different software algorithms (Sequest, Mascot, X!Tandem, and OMSSA) (14). Their results showed that for peptide identification the search algorithms provided relatively comparable results when consistent scoring procedures and false discovery rates were used for all the programs. Does this mean that each program identified exactly the same exact number of proteins? Of course not! Different software algorithms will sometimes identify a peptide from an MS^2 spectrum that another does not or even assign different peptide identifications to the same spectrum. Before the reader panics, however, keep in mind the overlap between the search algorithms is very high. Dr. Steve Gygi's group at Harvard showed that in the analysis of almost 50,000 MS^2 spectra acquired on a linear ion-trap mass spectrometer, the overlap in identifications between Mascot and Sequest was over 85% (15). Each algorithm produced about 15% of identifications that were unique.

There may still be investigators that are concerned that different algorithms do not produce exactly the same results starting with the exact same data. Admittedly, this would be a concern if the goal was to identify a highly purified protein and the algorithms produced a different result. This outcome is rarely, if ever, seen for a protein that provides a series of MS^2 spectra of reasonable quality. In global analysis, thousands of proteins are generally identified and the overall goal in these discovery driven experiments is to find a group of proteins that suggest some specific function relevant to the researchers' interest is activated. There should not be any real concern if either Sequest or Mascot missed 15% of the proteins that could have been identified. The amount of data provided is generally more than sufficient to formulate a hypothesis-driven study. For those who may still be unconvinced, it is always possible to analyze the data using multiple software packages and combine the results.

11.2.4 Peptide Spectral Libraries

Another bioinformatic method for identifying peptides from MS^2 data was proposed by Dr. John Yates in 1999 (16). This concept, known as spectral library searching, was based on the notion that peptide MS^2 spectra were reproducible and could be identified based on comparison of their profile to spectra within an annotated database that were acquired using the same conditions. Spectral libraries are searchable collections of spectra for which the identity of the originating molecule is known and have been employed for small molecules for decades. Spectral library searching is routinely used to identify small molecules detected using gas chromatography (GC)-MS or LC-MS (17). The National Institute of Standards and Technology/National Institutes of Health/Environmental Protection Agency (NIST/NIH/EPA) mass spectral library (http://www.nist.gov/srd/nist1a.cfm) contains over 200,000 entries of MS^2 spectra of primarily organic analytes. Unfortunately, at the time Dr. Yates proposed the concept of spectral library searching for proteomics, there was insufficient data to build the necessary database. Owing to the development and widespread use of shotgun proteomics leading to the rapid accumulation of peptide MS^2 spectra and the emergence of proteomic data repositories, several spectral searching methods were published in 2006–2007 including X!Hunter (18), Bibliospec (19), and SpectraST (20). In the same year, NIST expanded their mass spectral library to include peptides. The adoption of open and standardized data formats allowed data acquired by different laboratories to be effectively combined in growing libraries of peptide MS^2 spectra (21, 22). There are presently almost 1 million peptide reference spectra contained within 18 libraries of different organisms and biological samples that are publicly available in the NIST Libraries of Peptide Tandem Mass Spectra (www.peptide.nist.gov/). While still not widely used, in the future all peptides may be identified through direct comparison of experimental MS^2 spectra to those already contained within spectral library databases, in a manner analogous to the identification of small molecules.

11.3 PROTEIN QUANTITATION SOFTWARE

Besides identification, the major aim of proteomics experiments is the determination of relative protein abundances in different samples. As discussed in previous chapters, there are a myriad of different methods for comparing the relative abundances of proteins within proteomes. The four most popular methods for comparing protein abundances are 2D-PAGE, stable-isotope labeling, label free quantitation, and spectral counting.

11.3.1 SDS-PAGE

The first global method for characterizing proteomes was 2D-PAGE (23). Along with improvements in the ability to resolve complex mixtures, major advancements in software packages have enable the automated comparison of digitalized gel images. Quantitative analysis of gel images is a critical step in the identification of alterations in protein expression within a given biological system. The images are generally captured using laser densitometry, phosphor imaging, or by a charged coupled device camera.

Gel image analysis is performed using software packages that provide accuracy, reproducibility, sensitivity, and speed. These programs use a variety of different mathematical algorithms to perform spot recognition and quantitation of each spot prior correlating each position to a spot within a comparative gel image(s) (24). A majority of proteins (i.e., 70%) within a 2D-PAGE image will overlap with other proteins making the selection of software settings for determining the spot size, shape, and border of each protein critical to the overall results. Studies have shown that the selection of these parameters can result in significant error in the quantitative results (25, 26).

The biggest challenges in obtaining accurate quantitation using software based analysis comparison of 2D-PAGE gels include overlapping spots, weakly stained spots (e.g., low abundance proteins), and differential protein migration between gels resulting in mismatched spots. There are a large number of different software programs for quantitatively comparing 2D-PAGE gels including PDQuest, Progenesis, AlphaMatch 2D, Z3, Delta2D, ImageMaster, Melanie, and Proteome-Weaver (24). An important consideration when conducting long-term proteomic studies using 2D-PAGE is consistency in the analysis software. While software packages such as PDQuest and Progenesis will generally provide the same result for resolved protein spots, the measurements they each report for poorly resolved spots can vary (27).

11.3.2 Stable Isotope Labeling

There are a number of different algorithms for analyzing stable-isotope labeling data (Table 11.3) but they all work on the same premise of comparing the peak areas of peptides obtained from different proteomic sources. The different isotopic forms of a specific peptide will have identical elution profiles (except in some cases of deuterated

TABLE 11.3. Software Programs for Analysis of Isobaric[a] and Isotopic[b] Labeling Quantitative Proteomic Studies

Software	Label	Compatible Labels	URL (http://)
Multi-Q[a]	iTRAQ	Specific	ms.iis.sinica.edu.tw/Multi-Q
iTracker[a]	iTRAQ	Specific	www.cranfield.ac.uk/health/researchareas/ bioinformatics
Libra[a]	iTRAQ	Specific	tools.proteomecenter.org/wiki/
ProQuant[a]	iTRAQ	Specific	www.appliedbiosystems.com
ProteinPilot[a]	iTRAQ	Specific	www.appliedbiosystems.com
XPRESS[b]	ICAT	SILAC, ICPL	tools.proteomecenter.org/wiki/
ASAPRatio[b]	^{2}H	SILAC, ICPL, ICAT	tools.proteomecenter.org/wiki/
MSQuant[b]	SILAC	ICAT, ICPL	msquant.sourceforge.net
PeakPicker[b]	ICPL	Specific	www.appliedbiosystems.com
WARP-LC[b]	ICPL	Generic	www.bdal.com
STEM[b]	^{18}O	Generic	www.sci.metro-u.ac.jp/proteomicslab/ STEMDLP-0.html
ZoomQuant[b]	^{18}O	Specific	proteomics.mcw.edu/zoomquant

Abbreviations: iTRAQ: isobaric tag for relative and absolute quantitation; ICAT: isotope-coded affinity tags; SILAC: stable isotope labeling of amino acids in culture; ICPL; isotope-coded protein labeling.

standards); however, they are detectable by the predictable mass shift incorporated by the isotopic label. Once the peptides are recognized, a ratio between the "heavy" and "light" labeled peptide is calculated. If multiple peptides from a single protein are identified, their quantitative ratios can be averaged and their standard deviation calculated. Many software tools such as ASAPRatio (28), MaxQuant (29), Census (30), Vista (31), and UNiquant (32) have been produced to analyze stable isotope labeling-based quantitative proteomics data.

Isobaric labeling represents a slight twist on stable isotope labeling as reporter groups liberated during CID are used to quantitate differences in the abundance of proteins in up to 8 different samples concurrently (e.g., iTRAQ). There are a number of freely available software programs for analyzing this data including iTracker (33), Multi-Q (34), LTQ-iQuant (35), and the iTRAQ reporter ion counter (36). These software packages can utilize preprocessed MS^2 data from both Sequest and Mascot.

11.3.3 Label-Free Quantitation

Label-free quantitation is based on integrating the chromatographic peaks areas for any given peptide identified in an LC-MS^2 analysis (37). The peak area is considered as directly proportional to the concentration of the peptide, typically in the concentration range of 10 fmol-1000 pmol. The peak areas of a peptide measured in a series of LC-MS^2 experiments analyzing different samples are compared to calculate any change in its relative abundance. Although label free quantitation sounds straightforward, several parameters need to be considered to ensure reproducible and accurate quantitation when multiple samples are being compared. One of the foremost concerns is LC retention time, a major variable in proteomic analysis. Computational methods are required to properly align chromatograms so that the correct peptides are being compared between runs. If the peak alignment was done successfully, it will be confirmed by the MS^2 identifications. Another issue is peak abundance normalization. The finicky and sometimes erratic nature of the electrospray process or variability in column loading can cause global increases or decreases in ion signals. To correct for these phenomena, normalization software is applied to each LC-MS run.

11.3.4 Spectral Counting Quantitation

Spectral counting is another option for quantitating proteins across complex proteome samples (38). Spectral counting bases quantitation on the number of MS^2 identifications assigned to a protein. Spectral counting is probably the least software intensive method for conducting quantitative proteomics. Once the number of peptides per protein are identified and tabulated, these numbers can be compared using simple programs such as Microsoft Excel and Access.

A further refinement of using spectral count was developed by Dr. Matthias Mann (39). This refinement called exponentially modified protein abundance index (emPAI) takes into account not only the number of peptides identified per protein but also the MW of each protein. For example, when tryptically digested, larger proteins are more likely to provide more *observable* peptides in the preferred mass range for MS^2 analysis. The numbers of observable peptides per protein are calculated by conducting an *in silico* digest of the proteins within the targeted database and tabulating those that fall within the scan range of the mass spectrometer being used in the experiment. Peptides with extremely high hydrophilic or hydrophobic values are eliminated from the list of observable peptides. An in-house program was written in PHP to calculate the peptide number and was used to export all data to Microsoft Excel. A program is freely accessible to calculate the observable peptide number and export the data into Microsoft Excel (xome.hydra.mki.co.jp:8080/bitt/common/Menu). The number of experimentally observed peptides can be counted as either: (1) unique parent ions, (2) unique sequences, and (3) unique sequences that overlap owing to the presence of a missed tryptic cleavage (e.g., ASDGR and ASDGRR count as separate experimentally observed peptides). emPAI can be used for relative quantitation to compare protein abundances between different proteome samples but can also be used to rank the relative concentration of proteins within a single proteome sample.

In an application of emPAI, our laboratory worked with the laboratory of Dr. Jim Hartley, the inventor of Gateway® Recombinant Cloning for protein expression (40). In this application, a collection of 512 open reading frames (ORFs) from various human proteins were transfected into *Escherichia* (*E.*) *coli* cells (41). The cells were induced to express the various ORFs. The cells were collected and their proteomes extracted and analyzed using LC-MS^2. The relative levels of expressed human proteins in the *E. coli* proteome was calculated using emPAI. The purpose of the experiment was to quickly determine which ORFs expressed under the growth conditions utilized. The scores of the identified proteins were tabulated and human proteins that consistently gave high emPAI values were subjected to small-scale expression and purification validation. These clones also showed high levels of expression of soluble protein in individual experiments while those with low emPAI scores did not show any detectable expression. Overall, these results validated the ability of emPAI to provide a measure of the relative abundances of proteins in complex mixtures.

11.4   PROTEIN PATHWAY ANALYSIS

Now that all of the proteins have been identified and their relative quantities compared between proteomic samples, what do you do with all those numbers? One approach is to find an interesting protein in the list of those that show a change in abundance and use techniques such as Western blotting to validate the MS result. This approach is entirely appropriate for many proteomic experiments such as protein complex characterization and biomarker discovery. For global profiling experiments, however, the jump from a list containing thousands of proteins to the focus on a single protein seems too drastic. The amount of data that would be utilized in the proceeding experiments would typically be less than 0.1%.

An alternative is to use analysis tools that group the results into curated pathways of proteins that are known to interact. One of the most popular programs for assimilating proteomic data into protein pathways is Ingenuity Pathway Analysis (IPA) software (Ingenuity Systems, http://www.ingenuity.com/). Other popular software tools for this application include STRING (http://string-db.org/) GeneGo MetaCore (http://www

.genego.com/metacore.php) or Ariadne Pathway Studio Similar (http://www.ariad negenomics.com/products/pathway-studio/).

As an example of how these protein pathway programs function, I am going to use IPA. This software constructs hypothetical protein-interaction clusters from data present in the Ingenuity Pathways Knowledge Base (IPKB). The IPKB is a large curated, regularly updated, database that lists millions of individual relationships and interactions. The protein relationships are selected from published manuscripts (42, 43). Not only does IPA construct protein-interaction clusters, but it also groups the proteins into a wide range of different categories based on protein function, cellular localization, and small molecule and disease interrelationships. Protein pathways are displayed as nodes that represent individual proteins, and edges that represent the biological relationships between proteins. To input data into IPA, the data should list the proteins with their SWISS-PROT or NCBI number along with a ratio indicating the expression ratio of the proteins in the proteome samples being compared.

This input is used by IPA to build hypothetical protein pathways. Not every node within a pathway will be populated as the experimental proteomic data will not contain information for every protein in the curated cluster. For example, Figure 11.1 shows protein pathways curated using IPA populated using both proteomic and microarray data of osteoblast cells treated with inorganic phosphate. The color-coded change in abundance at the proteome level is shown on the left side of the oval, with the change in the corresponding transcript on the right side. As can be seen, there are several instances where transcript data were acquired without the corresponding protein being detected. The data are color coded to show the relative abundances changes in the apoptosis-specific proteins and transcripts resulting from inorganic phosphate treatment of the cells. In this example, the data from IPA were redrawn to visually enhance the presentation. Protein pathway generation is ranked by a score based on the inclusion of the proteins from the inputted expression profile. This *p*-score is calculated for each populated pathway. IPA computes a p-score for each possible network according to the fit of that network to the inputted proteins. The p-scores are generally used as an indication of the probability that proteins within a specific pathway are differentially abundant between the proteomes being compared. A p-score of ≥ 2 is considered to have at least a 99% confidence of not being generated by random chance alone. As shown in Figure 11.2, IPA can display a list of significantly altered functional pathways along with the specific proteins that were identified as being up- or downregulated in the pathway.

11.5 PROTEOME DATA REPOSITORIES

In Section 11.2.1 entitled "Protein Databases," I described some of the common databases that play an essential role in the analysis of raw MS data. These databases provide the foundation to which raw MS data are compared to provide protein identifications. Databases also play a critical role after protein identification by serving as repositories of data gleaned from proteomic experiments. The construction and public availability of proteomic data repositories has been critical to the development of proteomics enabling

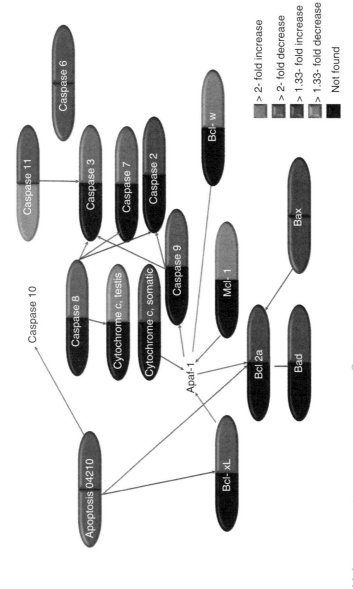

Figure 11.1. Protein pathway generation using IPA®. The IPA software is able to display quantitative proteomic (left side) and microarray (right side) data. Color codes are used to indicate the relative abundance changes in both types of molecules.

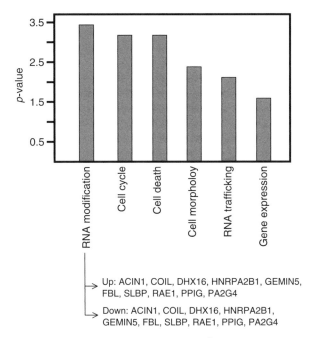

Figure 11.2. Example of data display provided by IPA®. The probability that the proteins within a curated protein pathway show a significant change in abundance based on the proteomic data input is displayed. The specific proteins that are up- or downregulated within each functional pathway can also be displayed.

investigators to share data and maximize the information contained within their own data sets. The main publicly available databases for proteomics data are the Global Proteome Machine Database (44), PeptideAtlas (45), the PRoteomics IDEntifications database (46), NCBI Peptidome (47), Human Proteinpedia (48), and Tranche (49).

These data repositories serve a major need in the proteomics community by allowing investigators to access data for other laboratories that are conducting similar studies. This capability can lead to the validation of important results and also allow errors in the experimental data to be quickly recognized. The repositories also serve as inventories of raw MS data for investigators interested in studying peptide fragmentation patterns or creating spectral libraries to develop novel methods for peptide identification.

Initially, the rate at which these data repositories grew was based on voluntary submission; however, a number of journals (e.g., *Molecular and Cellular Proteomics*) and funding agencies (e.g., NIH and The Wellcome Trust) now require investigators to deposit their proteomic data into these databases. These requirements are similar to those used to grow the Protein Data Bank, which contains three-dimensional structures of proteins determined primarily using X-ray crystallography and nuclear magnetic resonance spectroscopy. Obviously, these requirements have spurred the growth of these databases; however, they still do not approach the critical mass of data found in genomic and transcriptomic databases.

11.5.1   Information Storage in MS-Based Proteomics Databases

As the reader is undoubtedly realizing from perusing this book, MS is used for a wide variety of different experiments, each with a distinct goal. Some experiments are designed to identify large numbers of proteins within an entire cell lysate, while others aim to characterize a small subset of proteins. Not only is identification important, but many experiments focus on quantifying each protein or measuring sites that are posttranslationally modified. This variety of different MS experiments has an effect on the types of data that are generated and, therefore, the data formats. While repositories may have different requirements, there are three basic levels of information that must be captured: original raw MS data, results obtained from preliminary analysis of the raw data, and metadata.

The original raw MS data refer to the data recorded directly by the mass spectrometer. While the acquired mass spectra are typically stored originally in a binary instrument specific formatted file, most can be converted to open, XML-based file formats such as mzXML, mzData, or mzML format. Unfortunately, these XML files can be very large making text files that contain peak lists the most popular means of storing raw MS data. The results obtained from preliminary analysis of the raw data refer most commonly to the peptide/protein sequences determined using a specific search engine (e.g., Sequest or Mascot) and can include such things as posttranslational modifications. Depending on the search engine used to obtain the data, these results can be in many different formats depending on the search engine used. Fortunately, many of these formats are text based. The increasing move toward quantitation has added an additional challenge to data storage within repositories as many different quantitative methods exist and the reliability of some of these is not robust. At present, capturing qualitative data is much less of an issue than quantitative data.

The third critical piece of required information is metadata. This information typically contains the author's contact information, sample type, experimental protocol, type of mass spectrometer, data processing method, etc. Metadata are very important because these allow investigators the opportunity to determine how compatible the data within the repository are with their own data. Beyond ensuring that the same type of mass spectrometer was utilized, it is important to compare sample preparation protocols as these methods can be very specific to an experimental aim. Personally, I think that the author's contact information is the most important, and underutilized, piece of information within the metadata. In my opinion, anyone who utilizes or reanalyzes data within a repository should be required to contact the depositor of the data to obtain greater detail on how the data were generated and discuss their purpose of using the data. I also believe that if the data are used in any form of publication, the original depositor should be able to review the final manuscript prior to its publication and be included as an author.

11.6   CONCLUSIONS

I remember as an undergraduate student using scissors to carefully cut the peaks from a piece of recording paper generated by the detector within a GC system. These pieces were weighed on a balance and the resulting mass measurements extrapolated to determine the quantity of the compound I was measuring. While this method may seem laughable to many of the younger readers of this book, I am sure the more mature reader can relate. Now every measurement we take in the lab is driven by software. Although the Dutch theoretical biologist Paulien Hogeweg along with her colleague Ben Hesper coined the term bioinformatics in the early 1970s (50), I never heard the word bioinformatics as an undergraduate and graduate student during the years 1986–1994. In proteomics, the term "bioinformatics" is now as commonplace as the term "MS."

With the rapid advances in proteomics technologies, however, there is this constant race between the ability to gather data and the capability to analyze it. My first studies using MS involved measuring the calcium binding to a single isolated protein (i.e., calbindin D_{28K}) (51). While these studies required software to operate the mass spectrometer and deconvolute the spectra, most of the calculations could be done using "paper and pencil." The major trends in proteomics today, which include biomarker discovery and global quantitative comparisons, are absolutely dependent on bioinformatic tools. Even the identification of a single protein requires software to scan the available databases for the correct match between the experimental and *in silico* data. Considering state-of-the-art laboratories can generate data to quantitatively compare thousands of proteins, the bioinformatic needs are obvious. While there are a number of very useful programs available, none have yet reached the ability to utilize all of the data acquired in these global studies. In a majority of cases, the bioinformatics software is capable of providing enough information that leads investigators to the next correct steps in their research (e.g., pathway and network analysis). Even the best bioinformatics tools available, however, are still incapable of fitting *all* of the data acquired in global proteomic studies into a neat, complete package describing the dynamics that occur within a proteome. There are a number of factors that limit our ability to use all of the available data, with our lack of knowledge of the intricate dynamic interactions between genes, transcripts, proteins, and metabolites being the major limiting factor. The progress that has been made in such a short period of time makes it is easy to envision a point in the near future where bioinformatic tools will be able to take all the available data and provide an accurate working model of how cells and organisms respond to external and internal perturbations at a molecular level.

REFERENCES

1. The Rime of the Ancient Mariner. Samuel Taylor Coleridge. D. Appleton and Company. 346 and 348 Broadway, New York, NY. 1857.

2. Griss J, Côté RG, Gerner C, Hermjakob H, Vizcaíno JA. Published and perished? The influence of the searched protein database on the long-term storage of proteomics data. Mol. Cell. Proteomics 2011;10:M111.008490.

3. Magrane M, Consortium U. UniProt knowledgebase: A hub of integrated protein data. Database (Oxford) 2011;2011:bar009.

4. Sayers EW, Barrett T, Benson DA, Bolton E, Bryant SH, Canese K, Chetvernin V, Church DM, Dicuccio M, Federhen S, Feolo M, Fingerman IM, Geer LY, Helmberg W, Kapustin Y,

Krasnov S, Landsman D, Lipman DJ, Lu Z, Madden TL, Madej T, Maglott DR, Marchler-Bauer A, Miller V, Karsch-Mizrachi I, Ostell J, Panchenko A, Phan L, Pruitt KD, Schuler GD, Sequeira E, Sherry ST, Shumway M, Sirotkin K, Slotta D, Souvorov A, Starchenko G, Tatusova TA, Wagner L, Wang Y, Wilbur WJ, Yaschenko E, Ye J. Database resources of the national center for biotechnology information. Nucleic Acids Res 2012;40: D13–D25.

5. Kersey PJ, Duarte J, Williams A, Karavidopoulou Y, Birney E, Apweiler R. The international protein index: An integrated database for proteomics experiments. Proteomics 2004;4:1985–1988.

6. Pappin DJ, Hojrup P, Bleasby AJ. Rapid identification of proteins by peptide-mass finger-printing. Curr. Biol. 1993;3:327–332.

7. Sutton CW, Pemberton KS, Cottrell JS, Corbett JM, Wheeler CH, Dunn MJ, Pappin DJ. Identification of myocardial proteins from two-dimensional gels by peptide mass fingerprinting. Electrophoresis 1995;16:308–316.

8. McLafferty FW. Tandem mass spectrometry. Science 1981;214:280–287.

9. Roepstorff P, Fohlman J. Proposal for a common nomenclature for sequence ions in mass spectra of peptides. Biomed. Mass Spectrom. 1984;11:601.

10. Sanger F, Nicklen S, Coulson AR. DNA sequencing with chain-terminating inhibitors. Proc. Natl. Acad. Sci. U.S.A. 1977;74:5463–5467.

11. Eng JK, McKormack AL, Yates JR. An approach to correlate tandem mass-spectral data of peptides with amino-acid-sequences in a protein database. J. Am. Soc. Mass Spectrom 1994;5:976–989.

12. Yu LR, Conrads TP, Uo T, Kinoshita Y, Morrison RS, Lucas DA, Chan KC, Blonder J, Issaq HJ, Veenstra TD. Global analysis of the cortical neuron proteome. Mol. Cell. Proteomics 2004;3:896–907.

13. Perkins DN, Pappin DJC, Creasy DM, Cottrell JS. Probability-based protein identification by searching sequence databases using mass spectrometry data. Electrophoresis 1999;20:3551–3567.

14. Balgley BM, Laudeman T, Yang L, Song T, Lee CS. Comparative evaluation of tandem MS search algorithms using a target-decoy search strategy. Mol. Cell. Proteomics 2007;6:1599–1608.

15. Elias JE, Haas W, Faherty BK, Gygi SP. Comparative evaluation of mass spectrometry platforms used in large-scale proteomics investigations. Nat. Methods 2005;2:667–675.

16. Yates JR 3rd, Morgan SF, Gatlin CL, Griffin PR, Eng JK. Method to compare collision-induced dissociation spectra of peptides: Potential for library searching and subtractive analysis. Anal. Chem. 1998;70:3557–3565.

17. Dehaven CD, Evans AM, Dai H, Lawton KA. Organization of GC/MS and LC/MS metabolomics data into chemical libraries. J. Cheminform 2010 Oct 18;2(1):9.

18. Craig R, Cortens JC, Fenyo D, Beavis RC. Using annotated peptide mass spectrum libraries for protein identification. J. Proteome Res 2006;5:1843–1849.

19. Frewen BE, Merrihew GE, Wu CC, Noble WS, MacCoss MJ. Analysis of peptide MS/MS spectra from large-scale proteomics experiments using spectrum libraries. Anal. Chem. 2006;78:5678–5684.

20. Lam H, Deutsch EW, Eddes JS, Eng JK, King N, Stein SE, Aebersold R. Development and validation of a spectral library searching method for peptide identification from MS/MS. Proteomics 2007;7:655–667.

21. Riffle M, Eng JK. Proteomics data repositories. Proteomics 2009;9:4653–4663.

22. Vizcaíno JA, Foster JM, Martens L. Proteomics data repositories: Providing a safe haven for your data and acting as a springboard for further research. J. Proteomics 2010;73:2136–2146.

23. Cordwell SJ, Wilkins MR, Cerpa-Poljak A, Gooley AA, Duncan M, Williams KL, Humphery-Smith I. Cross-species identification of proteins separated by two-dimensional gel electrophoresis using matrix-assisted laser desorption ionisation/time-of-flight mass spectrometry and amino acid composition. Electrophoresis 1995;16:438–443.

24. Marengo E, Robotti E, Antonucci F, Cecconi D, Campostrini N, Righetti PG. Numerical approaches for quantitative analysis of two-dimensional maps: A review of commercial software and home-made systems. Proteomics 2005;5:654–666.

25. Wheelock AM, Buckpitt AR. Software-induced variance in two-dimensional gel electrophoresis image analysis. Electrophoresis 2005;26:4508–4520.

26. Damodaran S, Rabin RA. Minimizing variability in two-dimensional electrophoresis gel image analysis. OMICS 2007;11:225–230.

27. Arora PS, Yamagiwa H, Srivastava A, Bolander ME, Sarkar G. Comparative evaluation of two two-dimensional gel electrophoresis image analysis software applications using synovial fluids from patients with joint disease. J. Orthop. Sci 2005;10:160–166.

28. Li XJ, Zhang H, Ranish JA, Aebersold R. Automated statistical analysis of protein abundance ratios from data generated by stable-isotope dilution and tandem mass spectrometry. Anal. Chem. 2003;75:6648–6657.

29. Cox J, Mann M. MaxQuant enables high peptide identification rates, individualized p.p.b.-range mass accuracies and proteome-wide protein quantification. Nat. Biotechnol. 2008;26:1367–1372.

30. Park SK, Venable JD, Xu T, Yates JR 3rd. A quantitative analysis software tool for mass spectrometry-based proteomics. Nat. Methods 2008;5:319–322.

31. Bakalarski CE, Elias JE, Villén J, Haas W, Gerber SA, Everley PA, Gygi SP. The impact of peptide abundance and dynamic range on stable-isotope-based quantitative proteomic analyses. J. Proteome Res. 2008;7:4756–4765.

32. Huang X, Tolmachev AV, Shen Y, Liu M, Huang L, Zhang Z, Anderson GA, Smith RD, Chan WC, Hinrichs SH, Fu K, Ding SJ. UNiquant, a program for quantitative proteomics analysis using stable isotope labeling. J. Proteome Res. 2011;10:1228–1237.

33. Shadforth IP, Dunkley TPJ, Lilley KS, Bessant C. i-Tracker: For quantitative proteomics using iTRAQ. BMC Genomics 2005;6:145.

34. Lin WT, Hung WN, Yian YH, Wu KP, Han CL, Chen YR, Chen YJ, Sung TY, Hsu WL. Multi-Q: A fully automated tool for multiplexed protein quantitation. J. Proteome Res. 2006;5:2328–2338.

35. Onsongo G, Stone MD, Van Riper SK, Chilton J, Wu B, Higgins L, Lund TC, Carlis JV, Griffin TJ. LTQ-iQuant: A freely available software pipeline for automated and accurate protein quantification of isobaric tagged peptide data from LTQ instruments. Proteomics 2010;10:3533–3538.

36. Griffin TJ, Xie H, Bandhakavi S, Popko J, Mohan A, Carlis JV, Higgins L. iTRAQ reagent-based quantitative proteomic analysis on a linear ion trap mass spectrometer. J. Proteome Res. 2007;6:4200–4209.

37. Neilson KA, Ali NA, Muralidharan S, Mirzaei M, Mariani M, Assadourian G, Lee A, van Sluyter SC, Haynes PA. Less label, more free: Approaches in label-free quantitative mass spectrometry. Proteomics 2011;11:535–553.

38. Carvalho PC, Hewel J, Barbosa VC, Yates JR 3rd. Identifying differences in protein expression levels by spectral counting and feature selection. Genet. Mol. Res. 2008;7:342–356.

39. Ishihama Y, Oda Y, Tabata T, Sato T, Nagasu T, Rappsilber J, Mann M. Exponentially modified protein abundance index (emPAI) for estimation of absolute protein amount in proteomics by the number of sequenced peptides per protein. Mol. Cell. Proteomics 2005;4:1265–1272.

40. Hartley JL. Use of the gateway system for protein expression in multiple hosts. Curr. Protoc. Protein Sci. 2003;5:5–17.

41. Waybright T, Gillette W, Esposito D, Stephens R, Lucas D, Hartley J, Veenstra T. Identification of highly expressed, soluble proteins using an improved, high-throughput pooled ORF expression technology. Biotechniques 2008;45:307–315.

42. Raponi M, Belly RT, Karp JE, Lancet JE, Atkins D, Wang Y. Microarray analysis reveals genetic pathways modulated by tipifarnib in acute myeloid leukemia. BMC Cancer 2004;4:56.

43. Siripurapu V, Meth J, Kobayashi N, Hamaguchi M. DBC2 significantly influences cell-cycle, apoptosis, cytoskeleton and membrane-trafficking pathways. J. Mol. Biol. 2005;346:83–89.

44. Zhang CC, Rogalski JC, Evans DM, Klockenbusch C, Beavis RC, Kast J. In silico protein interaction analysis using the global proteome machine database. J. Proteome Res. 2011;10:656–668.

45. Deutsch EW, Lam H, Aebersold R. PeptideAtlas: A resource for target selection for emerging targeted proteomics workflows. EMBO Rep. 2008;9:429–434.

46. Martens L, Hermjakob H, Jones P, Adamski M, Taylor C, States D, Gevaert K, Vandekerckhove J, Apweiler R. PRIDE: The proteomics identifications database. Proteomics 2005;5:3537–3545.

47. Slotta DJ, Barrett T, Edgar R. NCBI peptidome: A new public repository for mass spectrometry peptide identifications. Nat. Biotechnol. 2009;27:600–601.

48. Prasad TS, Kandasamy K, Pandey A. Human protein reference database and human proteinpedia as discovery tools for systems biology. Methods Mol. Biol. 2009;577:67–79.

49. Smith BE, Hill JA, Gjukich MA, Andrews PC. Tranche distributed repository and proteomecommons.org. Methods Mol. Biol. 2011;696:123–145.

50. Hogeweg P. The roots of bioinformatics in theoretical biology. PLoS Comput. Biol. 2011;7:e1002021.

51. Veenstra TD, Johnson KL, Tomlinson AJ, Naylor S, Kumar R. Determination of calcium-binding sites in rat brain calbindin D28K by electrospray ionization mass spectrometry. Biochemistry 1997;36:3535–3542.

12

FUTURE PROSPECTS OF PROTEOMICS IN CANCER RESEARCH

12.1 INTRODUCTION

The era of proteomics erupted on the scene with great fanfare and expectations. Most of the expectations were centered on the protein identification rate of mass spectrometers. Therefore, the early part of this proteomics era was focused on how mass spectrometry (MS) could be used to obtain comprehensive analysis a cell's, tissue's, or organism's proteome. Along with MS, fractionation using two-dimensional polyacrylamide gel electrophoresis and liquid chromatography (LC) were the tools of choice. Mass spectrometers continued to improve in sensitivity and speed (1). Methods to quantitate proteins within proteomes, through the use of stable isotopes and label-free, were developed (2). Novel separation strategies were devised to fractionate proteomes prior to MS analysis (3). These separation methods not only provided higher resolution separation of entire proteomes, but also allowed tools to extract specific classes of proteins (e.g., subcellular fractionation of mitochondrial or membrane proteins) (4) or peptides (e.g., phosphopeptides) (5,6). Comparing the relative quantity of proteins in two samples was anticipated to reveal coordinated changes in protein abundances and make "systems

Proteomic Applications in Cancer Detection and Discovery, First Edition. Timothy D. Veenstra.
© 2013 John Wiley & Sons, Inc. Published 2013 by John Wiley & Sons, Inc.

biology" the wave of the future. This same analytical principal was also anticipated to begin a revolution in the discovery of biomarkers. MS was also going to provide the technology for mapping out the entire protein interactome. Obviously, these goals have not yet been fully achieved; however, tremendous technological strides have been made in just the past decade.

12.2 SIGNIFICANT DEVELOPMENTS OVER THE PAST DECADE

The speed at which proteins can be identified is the single biggest development in proteomics in the past decade. While there are a number of factors behind this increase, developments in MS technology has been the biggest contributor. Over the past decade, advances in MS technology have come along as rapidly as those for personal computers. It often seems that just as the laboratory is becoming comfortable with their latest MS purchase, an improved instrument is being released. Thinking back to the year 2000 when my group started our laboratory here at the National Cancer Institute, state-of-the-art proteomic laboratories were dominated by QqTOFs (7) and LCQ Deca ion traps (8). A few laboratories, such as Dr. Richard Smith at Pacific Northwest National Laboratories or Dr. Alan Marshall at the High Magnetic Field Laboratory, possessed Fourier transform ion cyclotron resonance (FTICR) mass spectrometers. Instruments with these capabilities, however, were definitely the exception and not the rule. While QqTOF's provided high resolution and mass accuracy for proteomics analysis, their duty cycle was slow. LCQ Deca's had a reasonable duty cycle, but their resolution and mass accuracy was poor. A major step forward was the introduction of the LTQ linear ion trap (9). This instrument not only had a rapid duty cycle but greater sensitivity than existing ion traps owing to the increased capacity of the ion trap. Arguably, the next great leap in making high performance MS available to the general community, was the release of FTICR mass spectrometers by Bruker and Thermo-Scientific. While Bruker's instrument was introduced earlier, Thermo-Scientific's release was more instrumental in bringing FTICR to laboratories on a broad scale. Subsequent to this advance, Orbi-trap mass spectrometers have brought capabilities beyond FTICR technology with less maintenance since it does not require cryogens. Nowadays, almost every MS laboratory has access to state-of-the-art MS.

While the past few years could be termed the "decade of the ion-trap," triple quadrupole MS development has also been impressive, primarily during the past 5 years. Triple quadrupole technology has been a cornerstone of analytical chemistry owing to its ability to quantitate specific molecules in complex mixtures. Triple-quadrupole MS has historically had the greatest impact in quantitating metabolites. As a glutton of potential biomarkers discovered using MS continues to grow; however, proteomic laboratories began looking for ways to verify or validate their efficacy. Therefore, techniques such as multiple reaction monitoring (MRM) for quantitating peptides using triple quadrupole mass spectrometers have become popular in the past few years (10). MS vendors responded in kind with the release increasingly sensitive instruments such as the TSQ Vantage and API 5500. By the time you read this book, mass spectrometers with even greater capabilities will have been commercially released.

12.3   THE MOVE FROM WESTERN TO MASS SPECTROMETRY

While instrumental improvements are always nice (e.g., most people would enjoy having a nicer automobile with more features), they are worth little unless they change how things are done (e.g., cannot legally drive a Lamborghini Aventador at 217 mph on American highways). One subtle change that has occurred over the past decade has been the movement from Western blotting to MS for protein identification and characterization (11). In the early part of this century, it was considered necessary to validate MS results using Western blotting. This requirement was often a challenge to MS labs whose expertise resided in instrumentation, not molecular biology. Over time the requirements started to change. MS labs started receiving requests from molecular biologists to confirm Western blot results. This request is particularly evident in the identification of posttranslational modifications (PTMs), where the specificity of MS is favored over that provided by antibodies (12, 13). While it has not completely replaced antibody-based molecular detection methods, MS's contribution to molecular biology has definitely increased as evidenced by the number of *Science*, *Nature*, and *Cell* manuscripts that currently rely on MS data for their conclusions.

12.4   THE FUTURE OF PROTEIN IDENTIFICATION

The future for protein identification is more, more, more. While this may sound contradictory to some other points made in this book and even this chapter, the trend is going to continue. There is no contradiction, however, if how the technology advances and how it will be applied are considered separately. MS technology will continue to increase in sensitivity, mass accuracy, and resolution, while possessing even shorter duty cycles. The net result will be an increase in the number of proteins identified within a proteome sample. While most reader's minds will automatically think about how this will affect biomarker discovery; that will not be the main advantage of obtaining deeper proteomic coverage. Increasing the efficacy of biomarker discovery will require much more than simply identifying a greater number of proteins. Faster and more sensitive protein identification will have the greatest impact on less complex samples, particularly the characterization of protein complexes. While the present strategy for characterizing protein complexes is not that much different than conducting biomarker discovery studies, identifying more proteins in a shorter time will increase the coverage per protein aiding in the ability to discriminate major and minor components.

Identifying more proteins will not only be realized through better mass spectrometers, but improved software and computers with faster processing speed. As mentioned in previous chapters in relation to data-dependent MS, proteomics is basically a "you-get-what-you-get" science. This limitation is not only related to the mass spectrometer operation but also the analysis of the data. Converting MS^2 (or MS^3) spectra into peptide identifications is still dominated by SEQUEST (14, 15) and MASCOT (16), which apply peptide collision-induced dissociation (CID) rules to correlate peaks to proteins found in databases. Do not get me wrong, both of these software algorithms have been enormously successful and proteomics would not be anywhere close to where it is without these programs. However, the limitations of the commonly used data analysis pipelines are very evident. Only about 20% of CID spectra result in a useful identification. A large percentage of spectra are not identified owing to poor signal-to-noise or a dearth of peaks; however, a sizable percentage of "good" spectra are never identified. If the CID of a peptide does not follow the established rules because of, for example, an unsuspected PTM, these popular software algorithms will not assign the spectra correctly. Even more of a limitation is if the peptide or protein is not within the genomic or proteomics database being searched, the MS^2 spectrum will not be identified. Another challenge is the identification of PTMs. In a typical experiment, the software is told to search a batch of CID spectra for a specific modification. For example, the software is told to include a dynamic modification of 79.99 Da on serine, threonine, and tyrosine residues in the search for phosphorylation sites. Finding additional modifications is done by researching the data for other residue-specific mass shifts due to the PTM of interest (i.e., acetylation and methylation).

Future peptide identification will be based on *de novo* sequencing of CID spectra that do not depend on an available database. In *de novo* sequencing, peptide sequences are interpreted by applying established fragmentation rules to the CID spectra of the peptide (17). Once established, the putative peptide sequences can be submitted for sequence alignment and similarity search using, for example, BLAST. There are a number of studies that already illustrate the potential value of *de novo* sequencing (18–20). Recently, Dr. Albert Heck's group developed a *de novo* pipeline (Figure 12.1) in which fragmentation spectra acquired of peptides produced from a Lys-N digest of a HEK293T proteome sample using both electron-transfer dissociation (ETD) and CID were analyzed (21). Noise is filtered from the spectra prior to any initial *de novo* interpretation. This initial *de novo* analysis resulted in 5.7 million possible peptide sequence solutions from over 11,000 ETD spectra. There are over 5.7 million solutions since every possible peptide sequence is considered at this stage of the analysis. To identify the best match, a database of the *de novo* sequence library along with a decoy was prepared and uploaded into Mascot to be used to match the ETD spectra to peptide sequences. The CID data were also searched using Mascot against the *de novo* and decoy database. The resulting matches were filtered so that only results for an ETD and its paired CID spectra obtained by Mascot originating from a *de novo* solution for the same spectrum were considered acceptable. In cases where multiple solutions match both the ETD and CID scans, Mascot scores for the same solution were combined and the highest ranking result was accepted as correct. The net result was the identification of 2744 unique nonredundant peptide sequences. While this approach relied to an extent on Mascot for peptide sequence confirmation, in the future *de novo* sequencing algorithms will stand on their own. These *de novo* approaches will be a valuable tool for filling in the gaps left by genomic analysis, such as the identification of PTMs, hypervariable regions within antibodies, and even characterization of proteomes from extinct species.

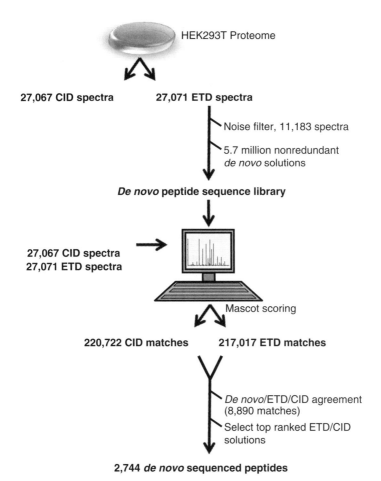

Figure 12.1. Schematic overview of a *de novo* peptide sequencing pipeline developing by the laboratory of Dr. Albert Heck. Peptides generated from HEK293T cells are sequenced using ETD and CID. After noise filtering, 11,183 spectra are subject to *de novo* interpretation resulting in a peptide sequence library containing about 5.7 million sequences. This library, along with a decoy library, is loaded into Mascot, which attempts to match the ETD and CID spectra. The resulting matches are filtered such that the ETD and CID spectra obtained by Mascot must agree with the result obtained for the same spectra via *de novo* sequencing. If more than one solution is found for the ETD and CID scans, the Mascot scores for the same solution are combined and the highest ranking result is considered correct. The entire process resulted in the identification of 2744 unique peptide sequences.

12.4.1   Top-Down Mass Spectrometry

Proteomic characterization is presently based almost solely on digestion of proteins into peptides and characterization of these surrogates via MS, MS^2, or MS^3. Unfortunately, this strategy is deficient in that it does not provide the direct evidence of important biological parameters such as alternative splice forms, diverse modifications, and variant sites of protein cleavages. To fully characterize the human proteome is going to require technologies that can detect mature forms of each protein. To differentiate various mature forms of very homologous proteins is going to require top-down MS. In top-down proteomics, intact protein ions, instead of their proteolytically created peptides, are characterized by MS and MS^2. Top-down proteomics is able to differentiate subtle differences in intact proteins such as those that have 100% sequence homology but different PTMs.

As mentioned, proteins are most commonly identified through peptides produced from enzymatically digesting the proteome. In a global proteomic analysis, these peptides are separated using multidimensional chromatography prior to MS analysis. The result is that peptides that originate from a single protein are scattered throughout a complex data set. It is literally impossible to put the peptide pieces back together again to form their parent protein with absolute certainty. Without any additional information, all peptides informatically linked to a protein sequence are assigned to a single protein. This data compilation does not always provide an accurate view of the sample. For example, proteins such as c-MET have multiple sites that can potentially be phosphorylated and play a role in the potential outcome of an individual with small cell lung carcinoma (22). Phosphorylation of a single site may activate the protein, while phosphorylation of multiple sites may cause deactivation (23). In a typical cell, these proteins will exist in both active and inactive forms. In a bottom-up approach all identified peptides will be grouped into a single protein sequence making it impossible to readily differentiate active and inactive populations (Figure 12.2).

In the future, top-down methods will dominate protein identification. Scientists such as Neil Kelleher and Roman Zubarev have pioneered the development and application of methods to obtain sequence information on intact proteins (24, 25). One of the primary advantages of top-down MS is the molecular weight of the intact protein is recorded. Once the protein is identified, any discrepancy between the calculated and measured mass of the protein immediately alerts the investigator to possible PTMs. Electron capture dissociation (ECD) has shown excellent promise in fragmenting proteins; however, it is not yet ready for prime time. Improving the sensitivity of top-down approaches is necessary. As top-down fragmentation produces a larger number of fragments, the number of each fragment type is relatively low compared to CID of peptides, in which fewer fragment types are possible. Top-down analysis presently requires between 1 and 2 orders of magnitude more material than required for typical peptide identification.

A number of developments, not all specifically related to the mass spectrometer, must be made for top-down proteomics to become the dominant method of protein identification in proteomics. Sample preparation is always to first step in proteomics. Conditions for cell lysis are well understood, both for preparing samples containing denatured and nondenatured proteins. What needs to be considered is keeping all

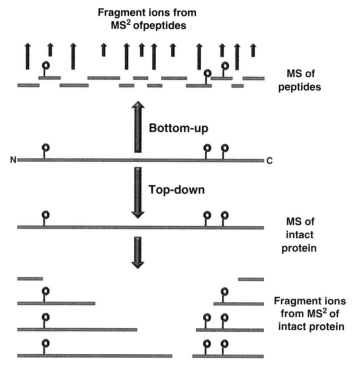

Figure 12.2. Schematic illustrating the advantages of top-down versus bottom-up approaches for identifying proteins. Using a bottom-up approach the proteins are digested into peptides that are individually identified. PTMs (represented by lollipops) are scrambled throughout this mixture of peptides. In a top-down approach, the intact protein is fragmented and PTMs remain associated with their protein of origin. Retaining the protein's entire sequence information enables confident assignment of the correct isoforms.

proteins in solution. Membrane proteins have much different solubility requirements than those within the cytosol. The wide range of protein isoelectric point (pI) values also needs to be considered when finding methods to keep intact proteins in solution. Chromatography needs to be improved and tailored to separate intact proteins. Peptides are readily fractionated using a variety of chromatographic techniques. In particular, peptides are soluble in the mobile phases and acidic conditions used for reversed phase chromatography. Unfortunately, a significant percentage of intact proteins will precipitate out under these conditions. Reversed-phase separations of intact proteins do not exhibit the high resolution that peptides do, nor do they give the same quality peak shape. Future advances in the separation of intact proteins, possibly using higher pressure LC systems or monolithic columns, will be critical for advancing top-down proteomics. I remember a conversation I had several years ago with Dr. Donald Hunt from the University of Virginia. We were talking about top-down proteomics and he made the comment that measuring intact proteins represented a more complex sample to separate than if the same sample was analyzed using a bottom-up approach. My initial reaction was puzzlement as conventional wisdom suggested that tryptic digestion of a proteome will increase its complexity since these proteins are all being chopped up into smaller pieces. He logically explained to me that the various isoforms and combinations of PTMs represent different entities when they are analyzed using top-down approaches. When all of these are considered, the intact mixture is more complicated and requires more sophisticated chromatographic fractionation than the digested mixture.

While top-down proteomics has existed for several years, novices in this area (such as myself) have often looked upon it as a technology that produces data that could identify a protein if you already knew it was in the sample. Most proteomic laboratories had a peripheral interest in it, but preferred to let other groups work on the developments required to bring top-down proteomics mainstream. While the prospects of bottom-up and top-down proteomics equally sharing the proteomics stage seemed to have waned recently, Dr. Kelleher recently published a study that may reinvigorate the enthusiasm toward top-down proteomics (26). Addressing the chromatographic concerns described above, Dr. Kelleher's group used a four-dimensional separation system, to identify over 3000 proteins within human cells via top-down proteomics. The intact proteins were initially fractionated by isoelectric point and size using solution isoelectric focusing followed by gel-eluted liquid fraction entrapment electrophoresis (27), respectively. These separated proteins were then analyzed via nanocapillary LC coupled directly to MS. The 3093 identified proteins originated from 1043 gene products, with many of the proteins originating from the same gene but having been processing via different PTMs, RNA splicing, and proteolysis events. Proteins over 100 kDa in molecular weight and membrane proteins containing up to 11 transmembrane helices were identified; illustrating the size and solubility range of the overall method. This study represented a greater than 20-fold increase in proteome coverage than had been seen previously using a top-down proteomics approach. The ability to measure over 3000 proteins in this study begins to bring top-down proteomics on-par with the numbers of proteins that can be routinely identified using bottom-up methods. Measuring intact proteins at this scale will enable the full, accurate blueprint encoded within the human genome to be elucidated at the protein level.

Continuing developments in novel spectral identification algorithms are going to be especially critical for top-down proteomics. This strong suit of peptide-based proteomics, however, is currently a weak suit of top-down proteomics. Examples of useful bioinformatic tools for analyzing top-down data have been published by Fagerquist et al. (28) and Neil Kelleher's group (ProSight PTM) (29). Another area of development that is going to be required to bring top-down proteomics into the mainstream is quantitation. It is conceivable that stable isotope labeling methods such as stable isotope labeling of amino acids in culture (SILAC), and peak intensity measurements could translate to intact proteins. While bottom-up approaches will continue to have a tremendous lead with respect to available bioinformatics and quantitation tools, top-down methods should continue to close this gap over the next few years. Indeed, there has been a recent upsurge in the demand for isotopically labeled intact proteins for quantitative proteomics purposes.

12.4.2 Bottom-Up Mass Spectrometry

As it exists today, bottom-up MS methods generate huge amounts of data. For most proteomic samples of low complexity, the coverage achievable today is already sufficient for answering most research questions. It is even doubtful if significantly increasing the number of identifications within complex proteomic samples will increase the information gleaned from these samples. Regardless, MS technology will continue to improve and an increase in the number of peptide identifications will naturally occur. While more peptides are identified per experiment than ever before, this is largely due to higher resolution chromatography coupled with the increased sensitivity of mass spectrometers and their faster duty cycles. In a vast majority of experiments, bioinformatic tools are absolutely critical for turning these data into information. As mentioned previously, about 20% of the recorded spectra are turned into a molecular identification. A recent study reported that of $>100,000$ features observed in a single LC-MS^2 analysis, only about 16% were targeted for MS^2 analysis and only about 10,000 proteins were subsequently identified (30). The inability to target every peak for identification was attributed to a lack of sequencing speed, sensitivity, and precursor ion isolation. While sequencing speed and sensitivity are obviously related to peptide identification, the role of precursor ion isolation is not so intuitive. The study found that the median intensity of the target peptide in the MS^2 selection window was only 14% of the total ion intensity. Decreasing the sample only had a minor effect on increasing the median intensity of the target peptide in the selection window. While faster sequencing speeds will undoubtedly increase the numbers of peptides that are identified in a complex proteome sample, this improvement by itself will not be enough to obtain complete proteomic coverage. In the future, multiplexed MS^2 analysis will become commonplace. In this format, multiple peptides are selected and subjected to CID (or ETD or ECD) resulting in sequence information for multiple peptides being contained within a single MS^2 spectrum. This data format will require a combination of higher resolution measurements and sophisticated data analysis software; however, many investigators have already made significant strides in this area (31–33).

12.4.3 Ionization

It is too easy to say that mass spectrometers will be more sensitive in the future. Everyone with knowledge of the field of proteomics would consider that statement to be completely obvious. Instruments with increased sensitivity, mass accuracy, and resolution will continue to be developed. Where the next revolution in MS will occur is within the ionization source. Presently, only a small fraction of the ions that are created in the ion source ever enter the mass analyzer region of the mass spectrometer. Continuing developments such as application of ion funnels will increase this ion transfer efficiency resulting in an immediate increase in sensitivity (34). More important than ion transfer, will be the design of novel ionization methods. Electrospray and matrix-assisted laser desorption/ionization (ESI and MALDI) have been utilized for the past 20 years in proteomics (35, 36). Their importance to biological sciences is underscored in the awarding of the Nobel prizes to Drs. John Fenn and Koichi Takashi for these inventions. Unfortunately, both methods cannot overcome the fact that all biological

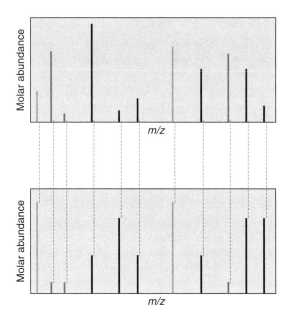

Figure 12.3. In the future an ionization source will be developed that ionize peptides based on their molar abundance. The result is a mass spectrum in which all peptides from a single protein will have equal intensities.

molecules do not possess the same physicochemical properties. The different ionization efficiencies of biological molecules (not just classes of molecules) results in signal intensities that do not correlate with their molar abundance. In addition, the ionization of a biological molecule is often dependent on other molecules within the environment during ionization. This "matrix-affect" can result in sample-to-sample variation in the signal provided by a biological molecule.

In the future, the signal detected by the mass spectrometer will be directly correlated to the molar equivalence of the molecule and the ionization source will label molecules based on their molar equivalence. For example, presently a tryptic digest of a protein produces a series of signals of varying intensity regardless of whether it is analyzed using ESI- or MALDI-MS (Figure 12.3). In the future, a single tryptically digested protein will produce a mass spectrum containing signals of equal intensity. The intensities of signals arising from different proteins will be directly correlated to their molarity allowing their relative abundance to be determined from the mass spectrum. The physicochemical properties or the matrix environment of the species being analyzed will not be a factor in the signal intensities.

How will this revolutionize proteomics? Technologies for identifying proteins are very good. Technologies for quantitating (absolute or relative) proteins are lagging behind. A variety of stable isotope labeling methods have been developed over the previous decade to improve quantitation of MS signals; however, the general lack of confidence in them is reflected in the fact that no single method has garnered favor over the others. Each has their specific advantages and disadvantages. While spectral counting, which does not require isotope labeling, is effective at comparing unlimited numbers of samples, its quantitative accuracy is lacking. Exponentially modified protein abundance index (emPAI) was developed several years ago as a method to measure the relative abundance of proteins in a complex mixture based on the number of identified peptides for each species (37). While it has shown some success, it cannot accurately quantify low abundant proteins that are generally identified by only one or two peptides (38). If the signal provided by a biomolecule correlated with its molar abundance, the stoichiometry of individual proteins in complexes could be calculated from the MS data. The stoichiometry of PTMs at specific sites within proteins could also be easily measured. The quantitative results from global proteomic studies would show less variation and maybe (just maybe) provide more confident leads in potential biomarkers.

12.5 GLOBAL QUANTITATIVE PROTEOMICS

Measuring differences in multiple proteomic samples is presently accomplished almost exclusively using relative quantitation. This type of measurement is not unique to proteomics; it is done in transcriptomics, Western blotting, global metabolomics, IHC, etc. Absolute quantitation is generally only performed using techniques that target a small number of analytes and incorporate internal or external calibration, such as ELISAs. In the future, quantitative comparisons of entire proteomes will be based on absolute quantitation. Being able to measure the absolute quantity of each protein and its potential myriad of isoforms will be a huge step in systems biology. Beyond being able to evaluate concordant changes of protein abundances within a single proteome, absolute quantitation enables comparisons of proteomic measurements made within many different laboratories under many different types of conditions. Much like analyzing massive amounts of data from sequencing large numbers of solid tumors has lead to the discovery of 296 tumor suppressors and 33 oncogenes that are actually involved in cancer (39), comparing a large, critical mass of absolute quantitation data from different proteomes will be required for recognizing trends and developing a predictive capability concerning how systems reacts to specific stimuli.

One attempt to provide absolute quantitation, combined data from three commonly used approaches—selected reaction monitoring (SRM) using a limited set of internal reference peptide standards, the average signal intensities provided by the top three peptides selected per protein, and weighted MS^2 spectral counts (Figure 12.4) (40). In the first phase of the project, the proteome of *Leptospira interrogans*, a spirochete responsible for leptospirosis was characterized via multidimensional fractionation and MS^2 analysis. In the second phase, 32 peptides corresponding to 19 proteins at different abundance levels were selected based on their spectral counts in the first phase. The absolute abundances of these proteins were determined by conducting SRM analysis of the samples containing known amounts of stable-isotope labeled peptides. By knowing the number of cells used to generate the sample and the amount of the heavy labeled peptides added, the copy number of these proteins was then calculated based

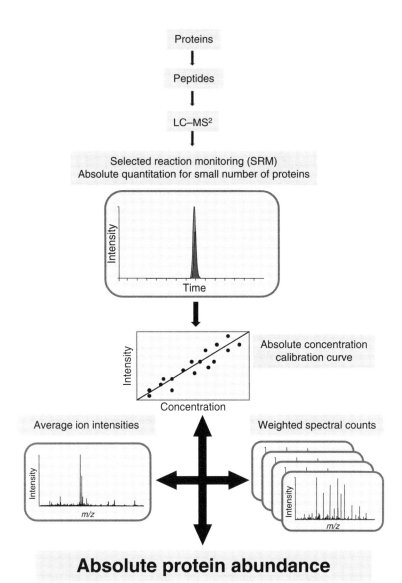

Figure 12.4. Protein absolute quantitation using SRM, ion intensities, and spectral counting. Proteins from the pathogen *Leptospira interrogans* were identified using a bottom-up LC-MS^2 approach. The absolute abundance of 19 proteins in this proteome was determined using SRM to construct an absolute concentration calibration curve. This calibration curve was used in conjunction with ion intensity and spectral count data to extrapolate the absolute abundance of 51% of the proteins predicted by this organism's genome.

on the SRM result. The abundance of these proteins ranged from 40 to 15,000 copies per cell.

The third phase took the extracted precursor ion intensities for peptides derived from LC-MS spectra and calculated the median total protein ion intensity of the three most intense peptides from a specific protein. The absolute concentration values of the 19 proteins measured using SRM were used as calibration points to extrapolate the absolute abundance measurements for the remaining proteins identified in the global analysis. The extracted ion intensities were used for absolute quantitation of 769 proteins while the absolute abundance of 1095 proteins was estimated by spectral counting. The total number of proteins whose absolute abundance was measured corresponded to 51% of the pathogens genome. The accuracy of this method was assessed to be less than threefold on average across the proteome.

While analysis such as the one presented above show potential, it is going to take more than existing technologies to obtain accurate measurements of protein absolute abundances across entire proteomes. The varying ionization efficiencies and matrix affects cannot be reliably compensated for across different peptides. Therefore, basing a proteins absolute abundance on the measurement of a single ion's intensity (or even the median of three peptides) is risky. Developing absolute abundance capabilities in the future will require a novel ionization technique that reflects a protein's true molar abundance in solution and is not influenced by its matrix. Also, absolute abundance measurements will be accomplished using top-down approaches so the any uncertainty of the origin of peptides that are measured using bottom-up methods is completely eliminated.

12.6   THE SHRINKING UNIVERSE OF PROTEOMICS

One general theme of proteomics in the past decade has been "bigger is better." As seen throughout this book, the proteome coverage attainable today is quite impressive compared to where we were 10 years ago. Unfortunately, this coverage is still a small percentage of the entire proteome. The proteome itself is a dynamic entity and does not have a defined endpoint; however, we know that we are currently interrogating only about 10% of it. Conventional wisdom would suggest that the field will continue on a trajectory of increasing the number of species that can be identified. For most simple proteomic samples, such as protein complexes, the depth of coverage presently attainable is adequate for most biological questions. Obviously, increasing the coverage for individual proteins would help ensure that more PTMs are identified. Even for complex proteome samples, the present coverage afforded by today's technology has been sufficient for generating hypothesis from the acquired data. Identifying all the detectable peptide features in a proteome sample may not dramatically improve biomarker discovery, for example.

In the future proteomics investigators will work "smarter not harder." In today's era of global proteomics where the focus may be on finding biomarkers or changes in the abundance of cellular proteins, upward of 90% of the proteins identified are not considered useful data that provide definite conclusions. We essentially create a thousand piece jigsaw puzzle and use only a few of the pieces to create a picture. Proteomics is very good at generating huge amounts of data hoping that it can be transformed into useful information. Unfortunately, this transformation rarely occurs to the extent required to the type of information that investigators are willing to risk additional resources in effectively pursuing. The literature is filled with studies in which 20–30% of the proteins compared between different types of samples show a significant change in abundance. With the capabilities of LC-MS^2 today, this can represent anywhere from 100 to 1000 proteins. The net conclusion of a vast majority of these studies will then focus on less than 10 proteins whose abundance change was correlated, by example, via ELISA or Western. What about the other 90–990 proteins? Do the data for these proteins not fit an existing model or is it untrustworthy? We do not presently have the knowledge, computational power, or confidence in the data to include all of the observed changes in a reasonable systems biology view of the cell.

To work smarter, proteomics needs to take a page from small molecule analytical chemistry. Historically, LC-MS has been used to measure targeted compounds in a complex sample that are known to be associated with a specific condition. In my opinion, one of the greatest clinical uses of LC-MS is for in-born errors of metabolism. Many of the initial studies in this area used LC-MS to screen newborns for phenylketonuria, tyrosinemia, maple syrup urine disease, and homocystinuria and other hypermethioninemias (41–43). Currently, this technology can screen for metabolic disorders in a single analysis using a dried blood spot. More than 3 million infants have been screened worldwide and more than 500 confirmed disorders detected using LC-MS. This assay takes the smart approach in that it measures known metabolites whose range of levels are known to be associated with a metabolic disorder. It does not simply rely on a single measurement; rather it uses a multiple-metabolite analysis for the detection of numerous metabolic disorders in a single LC-MS run.

The field of proteomics can learn a lot from the inborn errors of metabolism assay. I cannot even begin to estimate how much time and money has been wasted in using proteomics to find diagnostic and prognostic markers of cancer. Obviously, significant technological advances have been made, yet the return on biological advances has been poor. It seems that we do not learn much from previous studies, both our own and others. We often forget that knowledge we possess is much more powerful than our mass spectrometers. As mentioned in the chapter on protein arrays, we currently have a wealth of data on proteins that are involved in cancer. These data, accumulated primarily through genome sequencing, suggests that there are just over 300 tumor suppressors and oncogenes that are actually involved in cancer (39). In the future, MS assays will target known, critical proteins within biological mixtures instead of trying to identify as many as possible. While this strategy is currently being followed in the many single- and multiple-reaction monitoring (SRM and MRM) studies of peptides (44, 45), most of these still focus on proteins found in nonbiased discovery studies. The challenge is filling the MS chromatogram with only peptides of interest that are somehow connected via biology (e.g., protein pathway and patient health status).

Focusing only on proteins important in cancer will go a long way in bringing proteomics to the clinic. However, it is going to require a lot more than simply targeted MS to make proteomics a clinically useful technology. In my personal vision, as illustrated in Figure 12.5), tumor biopsies are acquired from the patient and laser capture

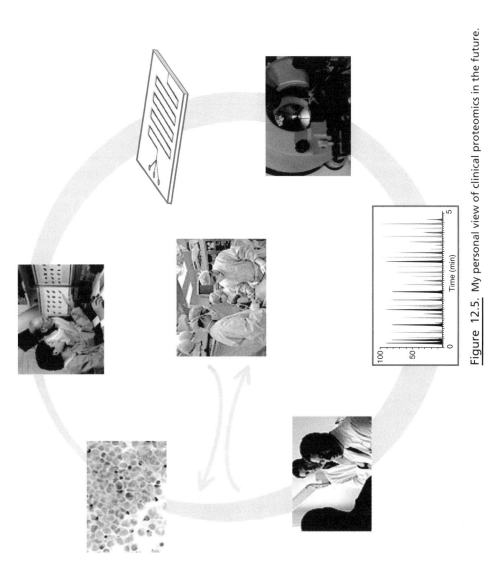

Figure 12.5. My personal view of clinical proteomics in the future.

microdissection (LCM) is used to acquire ∼100 cancer cells, thereby eliminating heterogeneity introduced by a combination of diseased and healthy cells. The cells are applied to a microfluidic device that conducts cell lysis, protein extraction, and fractionates the proteins. This device is coupled directly to the mass spectrometer, which has been configured to measure ∼100 proteins that are known to be related to the specific cancer suspected. Internal standards for these proteins are added to the sample after LCM enabling absolute quantitation of each protein from the patient. An algorithm trained on data from previous studies that measured the abundances of each protein in tumor cells from known types and grades of cancers is used to diagnose the current type and stage of the experimental sample. Measuring the protein abundances in absolute values is very important as the final output that the oncologist receives will list the amount of each protein within the tumor cells along with the expected range of abundances within normal cells. The ranges of these proteins can only be accurately determined if their absolute abundances are measured using a standard operating procedure (SOP). If a SOP and the proper reagents are used, these measurements will be comparable across laboratories and prospectively. Since separation technologies and MS duty cycles will be much faster in the future, these 100 analytes will be measured in less than 5 min. The report provided to the oncologist is summarized in such a manner that he/she is immediately able to make the correct diagnosis or determine the optimal treatment regiment for the patient. The key to the whole process is that it begins and ends with the patient.

12.7   OMIC DATA INTEGRATION

It is often mentioned that genomics and proteomics data need to be integrated to obtain a greater understanding of how the cell functions. Every biology and biochemistry student knows there is a close connection between genes and proteins. One does not exist without the other. Decades ago before the genomics and proteomics era, if a protein was analyzed using Western blotting, its mRNA was usually measured using Northern blotting. However, genomics and proteomics continue to exist in silos in the "omics" era. There are a number of reasons that genomics and proteomics have stayed in their respective silos. The throughput of genomic and proteomic measurements are not well matched. While we in proteomics used high throughput as an adjective to describe our science, it is not comparable to that attainable with modern gene sequencers. This discrepancy means that the genomic laboratory can analyze a lot more samples than the proteomics laboratory in the same time frame causing an imbalance in the amount of data being compared.

Another reason that genomics and proteomics remained silo-ed from each other is data coverage. The technology behind gene sequencing allows complete coverage of the genome; however, proteomic technologies only provide a small sampling of the proteome. This compromises the ability to make confident comparisons between events that occur at the gene and protein levels. There are numerous studies that demonstrate an overall lack of correlation between gene expression and protein abundances within the cell; however, there has to be some connection between the two. One of the challenges in finding this connection is that we are limited by only being able to produce static "snapshots" of a proteome and transcriptome. The proteome and transcriptome are very dynamic and capturing their fluctuations over time is going to be necessary if we ever hope to develop a coherent systems view of the cell. Acquiring this view is going to require collecting more data on fewer transcripts and proteins. Proteomic analysis is currently dominated by collecting little data on a lot of proteins. The first step toward true systems biology is going to require collecting a lot of data on a few proteins at many different time points. Top-down proteomics will be more effective at doing this than bottom-up approaches, simply because top-down methods provide direct evidence in changes to protein isoforms, which will be required to measure dynamic changes in PTM states of proteins.

Much of the present era of biomarker discovery has been consumed with following Einstein's Theory of Insanity. This theory states that insanity is "The endless repetition of the same experiments, in the hope of obtaining a different result" (46). Consider all of the published manuscripts on cancer biomarker discovery that end with a conclusion suggesting that the proteins identified in the study require further validation for clinical use (or something to that effect). How many of these investigators ever followed up any of these proteins as legitimate cancer biomarkers? No one would dispute that validating a cancer biomarker would be an extremely valuable premise. Any scientist that identified one would achieve great scientific fame and personal satisfaction knowing they made a huge positive impact. So why do so few ever put serious effort into validating potential markers? Much of it comes back to poorly thought out experiments. Before starting a biomarker discovery study, very few laboratories even consider potential validation studies. Our (and I am preaching to myself as well) general experimental designs go as far as discovery with the hope that we will publish a manuscript and then maybe someone else with the available resources will be interested enough to conduct the necessary validation studies using any of the potential markers. Biomarker studies need to follow the precepts of *The Oz Principle,* a book that describes the role of accountability in achieving results (47). The general mantra of the Oz Principle is See It, Own It, Solve It, Do It®. This principle challenges individuals and organizations to asked "What else can I do?" to achieve the desired result. Since the desired result is to find clinically useful biomarkers, laboratories need to design studies that incorporate both discovery and validation stages if we are ever going to achieve the desired result.

Integration of data at the systems level using all three "omics" technologies is required if we hope to increase our understanding of cancer biology to a point where intelligent decisions can drive the discovery and validation of diagnostic, prognostic, or therapeutic biomarkers. In the future, we must increase the number of samples analyzed via proteomics technologies or incorporate data concerning other biomolecules to recognize disease specific changes from normal interindividual variation (Figure 12.6). Since we have been unable to establish a "normal" proteome, future studies will incorporate genetic, proteomic, and metabolomic (more on that later) data to identify proteins that are dysregulated in cancer of can act as biomarkers. The assimilation of these data sets will provide the preponderance of evidence required to establish the biological importance of biomolecules in cancer.

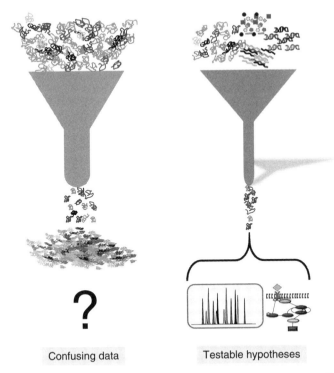

Confusing data   Testable hypotheses

Figure 12.6. Acquiring massive amounts of data only from proteomics studies will continue to create massive amounts of data but generate little information. To generate testable hypotheses, future studies will incorporate data from genomics, transcriptomics, proteomics, and metabolomics.

12.8   METABOLOMICS

While it is already here, the role of metabolomics in biomarker discovery and cell biology will increase exponentially in the future. The move toward metabolomics was in hindsight, quite obvious. In the most simplistic cellular model that every biology student is first taught, genes give rise to mRNA, which are translated into proteins, which proceed to carry out a myriad of functions within the cell including the metabolism of small molecules. With respect to technology, the development of metabolomics has come along at the perfect time. Metabolomics is a clear case of where technology drove the science. Both MS and nuclear magnetic resonance (NMR) spectroscopy, the two major technologies used in metabolomics, were quite mature when metabolomics began to pick up steam. These two technologies absolutely complement each other as the major weaknesses of MS are the major strengths of NMR spectroscopy and vice versa. NMR is inherently quantitative as the area of a peak in the spectrum is directly correlated to the concentration of specific nuclei (e.g., ^{1}H and ^{13}C). Peak area in a MS spectrum cannot be correlated directly to the abundance of a metabolite unless known amounts of internal standards are added. On the other hand, MS's major strength of sensitivity is NMR's major weakness. While it is orders of magnitude less sensitive than MS, cryo-genically cooled probe technology (48), higher field-strength superconducting magnets (49) [3], and miniaturized radiofrequency coils (50) [4] have dramatically improved NMR's sensitivity.

Metabolomics is in a unique position to impact discovery-driven science in a way that genomics, transcriptomics, and proteomics could not fully exploit. With the other three omic technologies being much further advanced, integrating metabolomics data into systems biology studies will be much more straightforward. This capability will be critical as the greatest chance to find truly useful disease biomarkers will be through studies that show correlation between biomolecules at all four levels. Years of working in technological "silos," particularly in genomics and proteomics, have proven to be an inefficient way to solve challenging biological problems.

The addition of metabolomics to the other "omic" technologies will finally be the ingredient that conquers the biomarker discovery challenge. On its own, metabolomics would probably be more successful than proteomics in finding biomarkers. However, on its own it will make many of the same mistakes as the field of proteomics. There will no doubt be a deluge of studies that publish hundreds of *potential* biomarkers that are never considered worth the effort of a validation study. As described above, integrating data from genomic and/or proteomic studies that corroborates metabolomics findings will provide the evidence needed to identify potential biomarkers that are worth attempting to validate. Validation in large cohorts is more straightforward than proteins since technologies (especially using MS) for measuring the absolute quantities of metabolites in clinical samples have been around for a long time. Although it is a younger science, metabolomics will become the dominant technology for producing clinically useful biomarkers for disease diagnosis and prognosis.

12.9 CONCLUSIONS

Writing this chapter was a lot more challenging than I thought it would be. I thought it would be the easiest chapter to complete as I envisioned having free license to speculate on what proteomics would look like a couple of decades from now. I thought I could be as ambitious (and ridiculous) as possible and predict such things as acquiring standard samples from colonies of isotopically labeled humans living on an island where they survive on diets of heavy lysine labeled plants and animals. While this whimsical approach would have been fun to write it would not be particularly useful. Hopefully, some of the predictions made in this chapter will stimulate the thinking of the readers and encourage them to pursue these directions or develop their own ideas on the future of proteomics. Most of the ideas in this chapter are already being pursued in one form or another and it will be interesting to pick up this book 20 years from now and see how far the science of proteomics has evolved.

Proteomic technology will continue to improve and new instrumentation will be introduced over the next couple of decades. These new capabilities will be great; however, they will not be what ultimately determines the success of proteomics in the fight against cancer. The key element will be our focus. I have posed the following question to a number of scientists over the years, "Imagine that you would spend the next 5 years of your career trying to identify a diagnostic biomarker for a cancer. During that time you would not publish any manuscripts, submit any abstracts, or attend any scientific conferences. After 5 years, I guarantee you that this biomarker would be clinically useful. Would you do it?" I have never had a scientist answer "No." We cannot escape the simple fact that identifying cancer biomarkers or developing a systems biology view of the cell is very, very, difficult. This era of proteomics has taken the field of protein science from analyzing data on a single protein to thousands of proteins per experiment. This extraordinarily giant leap did not result in the smooth landing we anticipated. Once our ability to comprehend this deluge of data catches up with our ability to collect data, proteomic's impact on cancer research and treatment will be invaluable.

REFERENCES

1. Zhou M, Veenstra T. Mass spectrometry: m/z 1983-2008. Biotechniques 2008;44:667–670.
2. Xie F, Liu T, Qian WJ, Petyuk VA, Smith RD. Liquid chromatography-mass spectrometry-based quantitative proteomics. J. Biol. Chem. 2011;286:25443–25449.
3. Hoffman SA, Joo WA, Echan LA, Speicher DW. Higher dimensional (Hi-D) separation strategies dramatically improve the potential for cancer biomarker detection in serum and plasma. J. Chromatogr. B Analyt. Technol. Biomed. Life Sci. 2007;849:43–52.
4. Lee YH, Tan HT, Chung MC. Subcellular fractionation methods and strategies for proteomics. Proteomics 2010;10:3935–3956.
5. McNulty DE, Annan RS. Hydrophilic interaction chromatography for fractionation and enrichment of the phosphoproteome. Methods Mol. Biol. 2009;527:93–105.
6. Zauner G, Deelder AM, Wuhrer M. Recent advances in hydrophilic interaction liquid chromatography (HILIC) for structural glycomics. Electrophoresis 2011;32:3456–3466.
7. Chernushevich IV, Loboda AV, Thomson BA. An introduction to quadrupole-time-of-flight mass spectrometry. J. Mass Spectrom. 2001;36:849–865.
8. Stafford G Jr. Ion trap mass spectrometry: A personal perspective. J. Am. Soc. Mass Spectrom. 2002;13:589–596.
9. Elias JE, Haas W, Faherty BK, Gygi SP. Comparative evaluation of mass spectrometry platforms used in large-scale proteomics investigations. Nat. Methods 2005;2:667–675.
10. Issaq HJ, Veenstra TD. Would you prefer multiple reaction monitoring or antibodies with your biomarker validation? Expert Rev. Proteomics 2008;5:761–763.
11. Yates JR, Ruse CI, Nakorchevsky A. Proteomics by mass spectrometry: Approaches, advances, and applications. Annu. Rev. Biomed. Eng. 2009;11:49–79.
12. Butt YK, Lo SC. Detecting nitrated proteins by proteomic technologies. Methods Enzymol. 2008;440:17–31.
13. Zee BM, Young NL, Garcia, BA. Quantitative proteomic approaches to studying histone modifications. Curr. Chem. Genomics 2011;5(Suppl 1):106–114.
14. Gatlin CL, Eng JK, Cross ST, Detter JC, Yates JR 3rd. Automated identification of amino acid sequence variations in proteins by HPLC/Microspray tandem mass spectrometry. Anal. Chem. 2000;72:757–763.

15. Link AJ, Eng J, Schieltz DM, Carmack E, Mize GJ, Morris DR, Garvik BM, Yates JR 3rd. Direct analysis of protein complexes using mass spectrometry. Nat. Biotechnol. 1999;17:676–682.

16. Perkins DN, Pappin DJ, Creasy DM, Cottrell JS. Probability-based protein identification by searching sequence databases using mass spectrometry data. Electrophoresis 1999;20:3551–3567.

17. Frank A, Pevzner P. PepNovo: De Novo peptide sequencing via probabilistic network modeling. Anal. Chem. 2005;77:964–973.

18. Zhang J, Xin L, Shan B, Chen W, Xie M, Yuen D, Zhang W, Zhang Z, Lajoie GA, Ma B. PEAKS DB: De Novo sequencing assisted database search for sensitive and accurate peptide identification. Mol. Cell. Proteomics 2012;11:M111.010587.

19. Ma B, Johnson R. De Novo sequencing and homology searching. Mol. Cell. Proteomics 2012;11:O111.014902.

20. Allmer J. Algorithms for the De Novo sequencing of peptides from tandem mass spectra. Expert Rev. Proteomics 2011;8:645–657.

21. Altelaar AF, Navarro D, Boekhorst J, van Breukelen B, Snel B, Mohammed S, Heck AJ. Database independent proteomics analysis of the ostrich and human proteome. Proc. Natl. Acad. Sci. U.S.A. 2012;109:407–412.

22. Arriola E, Cañadas I, Arumí-Uría M, Dómine M, Lopez-Vilariño JA, Arpí O, Salido M, Menéndez S, Grande E, Hirsch FR, Serrano S, Bellosillo B, Rojo F, Rovira A, Albanell J. MET phosphorylation predicts poor outcome in small cell lung carcinoma and its inhibition blocks HGF-induced effects in MET mutant cell lines. Br. J. Cancer 2011;105:814–823.

23. Miyata Y, Sagara Y, Kanda S, Hayashi T, Kanetake H. Phosphorylated hepatocyte growth factor receptor/C-Met is associated with tumor growth and prognosis in patients with bladder cancer: Correlation with matrix metalloproteinase-2 and -7 and Kelleher, N. L. top-down proteomics. Anal. Chem. 2004;76:197A–203A.

24. Zubarev RA. Electron-capture dissociation tandem mass spectrometry. Curr. Opin. Biotechnol. 2004;15:12–16.

25. Tipton JD, Tran JC, Catherman AD, Ahlf DR, Durbin KR, Kelleher NL. Analysis of intact protein isoforms by mass spectrometry. J. Biol. Chem. 2011;286:25451–25458.

26. Tran JC, Zamdborg L, Ahlf DR, Lee JE, Catherman AD, Durbin KR, Tipton JD, Vellaichamy A, Kellie JF, Li M, Wu C, Sweet SM, Early BP, Siuti N, LeDuc RD, Compton PD, Thomas PM, Kelleher NL. Mapping intact protein isoforms in discovery mode using top-down proteomics. Nature 2011;480:254–258.

27. Tran JC, Doucette AA. Gel-eluted liquid fraction entrapment electrophoresis: An electrophoretic method for broad molecular weight range proteome separation. Anal. Chem. 2008;80:1568–1573.

28. Fagerquist CK, Garbus BR, Williams KE, Bates AH, Boyle S, Harden LA. Web-based software for rapid top-down proteomic identification of protein biomarkers, with implications for bacterial identification. Appl. Environ. Microbiol. 2009;75:4341–4353.

29. Taylor GK, Kim YB, Forbes AJ, Meng F, McCarthy R, Kelleher NL. Web and database software for identification of intact proteins using "top down" mass spectrometry. Anal. Chem. 2003;75:4081–4086.

30. Michalski A, Cox J, Mann M. More than 100,000 detectable peptide species elute in single shotgun proteomics runs but the majority is inaccessible to data-dependent LC-MS/MS. J. Proteome Res. 2011;10:1785–1793.

31. Weisbrod CR, Eng JK, Hoopmann MR, Baker T, Bruce JE. Accurate peptide fragment mass analysis: Multiplexed peptide identification and quantification. J. Proteome Res. 2012;11:1621–1632.

32. Ledvina AR, Savitski MM, Zubarev AR, Good DM, Coon JJ, Zubarev RA. Increased throughput of proteomics analysis by multiplexing high-resolution tandem mass spectra. Anal. Chem. 2011;83:7651–7656.

33. Graichen AM, Vachet RW. Multiplexed MS/MS in a miniature rectilinear ion trap. J. Am. Soc. Mass Spectrom. 2011;22:683–688.

34. Kelly RT, Tolmachev AV, Page JS, Tang K, Smith RD. The ion funnel: Theory, implementations, and applications. Mass Spectrom. Rev. 2010;29:294–312.

35. Fenn JB, Mann M, Meng CK, Wong, SF. Electrospray ionization for mass spectrometry of large biomolecules. Science 1989;246:64–71.

36. Karas M, Bachmann D, Bahr U, Hillenkamp F. Matrix-assisted ultraviolet laser desorption of non-volatile compounds. Int. J. Mass Spectrom. Ion Processes 1987;78:53–68.

37. Ishihama Y, Oda Y, Tabata T, Sato T, Nagasu T, Rappsilber J, Mann M. Exponentially modified protein abundance index (emPAI) for estimation of absolute protein amount in proteomics by the number of sequenced peptides per protein. Mol. Cell. Proteomics 2005;4:1265–1272.

38. Waybright T, Gillette W, Esposito D, Stephens R, Lucas D, Hartley J, Veenstra T. Identification of highly expressed, soluble proteins using an improved, high-throughput pooled ORF expression technology. Biotechniques 2008;45:307–315.

39. http://news.sciencemag.org/scienceinsider/2010/04/a-skeptic-questions-cancer-genom.html

40. Malmström J, Beck M, Schmidt A, Lange V, Deutsch EW, Aebersold R. Proteome-wide cellular protein concentrations of the human pathogen leptospira interrogans. Nature 2009;460:762–765.

41. Chace DH, Human SL, Millington OS, Kahler SG, Adam BW, Levy HL. Rapid diagnosis of homocystinuria and other hypermethioninemias from newborns' blood spots by tandem mass spectrometry. Clin. Chem. 1996;42:349–355.

42. Chace DH, Millington DS, Terada N, Kahler SG, Roe CR, Hofman LF. Rapid diagnosis of phenylketonuria by quantitative analysis for phenylalanine and tyrosine in neonatal blood spots by tandem mass spectrometry. Clin. Chem. 1993;39:66–71.

43. Chace DH, Human SL, Millington OS, Kahler SG, Adam BW, Levy HL. Rapid diagnosis of maple syrup urine disease in blood spots from newborns by tandem mass spectrometry. Clin. Chem. 1995;41:62–68.

44. Schmitz-Spanke S, Rettenmeier AW. Protein expression profiling in chemical carcinogenesis: A proteomic-based approach. Proteomics 2011;11:644–656.

45. Makawita S, Diamandis EP. The bottleneck in the cancer biomarker pipeline and protein quantification through mass spectrometry-based approaches: Current strategies for candidate verification. Clin. Chem. 2010;56:212–222.

46. http://www.militaryphotos.net/forums/showthread.php?47335-Einstein-s-Theory-of-Insanity

47. Connors R, Smith T, Hickman C. The Oz Principle: Getting results through individual and organizational accountability. London, England: Penguin Books Ltd.; 2010.

48. Kovacs H, Moskau D, Spraul M. Cryogenically cooled probes-a leap in NMR technology. Prog. NMR Spec. 2005;46:131–155.

49. Felli IC, Brutscher B. Recent advances in solution NMR: Fast methods and heteronuclear direct detection. Chemphyschem 2009;10:1356–1368.

50. Kentgens AP, Bart J, van Bentum PJ, Brinkmann A, van Eck ER, Gardeniers JG, Janssen JW, Knijn P, Vasa S, Verkuijlen MH. High-resolution liquid- and solid-state nuclear magnetic resonance of nanoliter sample volumes using microcoil detectors. J. Chem. Phys. 2008;128:052202.

INDEX

Acetylation, 80, 89, 109, 110, 284
Acute lymphoblastic leukemia (ALL), 11, 156
Acute-phase protein, 27, 79, 207, 208
Aebersold, Ruedi, 82
Affymetrix, 10, 11
Alzheimer's disease, 187, 196
Anderson, Leigh, 178, 181, 253
Angiogenesis, 16, 18, 228
Antiproliferative factor (APF), 104, 105
Apoptosis, 16, 20, 21, 63, 68, 150–155, 158, 164, 229
Aptamer arrays, 226–228
Apweiler, Rolf, 263
Arkin, Adam, 263
ArrayCGH, 11

BCR-ABL, 156–159
Bench-to-bedside, 241
Biofluid collection, 173–175, 190
Biofluid preparation, 173–175, 190
Biofluid storage, 173–175, 190
Biomarker, 11, 27–29, 31, 34, 39, 44, 46, 48, 51, 52, 60, 62, 63, 65, 73, 76, 77, 79, 80, 82, 104, 108, 120, 146, 171–190, 195–197, 201, 203, 207–209, 211, 217, 220, 223, 226–229, 231, 245–247, 255, 263, 272, 277, 282, 283, 291, 293, 297–300
Blood, 43, 44, 51, 77, 79, 108, 171, 172, 174, 175–182, 183–186, 189, 195–197, 199, 204, 211, 242, 243, 255, 256, 258, 259, 294
Bottom-up mass spectrometry, 286–289, 292, 293, 297
Brent, Roger, 130

Calbindin D$_{28K}$, 2, 277
Capillary electrophoresis (CE), 31, 32
Caprioli, Richard, 13, 248
Capture arrays, 217, 220–227
Casein kinase-2, 151–155, 158

Caspase, 150–155, 158, 229, 274
Cell cycle, 16, 20, 68, 90, 164, 275
Cell division 5, 117, 118
Cerebrospinal fluid, 109, 175, 176, 185–187, 198
Chace, Donald, 196
Chromatin immunoprecipitation (ChIP), 11
Chronic myelogenous leukemia (CML), 156–158
c-Met, 148–150, 164, 165
Coleridge, Samuel Taylor, 263
Collisional-induced dissociation (CID), 34, 35, 43, 67, 70, 95–99, 107, 266, 271, 284–286, 289
CpG islands, 8, 12
Crux, 268

Dasatinib, 155–159, 163, 164
Data integration, 18, 19, 296, 297
De novo sequencing, 14, 264, 284, 285
Discovery-driven, 2, 3, 41, 42, 46, 48, 68, 89, 119, 120, 127, 128, 139, 140, 151, 153, 166, 299
Deoxyribonucleic acid (DNA), 1–11, 28, 31, 76, 81, 117, 121, 125–129, 131, 132, 198, 199, 217–219, 222, 224–226, 242, 267
DNA microarray, 10, 11
Dole, Malcolm, 29
Douglass, Frederick, 117
Driver kinase mutations, 158, 160, 163–165

EGFRvIII, 146–150, 223–225
Ekins, Roger, 216
Electron capture dissociation (ECD), 98, 99, 107, 108, 286, 289
Electron transfer dissociation (ETD), 98, 99, 107, 108, 284, 285, 289
Electrophoretic mobility shift assay (EMSA), 125–126, 128, 137
Electrospray ionization (ESI), 28, 29, 31, 32, 34, 37, 38, 41, 95, 252, 289, 290
Enzyme-linked immunosorbant assay (ELISA), 3, 51, 209, 216, 220, 246, 257–259, 291, 294

Proteomic Applications in Cancer Detection and Discovery, First Edition. Timothy D. Veenstra.
© 2013 John Wiley & Sons, Inc. Published 2013 by John Wiley & Sons, Inc.